PROGRESS IN POLYMER MATERIALS SCIENCE

Research, Development and Applications

About the Editors

Gennady E. Zaikov, DSc

Gennady E. Zaikov, DSc, is Head of the Polymer Division at the N. M. Emanuel Institute of Biochemical Physics, Russian Academy of Sciences, Moscow, Russia, and a professor at Moscow State Academy of Fine Chemical Technology, Russia, as well as a professor at Kazan National Research Technological University, Kazan, Russia. He is also a prolific author, researcher, and lecturer. He has received several awards for his work, including the the Russian Federation Scholarship for Outstanding Scientists. He has been a member of many professional organizations and on the editorial boards of many international science journals.

Oleg V. Stoyanov, DSc

Oleg V. Stoyanov, DSc, is Professor at the Kazan National Research Technological University, Kazan, Russia. He is a world-renowned scientist in the field of chemistry and the physics of oligomers, polymers, composites, and nanocomposites.

Elena I. Kulish, DSc

Elena I. Kulish, DSc, is Professor and Deputy Head of the Laboratory of Semenov at Bashkirian State University in Ufa, Russia. She is a specialist in the field of high-molecular compounds and chemical kinetics.

About the Editors

Apple Academic Press Inc.
3333 Mistwell Crescent
Oakville, ON L6L 0A2
Canada

Apple Academic Press Inc.
1613 Beaver Dam Road, Suite # 104
Point Pleasant, NJ 08742
USA

Exclusive worldwide distribution by CRC Press, a Taylor & Francis Group

International Standard Book Number: 978-1-926895-41-3 (Hardback)

Printed in the United States of America on acid-free paper

Library of Congress Control Number: 2012919713

Library and Archives Canada Cataloguing in Publication

Progress in polymer materials science : research, development and applications/edited by Gennady E. Zaikov, Oleg V. Stoyanov, and Elena I. Kulish.

Includes bibliographical references and index.
ISBN 978-1-926895-41-3
1. Polymers. 2. Polymers--Research. 3. Polymers--Industrial applications. 4. Materials science. I. Zaikov, G. E. (Gennadiĭ Efremovich), 1935- II. Stoyanov, Oleg V III. Kulish, Elena I

TA455.P58P76 2013 620.1'92 C2012-906367-3

Apple Academic Press also publishes its books in a variety of electronic formats. Some content that appears in print may not be available in electronic format. For information about Apple Academic Press products, visit our website at **www.appleacademicpress.com**

PROGRESS IN POLYMER MATERIALS SCIENCE

Research, Development and Applications

Edited by

Gennady E. Zaikov, DSc, Oleg V. Stoyanov, DSc and Elena I. Kulish, DSc

Apple Academic Press

TORONTO NEW JERSEY

Contents

List of Contributors .... *ix*
List of Abbreviations .... *xiii*
Preface .... *xvii*

1. **Update on Photoelectric Characteristics** **1**
D. J. Liaw, Y. C. Huang, B. M. Rumyantsev, V. I. Berendyaev, V. P. Zubov, A. A. Olkhov, G. V. Fetisov, G. E. Zaikov, and A. A. Ischenko

2. **Changes in the Polymer Molecular Mass During Hydrolysis** **15**
E. I. Kulish, V. V. Chernova, V. P. Volodina, S. V. Kolesov, and G. E. Zaikov

3. **Update on Enzymatic Destruction** **21**
E. I. Kulish, V. V. Chernova, V. P. Volodina, S. V. Kolesov, and G. E. Zaikov

4. **New Type of Bioadditives to Motor Fuel** **25**
V. B. Volieva, S. V. Usachev, I. S. Belostotskaya, N. L.Komissarova, A. V. Malkova, A. I. Nekhaev, A. L. Maksimov, G. G. Makarov, and S. D. Varfolomeev

5 **Interrelation of Viscoelastic and Electromagnetic Properties of Densely Cross-linked Polymers: Part I** **35**
T. R. Deberdeev, N. V. Ulitin, R. Ya. Deberdeev, and G. E. Zaikov

6. **Interrelation of Viscoelastic and Electromagnetic Properties of Densely Cross-linked Polymers: Part II** **49**
N. V. Ulitin, T. R. Deberdeev, R. Ya. Deberdeev, and G. E. Zaikov

7. **Practical Hints on Recovery of Strain Electromagnetic Susceptibility Relaxation** **61**
T. R. Deberdeev, R. Ya. Deberdeev, N. V. Ulitin, and G. E. Zaikov

8. **Susceptibility of Cross-linked Polymers: A Numerical Approach** **69**
T. R. Deberdeev, N. V. Ulitin, R. Ya. Deberdeev, and G. E. Zaikov

9. **Update on Cross-linked Polymers with Nanoscale Cross-site Chains** **75**
N. V. Ulitin, T. R. Deberdeev, R. Ya. Deberdeev, and G. E. Zaikov

10. **Role of Polymers in Technologies and Environment Protection** **83**
M. D. Goldfein and N. V. Kozhevnikov

11. **Update on Quantum-chemical Calculation** **95**
V. A. Babkin, D. S. Zakharov, and G. E. Zaikov

12. **Some Aspects of Silver Nanoparticles** **103**
N. I. Naumkina, O. V. Mikhailov, and T. Z. Lygina

13. **Mechanics of Textile Performance: Part I** 117
A. K. Haghi and G. E. Zaikov

14. **Mechanics of Textile Performance: Part II** 131
A .K. Haghi and G. E. Zaikov

15. **Some New Aspects of Polymeric Nanofibers: Part I** 141
A. K. Haghi and G. E. Zaikov

16. **Some new Aspects of Polymeric Nanofibers: Part II** 149
A. K. Haghi and G. E. Zaikov

17. **Mathematical Model of Nanofragment Cross-linked Polymers** 169
N. V. Ulitin, R. Ya. Deberdeev, E. V. Samarin, T. R. Deberdeev, I. I. Nasyrov, and G. E. Zaikov

18. **Internal Mechanical Properties of the Plain Woven Polymeric Fabrics: An Analytical Approach** 181
A. K. Haghi and G. E. Zaikov

19. **Effect of Insufficient Watering and Melamine Salt of Bis(Oxymethyl)phosphonic Acid (Melaphen) on the Fatty Acid Composition** 187
G. E. Zaikov, I. V. Zhigacheva, T. B. Durlakova, T. A. Misharina, M. B. Terenina, N. I. Krikunova, I. P. Generozova, A. P. Shugaev, and S. G. Fattakhov

20. **Meniscus Height Changing in the Capillary and Ultraviolet Spectra of Ascorbic Acid and Paracetamol High-diluted Solutions** 199
F. F. Niyazy, N. V. Kuvardin, E. A. Fatianova, and G. E. Zaikov

21. **Some Aspects of Chemical Physics in Polymers** 209
G. E. Zaikov, M. D. Goldfein, and N. V. Kozhevnikov

22. **A Note on Antioxidant Properties** 221
G. E. Zaikov, N. N. Sazhina, A. E. Ordyan, and V. M. Misin

23. **Antioxidant Activity of Various Biological Objects** 241
G. E. Zaikov, N. N. Sazhina, E. I. Korotkova, and V. M. Misin

Index 257

List of Contributors

V. A. Babkin
403343 SF VolgSABU, c. Mikhailovka, region Volgograds, Michurina 21.

I. S. Belostotskaya
Emanuel Institute of Biochemical Physics, Russian Academy of Sciences, Kosygin str. 4 Moscow-119334 Russia.
Topchiev Institute of Petrochemical Synthesis, Russian Academy of Sciences, Leninsky pr. 29, Moscow-119991 Russia.

V. I. Berendyaev
Institute of Chemical Physics, RAS, Moscow, Russia.

V. V. Chernova
Bashkir State University 32 Zaki Validy Str., Ufa, the Republic of Bashkortostan-450074, Russia.

R. Ya. Deberdeev
Kazan National Research Technological University.

T. R. Deberdeev
Kazan National Research Technological University

T. B. Durlakova
Emanual Institute of Biochemical Physics of the Russian Academy of Science.

E. A. Fatianova
Department General and Inorganic chemistry, South-West State University.

S. G. Fattakhov
Emanual Institute of Biochemical Physics of the Russian Academy of Science.
Arbuzov Institute of Organic and Physical Chemistry of the Russian Academy of Science.

G. V. Fetisov
Moscow Lomonosov State University, Chemistry Department, Moscow, Russia.

I. P. Generozova
Emanual Institute of Biochemical Physics of the Russian Academy of Science.
Timiryazev Institute of Plant Physiology of the Russian Academy of Science.

M. D. Goldfein
Saratov State University named after N. G. Chernyshevsky.

A. K. Haghi
University of Guilan, Rasht, Iran.

Y. C. Huang
Department of Chemical Engineering, National Taiwan University of Science and Technology, Taipei-10607, Taiwan.

A. A. Ischenko
Moscow Lomonosov State University of Fine Chemical Technology, Moscow, Russia.

S. V. Kolesov
The Institute of Organic Chemistry of the Ufa Scientific Center of the Russian Academy of Science 71 October Prospect, Ufa, the Republic of Bashkortostan-450054, Russia.

N. L. Komissarova
Emanuel Institute of Biochemical Physics, Russian Academy of Sciences, Kosygin str.4 Moscow-119334 Russia.
Topchiev Institute of Petrochemical Synthesis, Russian Academy of Sciences, Leninsky pr.29, Moscow-119991 Russia.

E. I. Korotkova
Tomsk Polytechnic University, 30 Lenin Street, 634050, Tomsk, Russia.

N. V. Kozhevnikov
Saratov State University named after N. G. Chernyshevsky.

N. I. Krikunova
Emanual Institute of Biochemical Physics of the Russian Academy of Science.

N. V. Kuvardin
Department "General and Inorganic chemistry", South-West State University.

E. I. Kulish
The Bashkir State University 32 Zaki Validy Str., Ufa, the Republic of Bashkortostan-450074, Russia.

J. Liaw
Department of Chemical Engineering, National Taiwan University of Science and Technology, Taipei-10607, Taiwan.

T. Z. Lygina
Central Scientific Research Institute of Geology Non-Ore Minerals, Zinin Street 4, 420097 Kazan, Russia.

G. G. Makarov
Emanuel Institute of Biochemical Physics, Russian Academy of Sciences,Kosygin str.4 Moscow-119334 Russia.
Topchiev Institute of Petrochemical Synthesis, Russian Academy of Sciences, Leninsky pr.29, Moscow-119991 Russia.

A. L. Maksimov
Emanuel Institute of Biochemical Physics, Russian Academy of Sciences,Kosygin str.4 Moscow-119334 Russia.
Topchiev Institute of Petrochemical Synthesis, Russian Academy of Sciences, Leninsky pr.29, Moscow-119991 Russia.

A. V. Malkova
Emanuel Institute of Biochemical Physics, Russian Academy of Sciences,Kosygin str.4 Moscow-119334 Russia.
Topchiev Institute of Petrochemical Synthesis, Russian Academy of Sciences, Leninsky pr.29, Moscow-119991 Russia.

O. V. Mikhailov
Kazan National Research Technological University, K. Marx Street 68, 420015 Kazan, Russia.

T. A. Misharina
Emanual Institute of Biochemical Physics of the Russian Academy of Science.

V. M. Misin
Emanuel Institute of Biochemical Physics Russian Academy of Sciences, 4 Kosygin Street-119334 Moscow, Russia.

I. I. Nasyrov,
Kazan National Research Technological University.

N. I. Naumkina
Central Scientific-Research Institute of Geology Non-ore Minerals, Zinin Street 4, 420097 Kazan, Russia.

A. I. Nekhaev
Emanuel Institute of Biochemical Physics, Russian Academy of Sciences, Kosygin str.4 Moscow-119334 Russia.
Topchiev Institute of Petrochemical Synthesis, Russian Academy of Sciences, Leninsky pr.29, Moscow-119991 Russia.

F. F. Niyazy
Department "General and Inorganic chemistry", South-West State University.

A. A. Olkhov
Moscow Lomonosov State University of Fine Chemical Technology, Moscow, Russia.

A. E. Ordyan
Emanuel Institute of Biochemical Physics Russian Academy of Sciences, 4 Kosygin Street, 119334 Moscow, Russia

B. M. Rumyantsev
Institute of Chemical Physics, RAS, Moscow, Russia.

E. V. Samarin,
Kazan National Research Technological University.

N. N. Sazhina
Emanuel Institute of Biochemical Physics Russian Academy of Sciences, 4 Kosygin Street, 119334 Moscow, Russia

A. P. Shugaev
Emanual Institute of Biochemical Physics of the Russian Academy of Science.
Timiryazev Institute of Plant Physiology of the Russian Academy of Science.

N. V. Ulitin
Kazan National Research Technological University.

S. V. Usachev
Emanuel Institute of Biochemical Physics, Russian Academy of Sciences,Kosygin str.4 Moscow-119334 Russia.
Topchiev Institute of Petrochemical Synthesis, Russian Academy of Sciences, Leninsky pr.29, Moscow-119991 Russia.

S. D. Varfolomeev
Emanuel Institute of Biochemical Physics, Russian Academy of Sciences,Kosygin str.4 Moscow-119334 Russia.
Topchiev Institute of Petrochemical Synthesis, Russian Academy of Sciences, Leninsky pr.29, Moscow-119991 Russia.

V. B. Volieva
Emanuel Institute of Biochemical Physics, Russian Academy of Sciences,Kosygin str.4 Moscow-119334 Russia.
Topchiev Institute of Petrochemical Synthesis, Russian Academy of Sciences, Leninsky pr.29, Moscow-119991 Russia.

V. P. Volodina
The Institute of Organic Chemistry of the Ufa Scientific Center of the Russian Academy of Science 71 October Prospect, Ufa, the Republic of Bashkortostan-450054, Russia.

G. E. Zaikov
Kazan National Research Technological University.
Institute of Biochemical Physics, Russian Academy of Sciences, 4 Kosygin Street, 117334 Moscow, Russia.
Saratov State University named after N.G. Chernyshevsky.

Emanuel Institute of Biochemical Physics Russian Academy of Sciences, 4 Kosygin Street-119334 Moscow, Russia
Department General and Inorganic chemistry, South-West State University.

D. S. Zakharov
403343 SF VolgSABU, c. Mikhailovka, region Volgograds. Michurina 21.

I. V. Zhigacheva
Emanual Institute of Biochemical Physics of the Russian Academy of Science.

V. P. Zubov
Moscow Lomonosov State University of Fine Chemical Technology, Moscow, Russia.

List of Abbreviations

NOMENCLATURES

Ef = Fabric modulus in warp direction (N/mm^2)
Ey = Modulus of opposed yarn (N/tex)
Eyf=Modified modulus of opposed yarn (N/tex)
F = Pullout force (N)
FS = Static friction force (N)
FD = Dynamic friction force (N)
F = Normalized pullout force per number of crossovers (N)
FN = Normal load at each crossover (N)
N = Number of crossovers in direction of the pulled yarn
M = Number of crossovers in opposite direction of the pulled yarn
Tf = Lateral force in fabric length direction (N)
Ty = Force propagated in the opposed yarn direction (N)
Tyf = Corrected force propagated in the opposed yarn direction (N)
H = Fabric height before pulling (mm)
h′ = Fabric height after pulling (mm)
L = Fabric length before pulling (mm)
L′ = Fabric length after pulling (mm)
p = Distance between two crossovers in opposed direction before pulling (mm)
p′ = Distance between two crossovers in opposed direction during yarn pulling (mm)
t = Fabric thickness before pulling (mm)
t′ = Fabric thickness after pulling (mm)
x = Length of yarns between two crossovers in opposed direction before pulling (mm)
x′ = Length of yarns between two crossovers in opposed direction during pulling (mm)
V = Sample volume before pulling (mm3)
V′ = Sample volume during pulling (mm3)
α = Fabric deformation angle
Δ = Displacement of fabric in the direction of pulled yarn
ΔS = Static displacement of fabric in the direction of pulled yarn
ΔD = Dynamic displacement of the fabric in the direction of the pulled yarn
εy = Yarn strain between two crossovers in opposed yarn direction (lateral strain)
μ = Yarn-to-yarn friction coefficient
ρ = Linear density of the opposed yarn (tex)
θ = Weave angle in the opposed direction, before pulling
θ' = Weave angle in the opposed direction, during pulling
Subscript S = Defines the parameters in maximum static situation
Subscript D = Defines the parameters in dynamic situations

AA	Acrylamide
ACN	Acetonitrile
AFD	Average fiber diameter
AH	1-aminohexane
AIBN	Azoisobutyronitrile
AlA	Allylacrylate
AN	Acrylonitrile
ANOVA	Analysis of variance
AO	Antioxidants
AOEM	Acryloxyethylmaleate
APS	Ammonium persulfate
ASSSC	Aqueous solution of sodium sulfocyanide
BA	Butylacrylate
BAS	Biological active substances
BDA	4,4′-bitetracarboxylic dianhydride
BET	Brunauer-emmett-teller
BP	Benzoyl peroxide
BTDA	4,4′-benzophenone tetracarboxylic dianhydride
CA	Contact angle
CCD	Central composite design
CHT	Chitosan
CTC	Charge transfer complexes
DGEBA	Diglycide ether of bisphenol-A
DMAc	*N,N*-Dimethylacetamide
DMF	N-N, dimethylformamide
DMFA	Dimethyl formamide
DPPF	1,1'-bis(diphenylphosphino)ferrocene
EA	Ethylacrylate
EPG	Electrophotographic
ER O2	Oxygen electroreduction
FAMEs	Fatty acid methyl esters
FATD	Field assisted thermo dissociation
FPU	Foam polyurethane
GA	Gallic acid
GC-MS	Chromato-mass-spectrometry
HEPA	High efficiency particulate air
HMDA	Hexamethylenediamine
HQ	Hydroquinone
IA	Itaconic acid
IP	Ion pairs
IS	Stearates of iron
ITO	Indium tin oxide,
IW	Insufficient watering
LPO	Lipid peroxidation
MA	Methylacrylate

MAA	Methacrylic acid
MAS	Methallyl sulfonate
MF	Melaphen
MFE	Mercury film electrode
MMA	Methylmethacrylate
NMP	N-methyl-2-pyrrolidinone
ODPA	4,4′-oxydiphthalic anhydride
PAA	polyacrylamide
PAN	Polyacrylonitrile
PES	Photoelectric sensitivity
PI	Polyimides
PLA	Polylacticacid
PSC	Photostimulated current
RCR	Respiratory control rate
RFBR	Russian Foundation of Basic Researches
RH	Relative humidity
ROS	Reactive oxygen species
RSM	Response surface methodology
SEM	Scanning electron microscope
SSD	Supersmall doses
SOD	Superoxide dismutase
TCQM	Tetracyanoquinodimethane
VA	Vinylacetate
VA-grams	Voltammograms
XRD	X-ray powder diffraction
ΔON	Octane number

Preface

This book, with chapters by the editors and other experts in the field of polymer science, covers a broad selection of important research advances in the field, including an update on photoelectric characteristics, a study on the changes in the polymer molecular mass during hydrolysis, an update on enzymatic destruction, a study on a new type of bioadditive for motor fuel, an exploration of the interrelation of viscoelastic and electromagnetic properties of densely cross-linked polymers, and much more.

We carefully selected papers on many important topics, such as a paper that offers practical hints on the recovery of strain electromagnetic susceptibility relaxation, a numerical approach to the susceptibility of cross-linked polymers, an update on cross-linked polymers with nanoscale cross-site chains, a paper addressing the role of polymers in technologies and environment protection, an update on quantum-chemical calculation, and a paper that covers some aspects of silver nanoparticles. Also included are chapters that discuss the problems of mechanics of textile performance, new aspects of polymeric nanofibers, a mathematical model of nanofragment cross-linked polymers, and much more.

Editors and contributors hope that you will find the information provided here enlightening and useful, and we will be happy to receive from readers their comments and insights that may be helpful to us in our future research.

— **Gennady E. Zaikov, DSc**

1 Update on Photoelectric Characteristics

D. J. Liaw, Y. C. Huang, B. M. Rumyantsev, V. I. Berendyaev, V. P. Zubov, A. A. Olkhov, G. V. Fetisov, G. E. Zaikov, and A. A. Ischenko

CONTENTS

1.1 Introduction 1
1.2 Experimental 2
  1.2.1 Synthesis of Homopolyimides 2
  1.2.2 Electrophotographic Study 3
1.3 Discussion and Results 4
1.4 Conclusion 8
Appendix 1 8
  Materials 8
Appendix 2 9
  Monomer Synthesis 9
Appendix 3 12
Acknowledgments 12
Keywords 12
References 13

1.1 INTRODUCTION

In the last two decades, p-conjugated polymers have attracted considerable interests because of their potential applications in electrochromics, [1-4] light emitting diodes, [5-9] organic thin film transistors, [10-14] photovoltaic's, [15-18] and polymer memories [19-20]. Fluorene and its analogous derivatives have drawn much attention of optoelectronics because they generally have good solubility, high luminescent efficiency, and very good charge-transfer mobility in both neutral and doped states [21-24]. However, it is also known that they have drawbacks such as unsatisfied thermal stability and excimers formation in the solid state [25, 26]. The applications for electrochromic conjugated polymers are quite diverse due to several favorable properties of these materials, like stable oxidation state, fast switching times, and excellent switching reproducibility [27]. These excellent properties led to the development of many technological applications such as self-darkening rear view mirrors, adjustably darkening windows, large-scale electrochromic screens, and chameleon materials [28-30]. Electron-rich triarylamines are known to be easily oxidized to form stable polarons and the oxidation process is always associated with a noticeable change of the coloration. Furthermore, triarylamine-based polymers are not only widely used as Hole transport layer in electroluminescent diodes but also show interesting electrochromic behavior [31-33]. A conjugated polymer derived from the Suzuki coupling reaction with a fluorene derivative was prepared, and its general properties such as thermal and optical properties as well as electrochemical and electrochromic property were investigated and discussed earlier [34].

The high photoelectric characteristics of polyimides (PIs) based on triphenylamine and its derivatives [35], and also their composites with the organic and inorganic semiconductors [36, 37] (photoelectric sensitivity (PES) S, the quantum yield of the charge carrier photogeneration β, their drift length l_D, collection coefficient C(Z)) are the basis of their applications in a whole series of optoelectronic devices: photovoltaic and electroluminescent cells [38, 39], photodetectors [40], and organic phototransistors [41, 42]. In the present work, the photoelectric characteristics of the newly synthesized PI [34, 43] are investigated by using the electrophotographic (EPG) method [42].

1.2 EXPERIMENTAL

1.2.1 Synthesis of Homopolyimides

The PIs were prepared in the procedures similar to one described in Appendix 1. One of the examples is described as follows. To the stirred solution of 0.6 g (0.883 mmol) of diamine (4, Appendix 1) in 5 ml of DMAc, 0.382 g (0.883 mmol) of 6FDA was gradually added. The mixture was stirred at room temperature for 4 hr under nitrogen atmosphere to form poly(amic acid). Chemical cyclodehydration was carried out by adding equal molar mixture of acetic anhydride and pyridine into the above mentioned poly(amic acid) solution with stirring at room temperature for 1 hr, and then treated at 100°C for 4 hr. The polymer solution was poured into methanol. The precipitate was collected by filtration, washed thoroughly with methanol, and then dried at 100°C under vacuum.

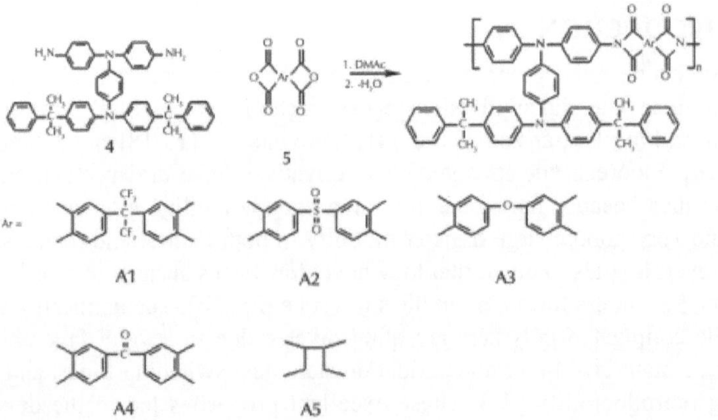

1.2.2 Electrophotographic Study

The EPG method involves the studying of the kinetics of the surface potential dark and photoinduced decay in polymer films, charged in the field of positive or negative corona discharge. The maximum potential of the charging V and the rate of the potential dark decay $(dV/dt)_D$ are determined by the dark conductivity of the filmsthe higher it is, the lower the V value and higher $(dV/dt)_D$ value. Potential photoinduced decay (1/I) (dV/dt) (where the I—intensity of excitation) is determined by the rate of the capacitor photo discharging formed by the ionic contact (aeroions on the film surface) as one electrode and glass conducting support (Indium Tin Oxide, (ITO)) as the second electrode with the induced charge of opposite sign on it. The potential photo discharging rate depends on the effective charge carrier photogeneration quantum yield in the film volume (xerographic output, β_{eff}), carrier collection efficiency on the electrodes C(Z) and the portion of the absorbed exciting light, P:

$$(1/I)\,[dV/dt - (dV/dt)_D)] = (ed/\varepsilon\varepsilon_0)\,\beta_{eff}\,P \qquad (1)$$

where $\beta_{eff} = \beta C(Z)$; C(Z)—charge carrier collection efficiency, $Z = \mu V\tau/d^2$ – ratio of the carrier drift length ($l_D = \mu V\tau/d$) to the film thickness d, μ = carrier drift mobility, τ = their lifetime, e = electron charge.

For the strong and weak absorption, the function C(Z) will be as follows:

for $Z < 1$, $C(Z) = Z$, and for $Z > 1$, $C(Z) = 1$ (strong absorption) and $C(Z) = 1/2$ for weak absorption.. For the polymers, β and C(Z) values usually are strongly depend on the field strength E = V/d [39]. The accuracy of the β measurement was determined by the accuracy of I, P, and film thickness d measurements and was estimated ~20%. The PES S (m^2/J) is defined as the reciprocal value of the half decay exposure time $t_{1/2}$ of the initial charging potential V:

$$S = (It_{1/2})^{-1} = (\beta_{eff}\,Pde)/[E(h\upsilon)V\varepsilon\varepsilon_0] \qquad (2)$$

where E(hυ) = excitation photon energy. The accuracy of the S value measurement was estimated ~10% and it was determined by the accuracy of excitation intensity I measurements. Thus, the EPG method makes it possible to obtain the following photoelectric characteristics of polymer samples: PES, the carrier photogeneration quantum yield (1), and from the field dependence of $\beta_{eff}(E)$ it is possible to estimate carrier drift length: at the field strength E_0, for which the change of C(Z) dependence from C(Z) = Z to C(Z) = constant is observed, drift length is equal to the film thickness, $l_D = \mu E_{0\tau} = d$.

Experimental setup makes it possible to determine both the optical density of the sample D_λ under monochromatic or integral excitation, and to measure also the influence of the ionic contact field on it. Knowing optical density, it is possible to estimate the portion of the absorbed excitation light energy:

$$P = 1 - \exp[-(D - D_0)] \qquad (3)$$

where D is the film optical density and D_0 = equivalent optical density, caused by the light reflection from the front and rear sample surfaces as well as by light scattering. The sign of major carriers can be determined *via* the comparison of PES values for the positive (S^+) and negative (S^-) corona charging of free surface under the heterogeneous excitation by the strongly absorbed UV light: with $S^+ > S^-$ major carriers are holes, with $S^- > S^+$–electrons. The PI films of 3 μm thickness were prepared by the cast of polymer solution in the chlorine containing solvents onto the conducting ITO glass supports and the subsequent drying under the ambient conditions at 50100°C. The PIs under study possess good solubility and excellent film forming properties.

1.3 DISCUSSION AND RESULTS

Study of the PES of the Pi films and Its connection with the charge transfer complex formation. The PES of the PI films is observed and the charge carrier photogeneration quantum yield is determined for the films of the new class of PI based on N, N', N'', and N''' substituted para phenylenediamine (electron donor fragment D) and dianhydrides of aromatic tetracarboxylic acids (electron acceptor fragment A). The PI series denoted as PI A1 to PI A5, see list of samples. A study of PES spectral dependence S(λ) shows that the highest sensitivity (up to 30 m^2/J) is observed at the UV region (200400 nm). In the visible region (400 700 nm) there is a PES band which collapses to the long wave edge (Figure1).

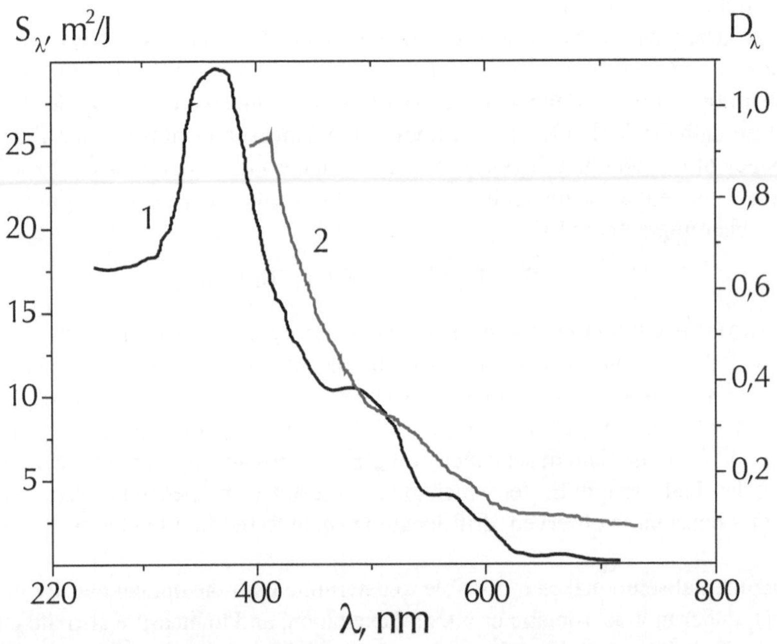

FIGURE 1 Spectra of the PES Sλ(1): (+ charging, V = 3036 V; the number of cycles N > 10) and the optical density Dλ(2) for the film PI A2 (thickness d = 3 μm).

Comparison of the PES spectral dependence with the absorption spectrum of the films evidences that PES in the visible region is due to the formation of weak electronic Donor Acceptor (DA) Charge Transfer Complexes (CTC) with absorption maxima in the region of 400600 nm [44]. The maxima and band absorption edge of the CTC are determined. The most clearly expressed CTC bands are observed for PI A2 and PI A3 films with flat absorption maxima in the 500–560 nm region. For PI A1, PI A4, and PI A5 films CTC bands are essentially weaker with flat maxima shifted to short wavelengths, 420–480 nm. The energy position of the long wavelength absorption band edge of the CTC and PES is determined, which is an analog of the band gap for semiconductors, E_g, that allows to estimate the relative affinity energy values for the acceptor fragments E_A (at the same values of the ionization potential of the donor fragments, $I_D = 7.0$, $P_h = 5.5$ eV, and P_h = polarization energy of holes) from the expression:

$$E_g = I_D - E_A - (P_h + P_e) \tag{4}$$

where P_e = polarization energy for electrons.

TABLE 1 E_g and $(E_A + P_e)$ values (in eV) for PI A1 – PI A5 film samples; $P_e = 1.5$ eV

PI A1	$E_g = 2.2^*$ $(E_A + P_e) = 3.3^*$
PI A2	1.9 3.6
PI A3	2.0 3.5
PI A4	2.6 2.9
PI A5	2.4 3.1

* Estimated average uncertainty for E_g and $(E_A + P_e)$ values is 0.1 eV.

As can be seen from Table 1, the highest E_A values are observed for PI A2 and PI A3 acceptor fragments (3.6 and 3.5 eV, respectively) which characterized by the lowest E_g (1.9 and 2.0 eV) and the most pronounced CTC band as well as high PES in the visible region (up to 520 m^2/J). The PI A1, PI A4 ,and PI A5 films possess the lower E_A values, weaker CTC bands and lower PES (of about 14 m^2/J).

These distinct spectral peaks (in the region of 440–480 nm, 540–560 nm, and 640–660 nm) are registered for PI A1, PI A3, PI A5 (absorption spectrum), and PI A2 (PES spectrum, Figure1). They are ascribed in this work to the formation and accumulation of the stabilized cation-radicals (D^+) (and perhaps anion radicals (A^-)) of polymer fragments arising in the PI as a result of the dark and photo processes [45]. Some evidence of this assumption is the PES found in the red region ($\lambda > 600$ nm, outside the CTC band) with a weak maximum in the absorption band of triphenylamine type cation radical (640–660 nm) due to its photostimulation [45].

The photo generation quantum yield for the UV (PI A1-PI A5) and visible spectrum (PI A1) is determined. On varying charging potential V a nonlinear field dependence of photo generation quantum yield $\beta(E) \sim E^n$ (Figure 2) is revealed. The exponent n increases with increasing excitation wavelength λ from $n \sim 1.2$ to $n \sim 1.8$ on changing λ from 257 to 547 nm, indicating that the photo generation occurs *via* the field assisted thermo dissociation (FATD) of ion pairs (IP), kinetically coupled with the excited states of the CTC: FATD

$$CTC + h\upsilon_{CT} \rightarrow CTC^* \leftrightarrow [D^+ A^-] \rightarrow \text{carriers (holes)} \quad (5)$$

E, T

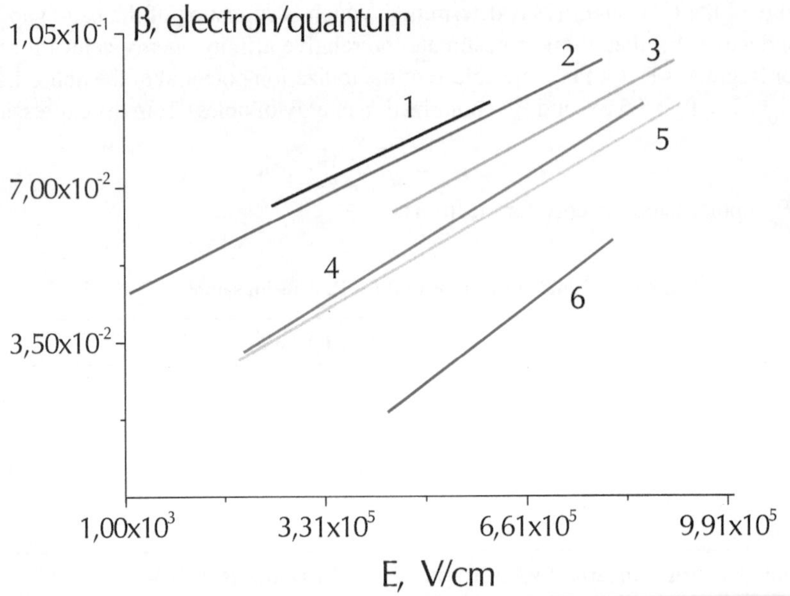

FIGURE 2 Field dependence of charge carrier photogeneration quantum yield β(E) for the PI A2 (1), PI A1 (2, 46), and PI A3 (3) films, under excitation by monochromatic light: 257 nm (2) and 365 nm (1, 3, and 4); 436 nm (5) and 547 nm (6). Positive charging (N > 10), the field changed by the time variation of corona. β in electrons/quanta, E in V/cm.

The highest β values in the UV region (up to 0.1 in a field $E = 5.7 \cdot 10^5$ V/cm) are obtained for films PI A3 and PI A2 ($\beta = 0.02$, $E = 10^5$ V/cm). Using the geminate recombination Onsager model [46, 47] to interpret the field dependence of β(E), it is possible to determine ion pair parametersthe initial yield φ_0 and initial separation r_0. For films PI A1 with increasing excitation wavelength (from 257 to 547 nm) value of r_0 is reduced from 3.64.5 nm to 2.0 nm, and the value of φ_0 increases from 0.2 to 0.7. However, under excitation in the red spectral region (640680 nm, outside of the CTC band), the value r_0 increases to 3.0 nm, which suggests that under photostimulation of stabilized cation radical IP_1 is formed that differs the IP in the Scheme (3). Comparison of the field dependences of S and β allows to conclude that for the PI films drift length of the generated carriers (holes) $l_D > d$ (3 μm) for $E > 10^5$ V/cm and hence to estimate $\mu\tau > 3 \cdot 10^{-9}$ cm²/V. Major carriers in the studied PI holes since under free surface excitation by strongly absorbed UV light, $S^+ > S^-$.

(2) Effect of stable cation-radicals accumulation during repeating charge discharge cycles on the photoelectric characteristics of the PI films; the observation of the photostimulated current (PSC). A strong dependence of the photoelectric characteristics of the samples (the potential of charging, PES in the red region, S_{red}) on the number of

charge discharge cycles N is found: when changing N from 1 to 10, V and S_{red} values significantly (of about several times) increase (Figure 3). In the UV region the rise only V is observed; the growth of PES is very small or completely absent.

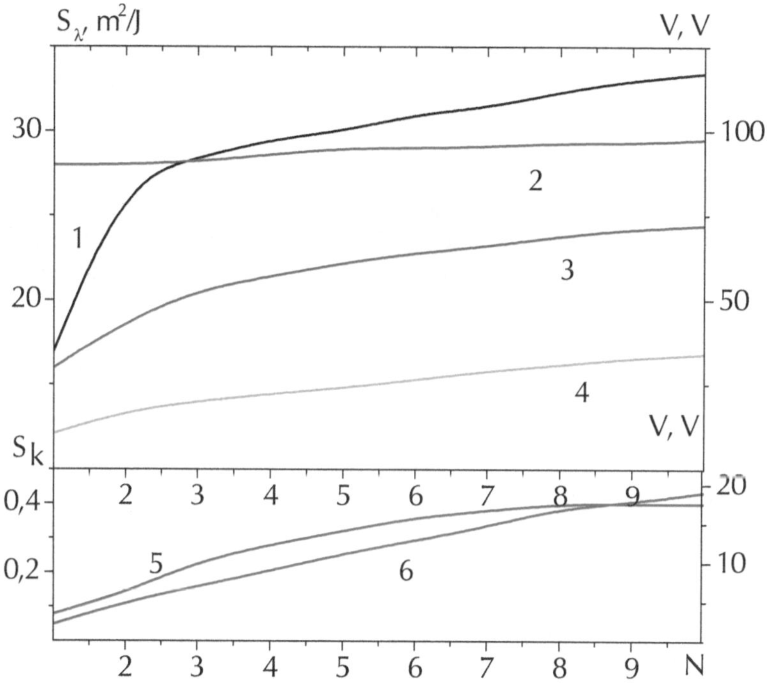

FIGURE 3 PES in the UV (λ = 365 nm) Sλ (1, 2) and red region (λ > 600 nm) Sred (5), as well as the maximal charge potential V (3, 4, 6) versus number of charge-discharge cycles N for films PI A5 (2, 3) and PI A2 (1, 46); + charging.

Usually the absence of surface charging associated with the dark injection of carriers from the electrodes into the bulk of the film (holes in case of positively charged free surface), leading to a sharp increase in dark conductivity. The increase in the value of V at the positive charge indicates blocking of the dark hole injection from the free surface when N > 34. The most probable reason is the appearance in the film bulk near the electrode positively charged layer of stabilized cation-radicals (electrode polarization). The latter are holes (h^+) trapped by deep centers:

$$\text{Dark injection Capture} + \text{Electrode} + D \rightarrow h^+ \text{ (mobile hole)} \rightarrow \text{cation-radical} \quad (6)$$

Accumulation and stabilization of the cation-radicals near the electrode leads not only to stop the dark hole injection but also to a drop of dark conductivity, increase the V value, as well as to the observation of PSCs, which manifest themselves as growth of PES in the red spectral range S_{red}, outside the absorption band of the CTC (absorption of cation-radicals) (Figure 3):

$$\text{FATD Cation-radical} + h\upsilon_1 \rightarrow IP_1 \rightarrow \text{curriers (PSC)} \quad (7)$$

The growth of $S_{red\ with}$ increasing N (Figure 3) is partly due to its field dependence caused by growth of the field strength E = V/d. Therefore, it was specially checked that in the red region the pure field dependence due to the ion pair IP_1 FATD (7) have the form: $\beta \sim E^n$ (n = 1.351.60) and by (2) $S \sim E^{n-1}$, that is weakly dependent on E, so that the growth of S_{red}(N) is partly due to the accumulation of the cation-radicals. Under conditions when V does not depend on N (for N > 10), the V value was changing by varying the time of corona discharge. In some cases (Figure 3) the growth of S_{red} was observed at constant V that points directly to the effect of the cation-radical accumulation.

The observation of PSC indicates a high cation-radical lifetime, $t > (\sigma I)^{-1}$ (σ = absorption cross section). If $\sigma = 10^{-17}$–10^{-16} cm^2 [5], I = 10^{15} cm^{-2} s^{-1}, an estimate of t > 10 s is obtained. It should be noted that the results of this study evidences that there is a range of the cation-radical states from labile (t = 1–10 s) to a fully stable ones (t > 10^3 s) which are involved in the process of photostimulated generation. Similar stabilized cation-radicals and related PSCs for the PI based on substituted triphenylamines resulting from irreversible photochemical transformation of free radical type with the halogen hydrocarbons are observed in [45].

1.4 CONCLUSION

A novel conjugated polymers, PI based on N, N', N", N"' substituted paraphenylenediamine and dianhydrides of aromatic tetracarboxylic acids, was successfully prepared *via* Suzuki coupling reaction. The polymer exhibits excellent solubility in common organic solvent, and has high thermal stability such as T_{d10} at 453°C in nitrogen atmosphere and T_g at about 140°C.

The PES of the PI films (3 μm thickness) is observed in the UV, and visible spectral regions, due to the interactions with charge transfer between donor and acceptor fragments of the PI chains (formation of CTC). Study of the photogeneration quantum yield field dependence gives the evidence that the photogeneration mechanism is a FATD of radical ion pairs kinetically associated with the excited CTC.

The second important mechanism of photogeneration is photostimulation of long-lived stable cation-radicals of the donor PI fragments, representing the hole (major carriers) captured by deep centers (PSC). Accumulation of the cation-radicals in the dark and photo processes leads to the dependence of photovoltaic characteristics on the number of charge discharge cycles of the sample.

APPENDIX 1

Materials

N,N,-bis(4-aminohenyl)-*N′,N′*-bis[4-(2-phenyl-2-isopropyl)phenyl]-1,4-phenylene diamine was synthesized by a well known synthetic route starting from bis(4-*tert*-butylphenyl)amine and p-fluoronitrobenzene as shown in Scheme 1 [43]. The synthetic details and the characterization data of this diaminemonomer have been described in [43]. Bis[4-(2-phenyl-2-isopropyl)phenyl] amine (OUCHI SHINKO), 4-fluoro-nitro- benzene (ACROS), cesium fluoride (ACROS), sodium hydride (95%;

dry; ALDRICH), 10% Pd/C (MERCK), and hydrazine monohydrate (MERCK) were used as received. *N,N*-Dimethylacetamide (DMAc; MERCK), dimethyl sulfoxide (DMSO), *N*-methyl-2-pyrrolidinone (NMP) (MERCK), and pyridine (MERCK), were dried over calcium hydride overnight, distilled under reduced pressure, and stored over 4 Å molecular sieves in a sealed bottle. Iodobenzene, bis(dibenzylideneacetone) palladium [Pd(dba)$_2$], 1,1'-bis(diphenylphosphino)ferrocene (DPPF), sodium *tert*-butoxide were purchased from ACROS. Commercially available aromatic tetracarboxylic dianhydrides such as 4,4′-hexafluoroisopropylidenediphathalic dianhydride (A1; CHRISKEV) (6FDA), 3,3′,4,4′-diphenyl sulfone-tetracarboxylic dianhydride (A2; TCI) (DSDA), 4,4′-oxydiphthalic anhydride (A3; TCI) (ODPA), 3,3′,4,4′-benzophenone tetracarboxylic dianhydride (A4; CHRISKEV) (BTDA) and 3,3′,4,4′-bitetracarboxylic dianhydride (A5; CHRISKEV) (BDA), and were purified by vacuum sublimation.

APPENDIX 2

Monomer synthesis [35]

Synthesis of bis[4-(2-phenyl-2-isopropyl)phenyl]-4-nitrophenylamine) (1). In a 500 ml three neck round-bottom flask was placed bis[4-(2-phenyl-2- isopropyl)phenyl] amine (20.0 g, 49 mmol), 4-fluoro-nitrobenzene (5.23 g, 49 mmol), sodium hydroxide (1.18 g, 49 mmol), and 120 ml DMSO. The mixture was heated with stirring at 120°C for 24 hr. The reaction mixture was cooled and then poured into 1 l methanol. The yellow precipitate was collected by filtration and dried under vacuum. The product was purified by silica gel column chromatography (*n*-hexane:dichloromethane = 2:1) to afford nitro compound 1 16.3 g in a 63% yield; mp 150–151°C by DSC (10°C/min).

IR (KBr): 1585, 1342 cm^{-1} (NO$_2$ stretch):

^{1}H NMR: (CDCl$_3$): δ(ppm) = 1.77 (s, 6H, H$_d$); 6.93–6.95 (d, 1H, H$_f$, J = 9.5 Hz); 7.12–7.14 (d, 2H, H$_e$, J = 10.1 Hz); 7.24–7.27 (m, 1H, H$_a$); 7.28–7.30 (d, 2H, H$_d$, J = 10.1 Hz); 7.35–7.36 (d, 2H, H$_b$, J = 5.0 Hz); 7.37–7.38 (d, 2H, H$_c$, J = 5.0 Hz); 8.05–8.08 (d, 1H, H$_g$, J = 15.0 Hz).

^{13}C NMR (CDCl$_3$): δ(ppm)=30.6 (C$_6$), 42.6 (C$_5$), 117.4 (C$_{12}$), 125.3 (C$_{13}$), 125.7 (C$_1$), 125.9 (C$_9$), 126.6 (C$_2$), 128.0 (C$_3$), 128.1 (C$_8$), 139.6 (C$_{14}$), 142.8 (C$_{10}$), 148.2 (C$_7$), 150.0 (C$_4$), 153.4 (C$_{11}$).

Anal. Calcd for $C_{36}H_{34}N_2O_2$: C, 82.10%; H, 6.51%; N, 5.32%. Found: C, 81.67%; H, 6.39%; N, 5.21%.

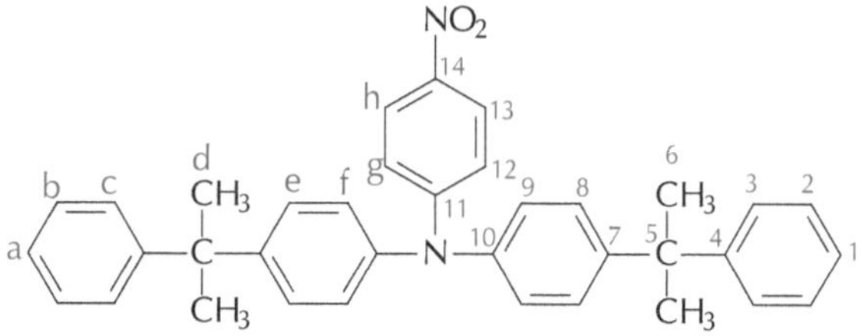

(1)

Synthesis of bis[4-(2-phenyl-2-isopropyl)phenyl]-4-aminophenylamine (2). In a 500 ml three neck round-bottom flask was placed nitro compound 1 (20 g, 38 mmol), Pd/C (0.4 g), ethanol 200 ml. After the addition of 14 ml of hydrazine monohydrate, the solution was stirred at reflux temperature for 12 hr. After the solution was cooled down to room temperature, the solution was filtered to remove the catalyst, and the crude product was recrystallized from ethanol yielded 12.5 g (yield: 66%) of the compound 2, mp 110-114°C by DSC (10°C/min).

IR (KBr): 3432, 3356 cm^{-1} (NH$_2$ stretch):
^{1}H NMR: (DMSO-d_6): δ(ppm)= 1.57 (s, 6H, H$_h$); 5.02 (s, 1H, NH$_2$); 6.56–6.58 (d, 1H, H$_g$, J= 8.5 Hz); 6.76–6.77 (d, 1H, H$_f$, J= 5.0 Hz); 6.77–6.79 (d, 2H, H$_e$, J= 10.0 Hz); 6.98–7.00 (d, 2H, H$_d$, J= 10.0 Hz); 7.09–7.12 (m, H, H$_a$); 7.19–7.20 (d, 2H, H$_b$, J= 5.0 Hz); 7.20–7.21 (d, 2H, H$_c$, J=5.0 Hz).

^{13}C NMR (DMSO-d_6): δ(ppm)= 30.3 (C$_6$), 41.7 (C$_5$), 114.9 (C$_{13}$), 120.73 (C$_9$), 125.3 (C$_1$), 126.2 (C$_3$), 126.9 (C$_8$), 127.8 (C$_2$), 127.9 (C$_{12}$), 135.2 (C$_{14}$), 142.5 (C$_{10}$), 145.4 (C$_7$), 145.9 (C$_{11}$), 150.3 (C$_4$).

E$_{LEM}$. A$_{NAL}$.Calcd. for $C_{36}H_{36}N_2$: C, 87.05%; H, 7.31%; N, 5.64%. Found: C, 86.9%; H, 7.13%; N, 5.61%.

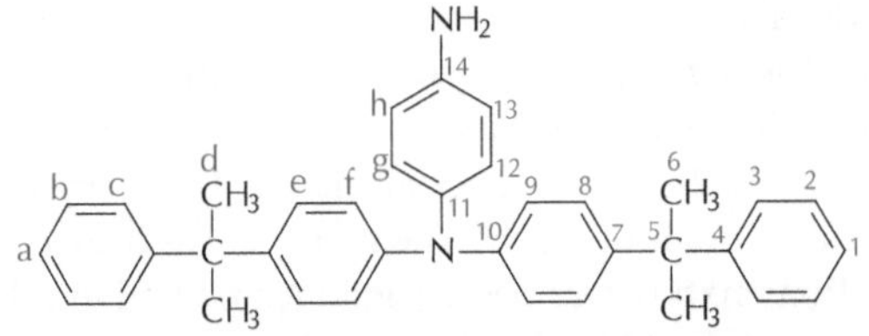

(2)

Synthesis of N,N,-bis(4-nitrophenyl)-N′,N′-bis[4-(2-phenyl-2-isopropyl) phenyl]-1,4-phenylene-diamine (3). In a 250 ml three neck round-bottom flask was placed bis- [4-(2-phenyl- 2-isopropyl)phenyl]-4-aminophenylamine (2) (7.26 g, 14.63 mmol), 4-fluoro-nitro- benzene (4.13 g, 29.27 mmol), cesium fluoride (4.41 g, 29.27 mmol), and 80 ml DMSO. The mixture was heated with stirring at 120°C for 24 hr. The reaction mixture was cooled and then poured into 500 ml methanol. The red precipitate was collected by filtration and dried under vacuum. The product was purified by silica gel column chromatography (n-hexane: dichloromethane = 1:1) to afford dinitro compound **(3)** in a 65% yield; mp 224–225 °C (by DSC; 10°C /min).

IR (KBr): 1580, 1341 cm^{-1}(NO$_2$ stretch):
^{1}H NMR (CDCl$_3$): δ(ppm)= 1.70 (s, 6H, H$_d$); 6.97-6.99 (d, 2H, H$_h$, J= 10.0 Hz); 7.04–7.05 (d, 2H, H$_f$, J= 5.0 Hz); 7.05–7.07 (d, 2H, H$_g$, J= 10.0 Hz); 7.15–7.17 (d, 2H, H$_e$, J = 10.0 Hz); 7.19–7.20 (d, 2H, H$_i$, J= 5.0 Hz); 7.20–7.21 (m, 1H, H$_a$); 7.28–7.29 (d, 2H, H$_c$, J= 5.0 Hz); 7.30–7.31 (d, 2H, H$_b$, J= 5.0 Hz); 8.14–8.17 (d, 2H, H$_j$, J= 15.0 Hz).

^{13}C NMR (CDCl$_3$): δ(ppm)= 30.7 (C$_6$), 42.5 (C$_5$), 121.8 (C$_{16}$), 123.2 (C$_{12}$), 124.3 (C$_9$), 125.4 (C$_{17}$), 125.6 (C$_1$), 126.6 (C$_3$), 127.7 (C$_8$), 127.9 (C$_2$), 128.0 (C$_{13}$), 137.2 (C$_{14}$), 142.3 (C$_{18}$), 144.3 (C$_{10}$), 146.1 (C$_7$), 147.0 (C$_{11}$), 150.4 (C$_4$), 151.7 (C$_{15}$).

E$_{LEM}$. A$_{NAL}$. Calcd. for $C_{48}H_{42}N_4O_4$: C, 78.03 %; H, 5.73%; N, 7.58%. Found: C, 77.57%; H, 5.63%; N, 7.45%.

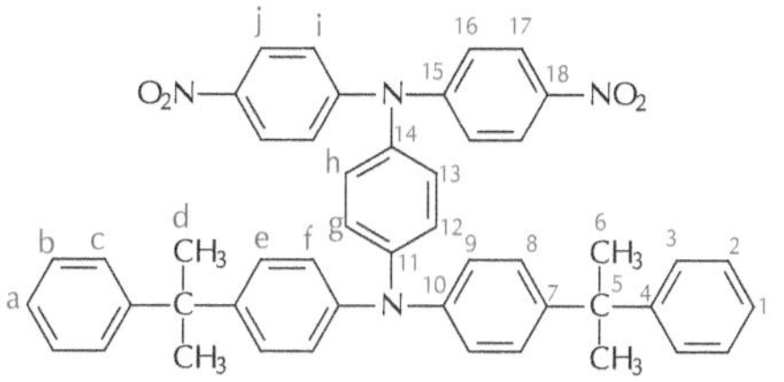

(3)

Synthesis of N,N,-bis(4-aminohenyl)-N′,N′-bis[4-(2-phenyl-2-isopropyl) phenyl]-1,4-phenylene-diamine (4). The dinitro compound (3) (5 g, 6.77 mmol), Pd/C (0.2 g), and 150 ml ethanol were taken in a three-necked flask and hydrazine monohydrate (10 ml) was added drop wise over a period of 30 min at 90°C. Upon completing the addition, the solution was stirred at reflux temperature for 12 hr. After the solution was cooled down to room temperature, the solution was filtered to remove the catalyst, and the crude product was purified by silica gel column chromatography (*n*-hexane: ethyl acetate = 2:1) to afford diamine monomer (4) 2.3 g (yield: 50%), mp 149–151°C by DSC (10°C/min).

IR (KBr): 3445, 3360 cm^{-1}(NH_2 stretch):

^{1}H NMR (DMSO-d_6): δ(ppm)= 1.56 (s, 6H, H_d); 6.53–6.55 (d, 2H, H_f, J = 10.1 Hz); 6.56–6.58 (d, 1H, H_h, J = 10.1 Hz); 6.75–6.77 (d, 1H, H_g, J = 10.1 Hz), 6.77–6.79 (d, 2H, H_i, J = 10.1 Hz,); 6.80-6.82 (d, 2H, H_e, J = 10.1 Hz); 6.99–7.01 (d, 2H, H_j, J = 10.1 Hz); 7.09–7.12 (m, 1H, H_a); 7.19–7.20 (d, 2H, H_c, J = 5.1 Hz); 7.21–7.23 (d, 2H, H_b J = 10.1 Hz).

^{13}C NMR (DMSO-d_6):δ(ppm)= 30.3 (C_6), 41.7 (C_5), 114.8 (C_8), 117.8 (C_{13}), 121.3 (C_{16}), 125.3 (C_1), 126.2 (C_3), 126.7 (C_{12}), 127.0 (C_9), 127.1 (C_{17}), 127.8 (C_2), 136.0 (C_{10}), 137.2 (C_{14}), 143.0 (C_{18}), 145.2 (C_{15}), 145.3 (C_7), 146.2 (C_{11}), 150.2 (C_4).

E_{LEM}. A_{NAL}. Calcd. for $C_{48}H_{46}N_4$: C, 84.92%; H, 6.83%; N, 8.25%. Found: C, 84.11%; H, 6.77%; N, 8.17%.

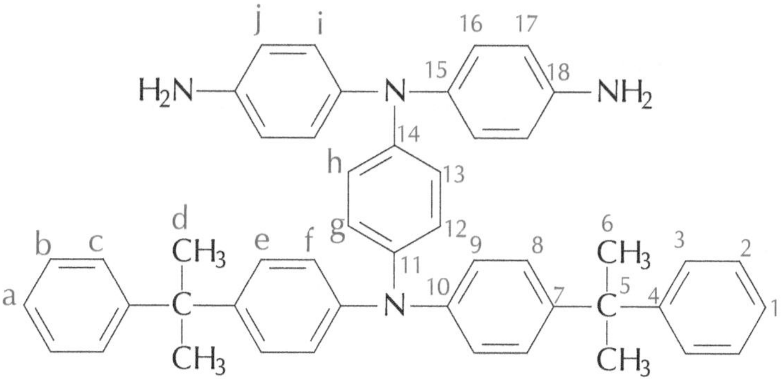

(4)

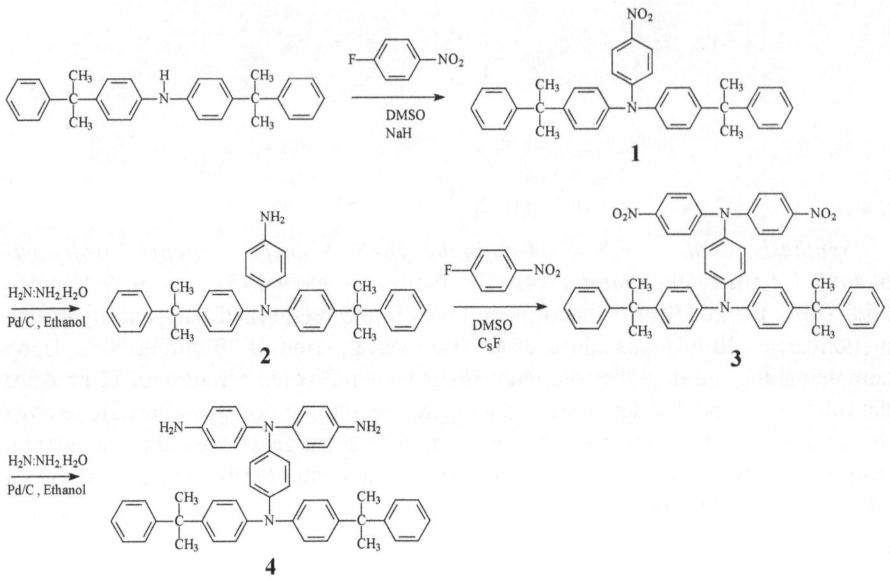

APPENDIX 3

The IR spectrum of A1 (film) exhibited characteristic imide absorption at 1779 (asymmetrical carbonyl stretching), 1726 (symmetrical carbonyl stretching) and 744 cm^{-1} (imide ring deformation).

^{1}H NMR ($CDCl_3$): δ(ppm)= 1.70 (s, 6H, H_d); 7.02–7.04 (d, 2H, H_f); 7.04–7.10 (d, 2H, H_h + H_g); 7.13–7.14 (d, 2H, H_e); 7.16–7.20 (m, 1H, H_a); 7.27–7.28 (d, 2H, H_i); 7.27–7.30 (d, 4H, H_b + H_c); 7.32–7.33 (d, 2H, H_j); 7.88–7.90 (d, 2H, H_l); 7.99 (s, 1H, H_k); 8.05–8.07 (d, 2H, H_m).

E_{LEM}. A_{NAL}.Calcd. for $(C_{69}H_{54}N_5O_4F_6)_n$: C, 74.18%; H, 4.87%; N, 5.01%. Found: C, 73.01%; H, 4.23%; N, 5.02%.

ACKNOWLEDGMENTS

This work supported by RFBR grants 10-02-92000-NNC_a, 11-02-00868-a, 11-02-12041-OFI-m-2011 and grant of the Ministry of Education and Science of the RF 1B-5-343.

KEYWORDS

- **Cation-radicals**
- **Electrophotographic method**
- **Photoelectric sensitivity**
- **Photogeneration**
- **Polyimides**

REFERENCES

1. Beaujuge, P. M., Vasilyeva, S. V., Ellinger, S., McCarley, T. D., and Reynolds, J. R. *Macromolecules*, **42**, 36943706 (2009).
2. Han, F. S., Higuchi, M., and Kurth, D. G. *Adv. Mater.*, **19**, 39283931 (2007).
3. Udum, Y. A., Yildiz, E., Gunbas, G., and Toppare, L. *J. Polym. Sci. Part A: Polym. Chem.*, **46**, 37233731 (2008).
4. Thompson, B. C., Kim, Y. G., McCarley, T. D., and Reynolds, *J. R. J. Am. Chem. Soc.*, **128**, 1271412725 (2006).
5. Michinobu, T., Kumazawa, H., Otsuki, E., Usui, H., and Shigehara, K. *J. Polym. Sci. Part A: Polym. Chem.*, **47**, 38803891 (2009).
6. Elschner, A., Heuer, H. W., Jonas, F., Kirchmeyer, S., Wehrmann, R., and Wussow, K. *Adv. Mater.*, **13**, 18111814 (2001).
7. Winter, A., Friebe, C., Chiper, M., Hager, M. D., and Schubert, U. S. *J. Polym. Sci. Part A: Polym. Chem.*, **47**, 40834098 (2009).
8. Forrest, S. R. *Nature,* **428**, 911918 (2004).
9. Liao, L., Cirpan, A., Chu, Q., Karase, F. E., and Pang, Y. *J. Polym. Sci. Part A: Polym. Chem.*, **45**, 20482058 (2007).
10. Yu, G., Gao, J., Hummelen, J. C., Wudl, F., and Heeger, A. J. *Science*, **270**, 17891791 (1995).
11. Kitamura, M. and Arakawa, Y. *Appl. Phys. Lett.*, **95**, 02350302503(3) (2009).
12. Dimitrakopoulos, C. D. and Malenfant, P. R. L. *Adv. Mater.*, **14**, 99117 (2002).
13. Liu, P., Wu, Y., Pan, H., Li, Y., Gardner, S., Ong, B. S., and Zhu, S. *Chem. Mater.*, **21**, 27272732 (2009).
14. Roberts, M. E., LeMieux, M. C., Sokolov, A. N., and Bao, Z. *Nano Lett.*, **9**, 25262531 (2009).
15. Durben, S., Nickel, D., KruË ger, R. A., Sutherland, T. C., and Baumgartner, T. *J. Polym. Sci. Part A: Polym. Chem.*, **46**, 81798190 (2008).
16. Segura, J. L., Martin, N., and Guldi, D. M. *Chem. Soc. Rev.*, **34**, 3147 (2005).
17. Chang, Y. T., Hsu, S. L., Su, M. H., and Wei, K. H. *Adv. Mater.*, **21**, 20932097 (2009).
18. Zhou, E. J., Tan, Z. A., He, Y. J., Yang, C. H., and Li, Y. F. *J. Polym. Sci. Part A: Polym. Chem.*, **45**, 629638 (2007).
19. Ling, Q. D., Liaw, D. J., Zhuc, C., Chanc, D. S. H., Kang, E. T., and Neoh, K. G. *Prog. Polym. Sci.* **33**, 917978 (2008).
20. Ling, Q. D., Liaw, D. J., Teo, E. Y. H., Zhu, C. Chan, D. S. H., Kang, E. T., and Neoh, K. G. *Polymer*, **48**, 51825201 (2007).
21. Scherf, U. and List, E. J. W. *Adv. Mater.*, **14**, 477487 (2002).
22. Naga, N., Tagaya, N., Noda, H., Imai, T., and Tomoda, H. *J. Polym. Sci. Part A: Polym. Chem.*, **46**, 45134521 (2008).
23. Wang, B., Shen, F., Lu, P., Tang, S., Zhang, W., Pan, S., Liu, M., Liu, L., Qiu, S., and Ma, Y. *J. Polym. Sci. Part A: Polym. Chem.*, **46**, 31203127 (2008).
24. Xu, Y., Guan, R., Jiang, J., Yang, W., Zhen, H., Peng, J., and Cao, Y. *J. Polym. Sci. Part A: Polym. Chem.*, **46**, 453463 (2008).
25. Ranger, M., Rondeau, D., and Leclerc, M. *Macromolecules,* **30**, 76867691 (1997).
26. Janietz, S., Bradley, D. D. C., Grell, M., Giebeler, C., Inbasekaran, M., and Woo, E. P. *Appl. Phys. Lett.*, **73**, 24532455 (1998).
27. Posadas, D. and Florit, M. I. *J. Phys. Chem .B*, **108**, 1547015476 (2004).
28. Sonmez, G. and Wudl, F. *J. Mater. Chem.*, **15**, 2022 (2005).
29. Rosseinsky, D. R. and Montimer, R. *J. Adv. Mater.*, **13**, 783793 (2001).
30. Durmus, A., Gunbas, G. E., Camurlu, P., and Toppare, L. *Chem. Commun.*, **31**, 32463248 (2007).
31. Ogino, K., Kanagae, A., Yamaguchi, R., Sato, H., and Kurtaja,. J. *J. Macromol. Rapid Commun.*, **20**, 103106 (1999).
32. Yu, W. L., Pei, J., Huang, W., and Heeger, A. *J. Chem. Commun.*, **8**, 681682 (2000).
33. Chou, M. Y., Leung, M. K., Su, Y. O., Chiang, S. L., Lin, C. C., Liu, J. H., Kuo, C. K., and Mou, C. Y. *Chem. Mater.*, **16**, 654661 (2001).

34. Wu, H. U., Wang, K. L., Liaw, D. J., Lee, K. R., and Lai, J. Y. *J. Polym. Sci.: Part A: Polym. Chem.*, **48**, 14691476 (2010).
35. Kotov, B. V., Berendyaev, V. I., Rumyantsev, B. M., Bespalov, B. P., Lunina, E. V., and Vasilenko, N. A. Molecular Design of Highly Sensitive Soluble Photoconductive Polyimides. Doklady RAS. *Physical Chemistry*, **367**, 183187 (1999).
36. Rumyantsev, B. M., Berendyaev, V. I., Tsegel'skaya, A. Yu., and Kotov, B. V. Molecular Aggregate Formation and Microphase Segregation Effects on the Photoelectrical and Photovoltaic Properties of Polyimide-Perylenediimide Composite Films. *Mol. Cryst. Liq. Cryst.*, **384**, 6167 (2002).
37. Rumyantsev, B. M., Berendyaev, V. I., Golub, A. S., Lenenko, N. D., Novikov, Yu. N., Krinichnaya, E. P., and Zhuravleva, T. S. Organic-Inorganic Polymer Nanocomposites for Photovoltaics. *J. High Energy Chem.*, **42**(7), 6163 (2008).
38. Mal'tsev, E. I., Berendyaev, V. I., Brusentseva, M. A., Tameev, A. R., Kolesnikov, V. A., Kozlov, A. A., Kotov, B. V., and Vannikov, A. V. Aromatic Polyimides as Efficient Materials for Organic Electroluminescent Devices. *Polym. International.*, **42**, 404 (1997).
39. Muhlbacher, D., Brabec, C. J., Sariciftsi, N. S., Kotov, B. V., Berendyaev, V. I., Rumyantsev, B. M., and Hummelen, J. C. Sensitization of Photoconductive Polyimides for Photovoltaic Applications. *Synth. Metals.*, **121**, 15501551 (2001).
40. Rumyantsev, B. M. and Berendyaev, V. I. Organic Polymer *p-n* Heterostructures for Optoelectronics. *J. Chem. Phys.*, (Russian) (in press).
41. Marjanovich, N., Singh, Th. B., Deunler, G., Gunes, S., Neugebauer, H., Sariciftsi, N. S., Schwodianer, R., and Bauer S. Photoresponse of Organic Field-Effect Transistors Based on Conjugated Polymer Fullerene Blends. *Organic Electronics*, **7**(4), 188194 (2006).
42. Grenishin, S. G. Electrophotographic process. *M. Science*, 374 (1970).
43. Chang, C. H., Wang, K. L., Jiang, J. C., Liaw, D. J., Lee, K. R., Lai, J. Y., and Lai, K. H. *Polymer*, **51**, 44934502 (2010).
44. Rumyantsev, B. M., Berendyaev, V. I., Vasilenko, N. A., Malenko, S. V., and Kotov, B. V. Photogeneration of Charge Carriers in Layers of Soluble Photoconducting Polyimides and Their Sensitization by Dyes. *Polymer Science Ser. A*, **39**(4), 506512 (1997).
45. Rumyantsev, B. M., Berendyaev, V. I., and Kotov, B. V. Photoconduction Kinetics and Nature of Intermediate Photogeneration Centers in Soluble Photoconductive Polyimides. Photochemistry and Magnetochemistry. *J. Phys. Chem.*, **73**(3), 538547 (Russian) (1999).
46. *Relaxation in Polymers*. T. Kobayashi (Ed.). World Scientific. Singapore, p. 329 (1993).
47. *Semiconductors and Semimetals*. Vol. 85. Quantum Efficiency in Complex Systems. Part II. From Molecular Aggregates to Organic Solar Cells. U. Wurfel, M. Thowart, and E. Weber (Eds.). Elsevier, Chapter 9, pp. 312, 341 (2011).

2 Changes in the Polymer Molecular Mass During Hydrolysis

E. I. Kulish, V. V. Chernova, V. P. Volodina, S. V. Kolesov, and G. E. Zaikov

CONTENTS

2.1 Introduction 15
2.2 Experimental 15
2.3 Discussion and Results 16
2.4 Conclusion 18
Keywords 19
References 19

2.1 INTRODUCTION

The problem of chitosan (CHT) hydrolysis, including enzymatic one, has been paid sufficiently much attention to in the recent years [1-6]. Many authors [4-6] choose the viscosimetric method as that of evaluating the change in the polymer molecular mass during hydrolysis. The method of viscosimetric evaluation of the molecular mass decrease makes it possible to establish the fact of the CHT hydrolysis process by the decrease in the intrinsic viscosity of the solutions. Meanwhile, the change in the polymer solution viscosity can reflect not only the change in the polymer molecular weight but also the possible rearrangement of the solution structural condition. In this connection, the aim of this investigation is the evaluation of correctness of using viscosimetric method in the studies on CHT enzymatic destruction in solution in the presence of some enzymes non-specific for CHT.

2.2 EXPERIMENTAL

The objects of investigation were food CHT produced by the company "Bioprogress" (Schelkovo) which was obtained by alkaline deacetylation of crab chitin (the deacetylation degree is ~83%, the molecular mass is 87 kDa) and enzymatic preparations food collegenase ("Bioprogress", Schelkovo) food pepsin ("Shako", Rostov-on-Don), lidase ("Immunopreparat", Ufa), and crystalline tripsin ("Microgen", Omsk). The CHT solution with the concentration of 2% mass was prepared by dissolving during 24 hr at room temperature. Acetic acid with the concentration of 1 g/dl was used as the solvent. The enzymatic preparation previously dissolved in a small quantity of water was introduced into the polymer solution in amount of 5% of CHT mass. Enzymatic destruction was carried out at 25°C for 10–180 hr. After exposure with enzyme the polymer was extracted from the solution, washed with distillated water and dried up to its constant weight. To determine the CHT intrinsic viscosity the 0.3% polymer solution in acetate buffer with pH = 4.5 was prepared. The definition of the reducing carbohydrates concentration was carried out by the ferricyanide method [7].

2.3 DISCUSSION AND RESULTS

During the investigation conducted by the authors two facts have been established which seem to be mutually eliminating. On the one hand, the standing of CHT solutions with enzymes non-specific for it trypsin, lidase, collagenase, and pepsin is accompanied by the decrease in the value of the intrinsic viscosity of its solutions (Figure. 1, curves 14). This fact was explained in terms of the process of CHT main chain destruction taking place under the action of enzymes. On the other hand, even if enzyme is absent CHT dissolving in acetic acid is also accompanied by a sharp one-fold change (decrease) in the value of its intrinsic viscosity. So, the value of intrinsic viscosity of CHT which did not undergo the stage of dissolving in acetic acid, determined in acetate buffer is [η] = 4.2 dl/g. If we hold CHT in 1% acetic acid during 2500 hr (Figure 1, curve 5), extract it from the solution and determine its intrinsic viscosity in acetate buffer the obtained value [η] is 3.1 dl/g, however, it does not depend on the time of standing of the initial polymer solution. The fact of CHT viscosity decrease in acetic acid solutions has been discussed in the literature and explained in terms of rearrangement of its supermolecular structure [5, 10]. Really, the change in the polymer solution viscosity can reflect not only the change in the polymer molecular weight but also a possible rearrangement in the solution of macromolecules aggregates.

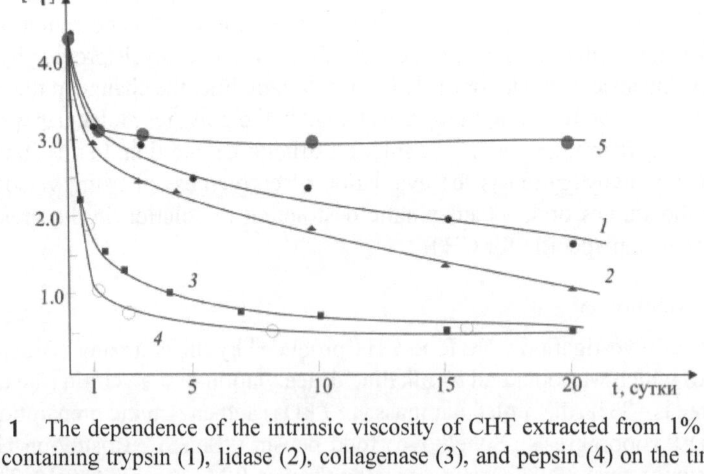

FIGURE 1 The dependence of the intrinsic viscosity of CHT extracted from 1% acetic acid solution containing trypsin (1), lidase (2), collagenase (3), and pepsin (4) on the time of CHT standing in solution.

Thus, one and the same experimental fact the decrease in the intrinsic viscosity of CHT specimens obtained under similar experimental conclusions can be considerable in these two cases in different ways–in the presence of enzymes one can speak about the hydrolysis process, in the absence of enzymes about structural rearrangements in solution. Thus, the conclusion about CHT destruction taking place under the action of non-specific enzymes can be ambiguous if made according to the viscosimetric data. The dependences [ŋ] = f(t) can be in favor of destruction. The intrinsic viscosity is seen to decrease regularly with the increase of the time of exposure of the polymer enzyme system.

According to the scheme of the hydrolytic rupture of the glycoside bond (Figure 2) the resulting destruction products are carbohydrates having reducing capacity.

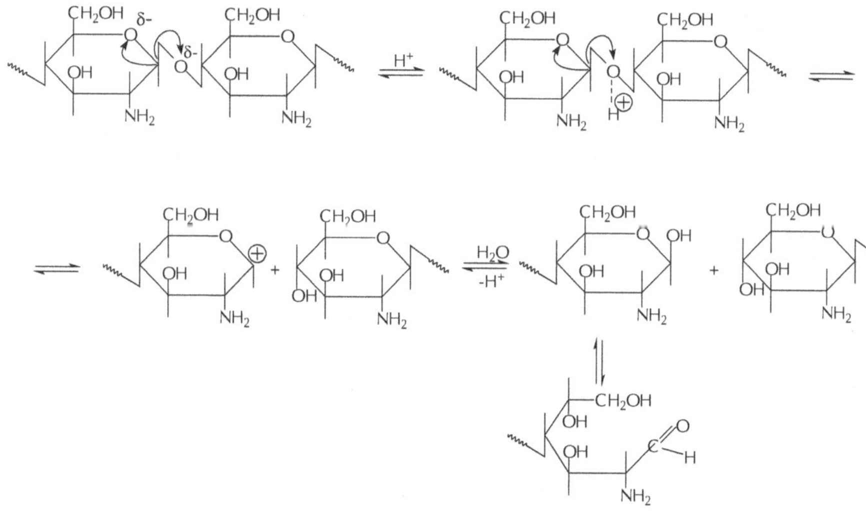

FIGURE 2 The scheme of hydrolytic rupture of the glycoside bond.

Direct determination of the reducing carbohydrates concentration during the experiment showed that a prolonged (for 20 days) standing of CHT solutions in acetic acid in the absence of enzymes is not accompanied by any change in their concentration. On the contrary, the standing of CHT solutions in acetic acid in the presence of enzymatic preparations results in a regular increase in reducing carbohydrates concentrations (Figure 3).

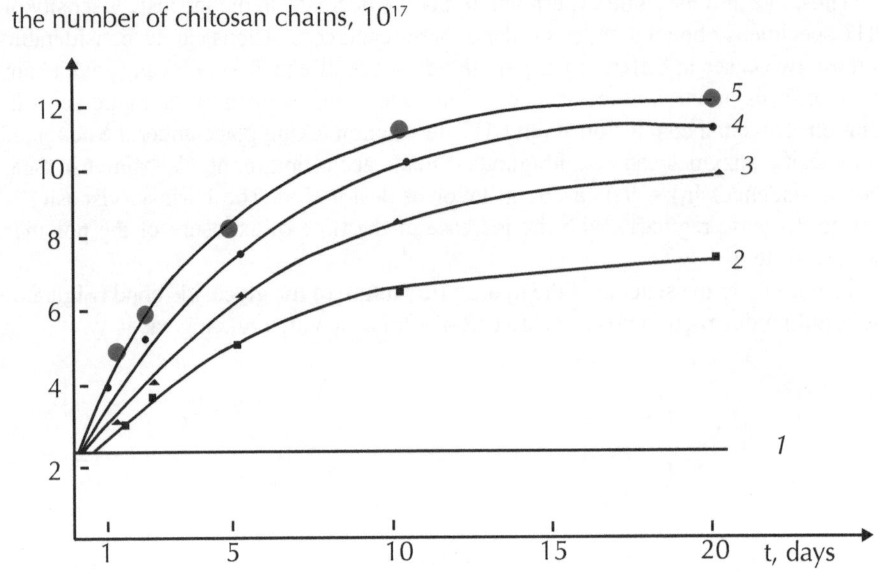

FIGURE 3 The dependence of the number of CHT chains placed in 0.1 ml of 0.2% CHT solution in acetic acid (1) in the presence of enzymes trypsin (2), lidase (3), collagenase (4), and pepsin (5) on the time of solution standing.

As this takes place, the type of corresponding kinetic dependences does not coincide with the dependences given in Figure 1 both in the course of time and in enzymes efficiency?

In the case of CHT specimens having no contact with enzymes the decrease in intrinsic viscosity observed after its reprecipitation from acetic acid solutions is seen not to result from the hydrolysis process, which is indicated by the time constancy of the reducing carbohydrates concentration. This is most likely to be connected with the destruction of the initial supermolecular structure of the polar CHT disposed to forming intermolecular hydrogen bonds. Correspondingly, the viscosimetric applied to CHT gives only a qualitative idea how destructive transformations take place. In this connection, the direct method of determining the reducing carbohydrates concentration seems to be more correct at investigation of the CHT enzymatic destruction.

2.4 CONCLUSION

Thus in the course of the investigation it has been established that:

1. In the absence of enzymatic preparations no accumulation of the reducing carbohydrates amount takes place. The observed changes in the values of intrinsic viscosity of CHT which underwent the stage of dissolving in acetic acid are evidently due to rearrangement of the supermolecular polymer structure;
2. In the presence of enzymatic preparation there occurs the accumulation of the reducing carbohydrates amount caused by the CHT main chain hydrolysis taking place under the action of enzymes.

The work has been carried out due to the financial support of the FAP "Research and scientific pedagogical personnel of the innovational Russia" (g/k №02.740.11.0648).

KEYWORDS

- **Chitosan hydrolysis**
- **Ferricyanide method**
- **Supermolecular structure**
- **Viscosimetric method**
- **Viscosimetric**

REFERENCES

1. Ilyina, A. B. and Varlamov, V. P. In the book *Chitin and chitosan: Obtaining, properties and application. red.* K. G. Scryabin, G. A. Vikhoreva, V. P. M. Varlamov (Eds.) Nauka, pp. 7990 (2002).
2. Panteleone, D. and Yalpani, M. In *Carbohydrates and carbohydrate polymer Analysis, Biotechnology, Modification, Antiviral, Biomedical, and other application.* M. Yalpani (Ed.). ATL Press, p. 44 (1993).
3. Yalpani, M. and Panteleone, D. *Carbohydr. Res.*, **256**, 159 (1994).
4. Mullagaliev, I. R., Aktuganov, G. E., and Melentyev, A. I. Proceedings of the 7th International Conference "Modern prospects in the investigation of chitin and chitosan". M. VNIRO, p. 10791084 (2006).
5. Vikhoreva, G. A., Rogovina, S. Z., Pchelenko, O. M., and Golbraikh L. S. *Vysocomolec. Soed. S. B*,. **43**(6), 10791084 (2001).
6. Fedoseeva, E. N., Semchicov, Yu. D., and Smirnova, L. A. *Vysocomolec. soed. S. B.*, **48**(10), 19301935.
7. Practical studies on biochemistry. Red. S. E. Severina, G. A. Solovyeva. MGU, Moscow. p. 509 (1989).
8. Chernova, V. V., Kulish, E. I., Volodina, V. P., and Kolesov S. V. Proceedings of the 9th International Conference "Modern prospects in the investigation of chitin and chitosan". M. VNIRO, p. 234 (2008).
9. Kulish, E. I., Chernova, V. V., Volodina, V. P., Torlopov, M. A., and Kolesov, S. V. Proceedings of the 10th International Conference "Modern prospects in the investigation of chitin and chitosan". M. VNIRO, p. 274 (2010).
10. Sklyar, A. M., Gamzazade, A. I., Rogovina, L. Z., Tytkova, L. V., Pavlova, S. A., Rogozhin, S. V., and Slonimskiy, G. L. *Vysocomolec. soed. S. A.*, **23**(6), 13961403 (1981).

3 Update on Enzymatic Destruction

E. I. Kulish, V. V. Chernova, V. P. Volodina, S. V. Kolesov, and G. E. Zaikov

CONTENTS

3.1 Introduction 21
3.2 Experimental.. 21
3.3 Discussion and Results 22
Keywords.. 24
References.. 24

3.1 INTRODUCTION

It has been established that the process of enzymatic destruction of chitosan (CHT) film specimens obtained from acetic acid solution experiences the influence of the prehistory of film formation, including the concentration of acetic acid used for films preparation and the thermal film modification.

One of the advantages of using CHT for obtaining various medicinal materials-films, threads, coatings, gels, is its capacity for enzymatic destruction under the action of enzymes of a living organism [1].

However, it should be taken into consideration that the reactions involving polymers can be controlled in many cases by supermolecular polymer structure and by the density of packing of macromolecular chains [2] which are laid at the stage of obtaining polymeric material. The present work reveals the interaction between the degree of enzymatic destruction of CHT film coating and the prehistory of film formation.

3.2 EXPERIMENTAL

A CHT specimen (company "Chimme", Russia) obtained by alkaline deacetylation of crab chitin was chosen as the object of investigation. To prepare CHT film specimens semi-diluted (2g/dl) solutions were made by dissolving a dry polymer weight at room temperature for 8–10 hr. Acetic acid with the concentration of 170 g/dl was used as the solvent.

The CHT film specimens with the thickness of 0.1 mm were prepared by casting the polymer solution onto glass surface. Viscosimetric studies were carried out according to the standard technique on the Webblode viscosimeter at 25°C. For this purpose the powder of the initial CHT specimen was dissolved in a number of acetate buffers with pH = 3.6, 4.0, 4.5, and 4.8.

To study the process of enzymatic transformation the CHT films extracted from acetic acid solutions of different concentration were put on the base moistened by enzymatic preparation solution and were kept at 35°C for a definite period of time.

The extent of enzymatic destruction of films was estimated by the difference between the value of initial intrinsic viscosity of CHT solutions in acetate buffer with pH = 4.5 obtained from film specimens and that of intrinsic viscosity $[\eta]_{dest}$ of solutions from film specimens subjected to enzymatic destruction for 1 hr. Food collagenase (Bioprogress, Schelkovo) was used as an enzymatic preparation. The enzymatic preparation concentration on the base was 5% mass of the CHT mass.

3.3 DISCUSSION AND RESULTS

At describing the viscosity properties of diluted solution one usually proceeds from the linear dependence of an increment in viscosity on the polymer solution concentration. However, in the case of polar polymers to which CHT belongs there is a possibility of the occurrence of reversible agglomeration process which can take place not only in the area of semi-diluted solutions but even in the area of diluted ones. In this case the contribution to viscosity is made not by separate particles with V_0 volume but by their aggregates whose volume V(n) depends not only on the number of particles constituting it, but also on their density characterized by fraction dimensions D [3]:

$$V(n) = V_0 n^{3/D} \tag{1}$$

Due to non-dense particles packing in the aggregate their contribution to viscosity begins to depend on the concentration in a non-linear way:

$$\Delta\eta \sim \eta_0 c^{\delta}, \delta > 1 \tag{2}$$

Consequently, any deviation of the δ index from one testifies to the fact that this system is structurized.

The processing of experimental dependences of specific viscosity of CHT solutions on the concentration in acetate buffer solutions with different pH in double logarithmic coordinates makes it possible to determine the degree index δ in relationship (2) whose corresponding values are given in Table 1. The table also demonstrates the values of initial intrinsic viscosity, Huggins constant, and the concentration of crossover point C which allow us to judge the quality of the solvent used. The analysis of these data makes it possible to speak about the deterioration of the solvent quality and the decrease in the size of the coils with increasing pH of the solvent. The comparison of the obtained values shows that in all the cases under consideration the values of δ index are higher than 1, even in the area of concentrations up to the crossover point, which unambiguously indicates the fact that aggregation processes in a polymer solution begin in the area of diluted solutions. The given experimental fact is confirmed by rather numerous investigations of the properties of CHT solution in acetic acid [4-6] and allow us to speak about their being structurized systems.

TABLE 1 Physico chemical characteristics of solutions of the CHT specimen used 1solutions obtained from initial CHT powder, 2solutions obtained from the film isolated from 1% acetic acid.

Type of the used Chitosan material	pH of the used acetate buffer	$[h]_{исх}$	K_x	C^*	d
1	3.6	3.72	0.36	0.22	1.25
1	4.0	3.65	0.44	0.23	1.28
1	4.5	3.45	0.47	0.24	1.31
2	4.5	2.78	0.32	0.22	1.18
1	4.8	3.09	0.49	0.25	1.32

At extracting CHT film specimens from acetic acid solutions both the supermolecular CHT structure and the extent of its structurization (aggregation) change. This is demonstrated not only by the decrease in the intrinsic viscosity of the film specimen but also by considerable reduction of the δ index (Table 1). However, it is of fundamental importance that the prehistory of film obtaining affects both the extent of δ system structurization and the extent of enzymatic destruction of films. The variation of concentration of acetic acid used for films preparation changes not only the thermodynamic quality of the solvent relative to CHT and the supermolecular solution structure [4] but also the supermolecular structure of the films extracted from the solution. As a result, the CHT solutions in acetate buffer obtained due to films dissolving are characterized by different extent of enzymatic stability (Table 2). However, increasing the extent of films structurization by means of their thermo-modification [7] it is possible to increase significantly their stability to enzymatic destruction.

As seen from the data in Table 2, the increase of the modification period is accompanied by the regular increase of the δ index and by corresponding decrease of the degree of films enzymatic destruction.

TABLE 2 Physico-chemical characteristics of CHT film specimens.

The concentration of the acetic acid (g/dl) used for films preparation	The time of termomodification of films, min	The extent of aggregation of diluted CHT solution in acetate buffer with pH = 4.5	The degree of enzymatic destruction of films Δη
1	0	1.18	1.08
5	0	1.13	1.38
5	30	1.17	1.29
5	60	1.22	1.25
10	0	1.14	1.36
10	60	1.20	1,13
20	0	1.17	1.09
50	0	1.10	1.58
50	60	1,21	1,22
70	0	1.09	1.59
70	60	1.22	1.10

Thus, the pre-formation of the film compositionthe variation of the concentration of acetic acid used as the solvent as well as conducting the thermo-modification of films, affects the extent of film enzymatic destruction since in the process of preparing and modifying film specimens it is possible to form systems with different structurization degree.

KEYWORDS

- **Chitosan**
- **Enzymatic destruction**
- **Macromolecular chains**
- **Thermo-modification**
- **Webblode viscosimeter**

REFERENCES

1. Skryabin, K. G., Vikhoreva, G. A., and Varlamov, V. P. Chitin and Chitosan. *Obtaining, properties and application*. Nauka, Moscow, p. 365 (2002).
2. Kulish, E. I., Volodina, V. P., Fatkullina, R. R., Kolesov, S. V., and Zaykov, G. E. *Vysokomolec. soed.*, **50**(7), 1277–1280 (2008).
3. Arinshtein, A. E. *J. Experiment and theor. phys.*, **101**(4), 1209–1215 (1992).
4. Vikhoreva, G. A., Rogovina, S. Z., Pchelko, O. M., and Galbraikh, Z. S. *Vysokomolec. soed.*, **43**(6), 1079–1084 (2001).
5. Nudga, L. A., Petrova, V. A., Bochek, A. M., Kallistov, O. V., Petrova, S. F., and Petropavlovskiy, G. A. *Vysokomolec. soed.*, **39**(7), 1232–1236 (1997).
6. Gamzazade, A. N., Sklyar, A. M., Palvova, S. A., and Rogozgin, S. V. *Vysokomolec. soed.*, **23**(3), 594–597 (1981).
7. Zotkin, M. A., Vikhoreva, G. A., and Kechekyan, A. S. *Vysokomolec. soed.*, **46**(2), 359–363 (2004).

4 New Type of Bioadditives to Motor Fuel

V. B. Volieva, S. V. Usachev, I. S. Belostotskaya, N. L.Komissarova, A. V. Malkova, A. I. Nekhaev, A. L. Maksimov, G. G. Makarov, and S. D. Varfolomeev

CONTENTS

4.1 Introduction 25
4.2 Experimental 26
 4.2.1 4-Hydroxymethyl -2,2-Dimethyl -1,3-Dioxolane (I): 26
 4.2.2 4-Acetoxymethyl -2,2-Dimethyl -1,3-Dioxolane (IV): 28
 4.2.3 4-Hydroxymethyl-2-Methyl-2-Ethyl-1,3-Dioxolane (II): 28
 4.2.4 4-Hydroxymethyl -2,2-Pentamethylene -1,3-Dioxolane (III): 28
 4.2.5 2,2-Dimethyl-1 ,3-Dioxolane (V): 28
4.3 Discussion and Results 29
4.4 Conclusion 33
Keywords 34
References 34

4.1 INTRODUCTION

The reduction of petroleum supplies, deterioration of its quality, and more expensive processing necessitate the search for alternative motor fuels derived from sources other than petroleum or reducing consumption of petroleum fuels. Thus, the production and consumption of ethanol-gasoline blends has been licensed in Russia recently. A shortcoming of such blends is a tendency to watering and subsequent phase separation under conditions of low temperatures. In the Emanuel Institute of Biochemical Physics of the Russian Academy of Sciences, a new type of additives to alcohol-gasoline blends is developed reducing additionally the content of hydrocarbons with a significant boost in the octane index and phase stability of fuel compositions in a wide range of temperature [1]. These additives are derivatives of polyatomic alcohols with completely or partly substituted hydroxyl groups. They are produced from renewable natural sources and farm wastes (straw, vine trimming, etc.), most of which are accumulated or simply destroyed polluting the environment. Besides polyatomic alcohols, commercially available carbonyl compounds (lower ketones and aldehydes, mainly acetone) are required. At present, convenient synthesis methods applicable in a large-scale production of these additives are developed. The antiknock efficiency of the additives is evaluated using the motor and research methods and express methods equivalent to the standard ones. A relationship is found between the octane index increase and additive structure and concentration in the range from 1 to 15%. The ability to phase homogeneity stabilization is characterized by the dependence of cloud and separation temperatures of fuel compositions on the additive structure and concentration. These dependencies allow variations in structures and concentrations of the additives along with economic value estimation. Research is underway to study other characteristics of new fuel compositions (gumming, detergent and corrosive characteristics, tendency to pre-ignition, storage stability, etc.). It is shown that the additives are not toxic and, therefore, are environmentally friendly. The proposed additives are a new economic type of bifunctional additives reducing the consumption of hydrocarbons with a significant boost in the octane index and phase stability. Solketal (4-hydroxy-2,2-dimethyl-1,3-dioxolane, I) the product of glycerol and acetone condensationhas been known for many years. It is used as universal nontoxic solvent in many areas of organochemical production, in perfumery and cosmetic compositions, in the creation of dosage forms. Now solketal along with related cyclic ketal has attracted particular attention due to the possibility of using it as a stabilizing and high octant additives to motor fuels. It is important that glycerol produced as a byproduct in the biodiesel production can be used in the synthesis of such additives. The growing scale of the production overtakes the commercial demand for glycerol. Existing powers for its processing into acrolein [2-5], glyceric acid and dihydroxyacetone [6], 1,2- and 1,3-propanediols [7-8] do not provide its complete utilization. Therefore, a substantial part of glycerol from biodiesel still regarded as a waste. The use of glycerol in the synthesis of fuel additives might be an optimal way to modernize the production of biodiesel making it more economic and environmentally friendly. In this regard, it becomes urgent to develop alternative to traditional methods for the synthesis of cyclic ketals corresponding to the latest environmental requirements and suitable for realization on a large scale.

4.2 EXPERIMENTAL

The acid catalyzed condensation of glycerol and acetone was performed by the following procedures:

4.2.1 4-hydroxymethyl -2,2-dimethyl -1,3-dioxolane (I):

(a) A mixture of glycerol (9.2 g, 0.1 M), acetone (7 g, 0.12 M), and Amberlyst or KU2 (1 g) was boiled for 3 hr while stirring with a magnetic stirrer. The catalyst was separated by filtration. The reaction products were analyzed by GLC (see below). Acetone was distilled off at atmospheric pressure, and 2,2-dimethyl-1,3-dioxolane-4-methanol (solketal) (5 g, 38%) was separated from the residual mixture by vacuum distillation (10 Torr), b.p. = 89–90°C. Spectrum NMR ^{1}H, d, ppm: 4.18 m (1H, CH), 3.98, 3.72 dd (2H, CH_2, J 8.3,

6.8 Hz), 3.65, 3.55 dd (2H, CH_2OH, J 12.0, 6.1, 5.5 Hz), 2.50 t (1H, OH, J 5.5Hz), 1.40, 1.34 s (3H, CH_3).

(b) AcOH (1 ml) was added while stirring to a mixture of glycerol (9.2 g, 0.1 M) and acetone (7 g, 0.12 M). A homogeneous solution formed in 4 or 5 min. Acetic anhydride (10 g, 0.1 M) was added to the solution. The mixture was boiled for 3 hr and then fractionated by vacuum distillation (10 Torr). Two fractions were collected: AcOH (11 g, below 70°C) and solketal (12.3 g, 93%, 89–91°C). The TLC analysis of the product on Silufol in a 1:1 hexane–ether mixture revealed a small admixture of ketal acylated at the hydroxymethyl group. The standard acetate was obtained via the reaction of solketal with acetic anhydride (technique 2).

(c) A mixture of glycerol (18.4 g, 0.2 M), acetone (14 g, 0.24 M), and Amberlyst or KU 2 (2 g) was boiled for 25–30 min before homogenization. A capsule of porous glass or fiber filled with a drying agent (25 g) was then suspended in the solution and heating continued for 2.5 hr. The capsule was removed and washed with acetone, and acetone was added to the reaction mixture. The catalyst was separated by filtration. Acetone was distilled off and solketal (24 g, 90%) was isolated from the residue by vacuum distillation (10 Torr); b.p = 89–91°C. At a low glycerol contents, the mixture was efficiently separated into components by passing the acetone solution through a silicagel layer (after separating the catalyst). Ketal was separated first and glycerol second. The separation was monitored by TLC on Silufol UV-254 in a 1:1 hexane–ether system. Solketal–glycerol mixtures with high glycerol contents can be separated by solketal extraction with chloroform. According to TLC, glycerol was not transmitted to the extract. The chromatograms were developed in iodine vapors. The catalysts used in this study included Amberlyst; cationite in the form of KU-2 acid (average particle size, 1–2 mm; acidity, 1.4 mmol/g); fluorinated sulfocationite in the form of S_4SF acid (acidity, 1 mmol/g); a catalyst based on the acid form of β-zeolite (cylinder, ≈1.5 mm in diameter; 70% zeolite, $SiO_2/Al_2O_3 = 30$); TsVM zeolite in acid form; β_zeolite in acid form (Zeolyst, $SiO_2/Al_2O_3 = 40$); and zeolite X (the sodium form of the zeolite was subjected to three exchanges with ammonium nitrate and thermally treated at 550°C). Magneto sensitive sulfocationite containing metal (mainly iron) particles in its grains was prepared from polystyrene using an ethyleneglycol dimethacrylate cross-linking agent. The sulfo groups were introduced in the grains of the intermediate composite copolymer by low temperature (40°C) sulfonation with sulfuric acid in the presence of silver sulfate. All other procedures were standard techniques for the preparation of sulfocationites described, for example, in [9].Sulfated oxides were synthesized by the procedure described in [10]. TiO_2 (Sigma–Aldrich), γ Al_2O_3 (Fluka, pure), and silcagel KSK (pore diameter, 100 nm; specific surface area, 150 m/g) were used as supports. A fraction with sizes of 1.02–1.20 mm was isolated from the supports and impregnated with sulfuric acid (10 wt %). The catalyst was dried for 1 day at 100°C and for 1 day at 500°C in an air flow. Sulfated zirconium oxide was synthesized by the following procedure: Zirconium hydroxide was prepared by precipitation from 25% aqueous ammonia (pH below 10). The precipitate was stored for 24 hr and then separated, washed, and dried for 72 hr at 110°C. For sulfation, we used a 0.5 M solution of sulfuric acid, adding H_2SO_4 (2.1 ml) per 1 g of the product. The precipitate was filtered off and dried for 12 hr at 110°C; then it was calcinated for 5 hr at 550°C.The products of glycerol ketalization were analyzed by GLC on a Kristall Lux 2000 chromatograph equipped with a flame ionization detector having an SPBTM Octyl L (Supelco) column (30 m, 0.25 mm, df 1.00 μm). The temperature of the evaporator and detector was 300°C. The analysis was performed in the temperature programming mode; the initial column temperature was 60°C (3 min), the heating rate was 20°C/min, and the final column temperature was 240°C (3 min).

4.2.2 4-acetoxymethyl -2,2-dimethyl -1,3-dioxolane (IV):

A mixture of solketal (3.3 g, 2.5, 0.01 M) and Ac_2O (2.5 мл 2.5, 0.01 M) was heated during 3 hr while stirring at 50ºC. The reaction mixture was distillated. 1.6 g (36%) of ketal IV was obtained, b.p. = 8992ºC (18 Torr). Spectrum NMR ^1H, d, ppm: 4.28 m (1H, CH), 4.07, 4.15 dd (2H, CH_2, J 11.4, 4.6 Hz), 3.76, 3.78 d (2H, CH_2OCOCH_3, J 6.4 Hz), 2.08 s (3H, $COCH_3$), 1.36 s (3H, CH_3), 1.42 s (3H, CH_3). Found, %: C 55.30; H 8.21. $C_9H_{16}O_3$.Calculated, %: C 55.16; H 8.10.

4.2.3 4-Hydroxymethyl-2-methyl-2-ethyl-1,3-dioxolane (II):

A mixture of solketal (13.2 g, 0.1 M), methylethylketone (8 g, 1.2×10^{-1} M), and Amberlyst (1 g) was heateded £56˚C for 20 min while stirring and acetone distillating. The catalyst was separated by filtration. The filtrate was distillated off in vacuum and 12.9 g (88%) of ketal II was obtained, b.p. = 9091ºC (6 Torr). Spectrum NMR ^1H, d, ppm: 4.20, 4.27 m (1H, CH), 4.04, 3.75 dd (2H, CH_2, J 12.0, 8.0 Hz), 3.73, 3.60 m (2H, CH_2OH) 2. 08 s (1H, OH) 1.70, 1.65 q (2H, CH_2, J 8.0 Hz). 1.37, 1.32 s (3H, CH_3) 0.95, 0.93 t (3H, CH_3, J 8.0 Hz).

4.2.4 4-Hydroxymethyl -2,2-pentamethylene -1,3-dioxolane (III):

A mixture of solketal (13.2 g, 0.1 M), cyclohexanone (11.8 g, 1.2×10^{-1} M), and Amberlyst (1 g) was heateded £56˚C for 20 min while stirring and acetone distillating. The catalyst was separated by filtration, the reaction product was distillated off in vacuum and 15.5 g (90%) of ketal III was obtained , b.p. = 121123ºC (5 Torr). Spectrum NMR ^1H, d, ppm: 4.03 s (1H, CH_2), 1.401.60 m (10 H, C_6H_{10}). Found, %: C 62.98; H 9.56. $C_9H_{16}O_3$. Calculated, %: C 62.84; H 9.38.

4.2.5 2,2-dimethyl-1 ,3-dioxolane (V):

A mixture of solketal (13.2 g, 0.1 M), ethyleneglycole (1.74 g, 1.2×10^{-1} M), and Amberlyst (1.5 g) was heateded to ≈60˚C for 20 min while stirring. Then temperature was increased to the start of boiling of ketal, b.p. = 8991ºC. The yield of the product is 9.7 g (95%). Spectrum NMR ^1H, d, ppm: 3.94 s (2H, CH_2,), 1.38 s (3H, CH_3).

4.3 DISCUSSION AND RESULTS

Natural polyols are incompatible with nonpolar hydrocarbon media. They can be chemically modified into a lipophilic form by transforming them into cyclic ketals (dioxolanes) by condensation with carbonyl compounds (acetone and other ketones and aldehydes):

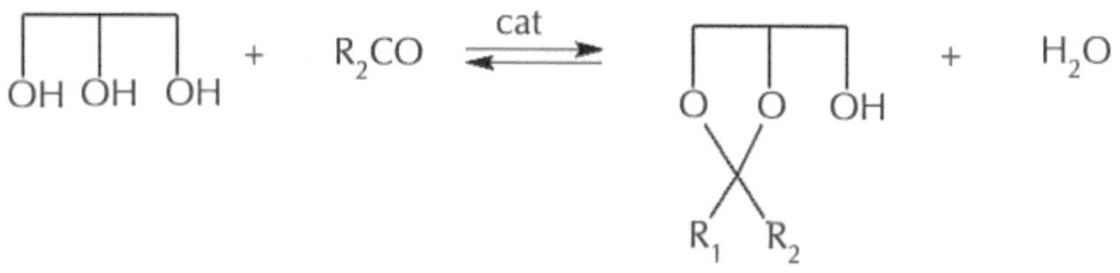

I-III

I, $R_1 = R_2 = CH_3$;
II, $R_1 = CH_3$, $R_2 = C_2H_5$; Cat = H_2SO_4, HCl, TsOH and al.
III, R_1, $R_2 = (CH_2)_5$.

Ketals are well compatible with hydrocarbon fuels and can raise the octane number and stabilize the phase homogeneity of composite fuels over a wide range of temperatures when combined with alcohols (Tables 1 and 2).

TABLE 1 Variation in the octane number (ΔON) of hydrocarbon fuel compositions with ketals and ethanol.

I Ketal	Ketal and alchol contents (vol %) in the composition		DON
	Ket ketal	alchol	
I (glycerol-acetone)	0	10	4
	10	0	1.7
	10	10	9.4
	15	15	14.8
II (glycerol-methyl-ethylketone)	10	0	0.9
	10	10	8.5
	10	0	0.5
III (glycerol-cyclohexanone)	10	10	6.6

TABLE 2 Stabilizing properties of ketal additives to alcohol-gasoline composition with 10% alcohol.

Ketal	Additives (vol %) to the composition	$T_{separation}$ (°C)
I (glycerol-acetone)	0	−10.4
	1	−10.6
	10	<−30
II (glycerol-methyl-ethylketone)	1	−16.2
	10	<−30
III (glycerol-cyclohexanone)	1	−16.5
	10	<−30

The formation of ketals is a reversible acid catalyzed process. The development of the technological basis for their synthesis is therefore aimed at creating an effective catalytic system that satisfies modern ecological requirements.

A number of homogeneous and heterogeneous acid catalytic systems were studied, using the eminently practical condensation of glycerol and acetone with the formation of solketal as an example. The first group of catalytic systems included complex compounds of glycerol formed in reactions with H_3BO_3, B_2O_3, $CaCl_2$, P_2O_5; acetic acid, and its anhydride. This choice was dictated by their rather high acidity and their ability to absorb or bind water excluded during condensation. Note that these systems are actually two-phase (glycerol is mixed with acetone in limited amounts) and homogenization is observed when a certain concentration of ketal is attained. A series of experiments were performed with different amounts of catalyst (from 5 to 10% of the mass of glycerol) and ratios glycerol: acetone = 2:1, 1:1, and 1:2, temperature modes 20–56°C and different reaction times. The conversion of glycerol was up to 45–55% (GLC). That is corresponded to the condensation equilibrium under the chosen conditions in all cases. The yield of solketal was almost quantitative based on the altered glycerol. It was shown that bottom glycerol with catalyst could be recycled, making the process wasteless. Ketalization with 5–50 vol% ACOH as catalyst proceeded in a homogeneous reaction mixture, but this did not substantially affect the conversion of glycerol (≈40%). The yield of solketal was increased to 90% in one cycle by using an equivalent amount of anhydride Ac_2O, which chemically bound the eliminated water. An inappreciable amount of ketal acetate (≈5%) was formed, but this did not deteriorate the octane index of the fuel composition. The necessity of vacuum rectification for isolating the product from catalyst is a major disadvantage of homogeneous catalytic ketalization.

In the series of heterogeneous acid catalysts, we tested specially prepared aluminum, titanium, zirconium, and silicon oxides; phosphoric acid on silica gel subjected to high temperature treatment; and polymeric sulfoacids (commercially accessible Amberlyst, Nafion, KU-2). The conversion of glycerol was 23–55% with a double excess of acetone. The catalyst must have medium acidity for the process to occur selectively without acetone autocondensation. The optimum catalysts were cation exchange resins in the form of KU-2 acid and Amberlyst. The yields of ketal were more lower (≈20%) in the case of highly acid catalysts such as sulfated alumina or zirconia.

To make ketalization more economical, it is necessary to shift the equilibrium toward the product. This can be done by using either substantial quantities of a dehydrating catalyst or a large excess of acetone. Glycerol ketalization occurred with acceptable yields (≈80% for 3 hr according to GLC) with a fourfold excess of acetone at 65–75°C. To obtain a significant yield of product, the catalyst to substrate mass ratio must be 1:5.

When polymeric sulfoacids were used, the process was not complicated by side transformations. The yield of solketal was ≈45% and did not increase when the process was prolonged under identical conditions (glycerol:acetone = 1:1.2, 10 wt% catalyst, t = 56°C, stirring). This was used as the basis for developing an improved catalytic system for ketalization in order to obtain a nearly quantitative yield of the desired product and simplify its isolation. A new catalyst was synthesized under the laboratory conditions. This was a modified analog of Amberlyst containing magnetosensitive metal (mainly iron) particles. The catalyst was synthesized by the suspension (granular) polymerization of styrene in the presence of the metal powder. Ethyleneglycol dimethacrylate was used as a cross-linking agent, yielding a reticular porous structure. This catalyst can readily be separated from the reaction mixture with a magnet. The synthetic procedure afforded grains with a more developed surface by etching the metal particles out of the product by washing it with hydrochloric acid. The amount of catalyst used in one cycle was thus reduced to 5–7%, while the time after which equilibrium concentration of ketal was reached remained the same (3 hr, 45%). It is obviously necessary to remove water from the reaction system in order to appreciably increase the ketal yield. A water removal technique was implemented in a pilot unit and ensured the passing of boiling acetone vapor together with eliminated water vapor through a column with an adsorbent dehydrator and the return of the acetone condensate to the reactor. The installation was a closed circuit of three units: a reactor (working volume 7.5 l), refrigerator, and dehydrator mounted on a main frame and equipped with the necessary communications and automation devices. The temperature in the reactor was maintained automatically with a sensor and a temperature relay. Unit operation time (150–180 min) was set in a program time relay and changed after the dehydrator adsorbed water. The unit design provided the possibility of upgrading the control for fully automated operation.

Water could also be removed, together with acetone vapor, by bubbling acetone vapor through glycerol heated up to 70–75°C in the presence of the catalyst. Mechanical stirring was then unnecessary, and the solketal yield was 65% after 3 hr of bubbling.

The new version of solketal synthesis based on the use of ethanol, can raise the yield to a practically quantitative, because ethanol is more effectively involved in the azeotropic removal of water and it also alters the chemical nature of the reaction medium by semiketal formation in interaction with acetone. It is known that the interaction of nonhindered lower alcohols and ketones gives almost quantitative formation of semiketals. This is detected by the change of physical parameters UV-absorbtion, density, refractive index. Due to semiketal formation boiling point increases what allows raising the temperature of the reaction mass. Most importantly, however, that semiketal can serve as a carrier of isopropenilic group on glycerol, that is, acts as an intermediate in the stepwise process of the solketal formation with crossketalization mechanism. Ethanol is not consumed, so it can be considered as co-catalyst in the process.

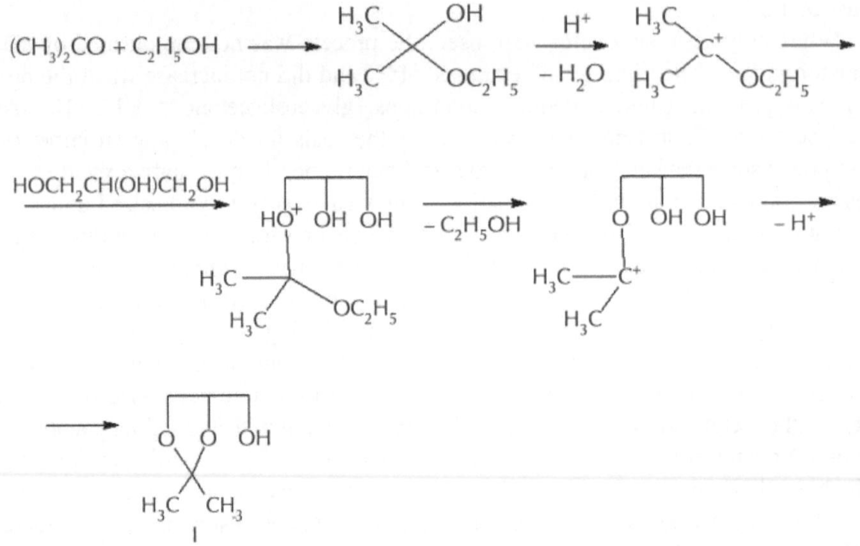

I

Near the same resultsolketal yield near 90% (3 hr) was obtained in a catalytic system with a composite catalyst whose components performed the functions of an acid and dehydrating agent (alkaline earth salts). In the reactor dehydrating agent was separated by a membrane (porous glass, textile) permeable to a glycerol–acetone mixture. This prevents deactivation of the acid catalyst surface by dehydrator during mechanical mixing. Importantly, the coordination sphere of the dehydrator salt absorbs not only water, but also glycerol. As mentioned above, these glycerol complexes can also perform a catalytic function during ketalization, enhancing the action of the main (sulfonate) catalyst. The system provides a double catalytic effect with the simultaneous binding of water. The efficiency of this system is obviously determined by increasing of alkaline earth salts ability to bind water due to their complexation with glycerol. It must be taken into account in developing of technological scheme for the ketalyzation process. Solketal, having a rather high boiling point, can serve as a starting compound for exchange reactions leading to ketals with other structural parameters. It is possible to exchange both a diol and a carbonyl component. Thus, the interaction of solketal with methyl ethyl ketone and cyclohexanone in the presence of Amberlist gives rise to 4-hydroxymethyl-2-methyl-2-ethyl-1,3-dioxolane (II) and 4-hydroxymethyl-2,2-pentamethylene-1,3-dioxolane (III) respectively in high yields.

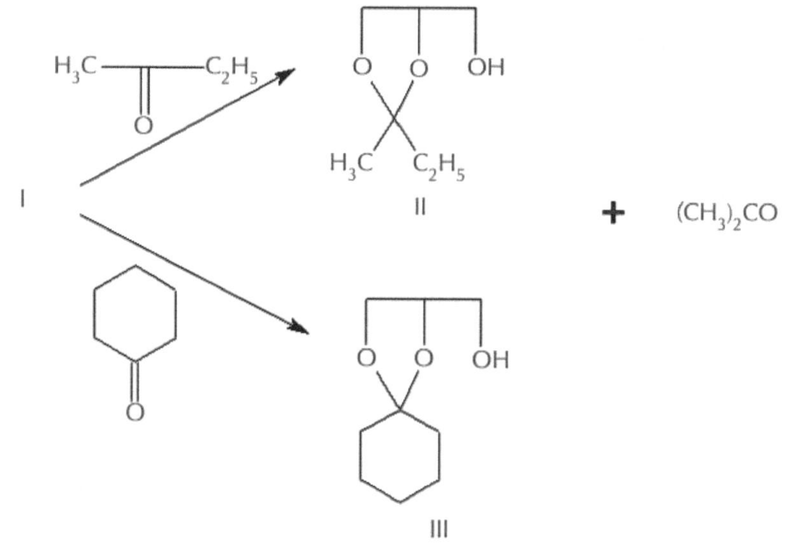

Crossketalization of solketal with ethylene glycol proved to be an express way to obtain 2,2-dimethyl-1 ,3-dioxolane (V), characterized by high yield and purity of the product.

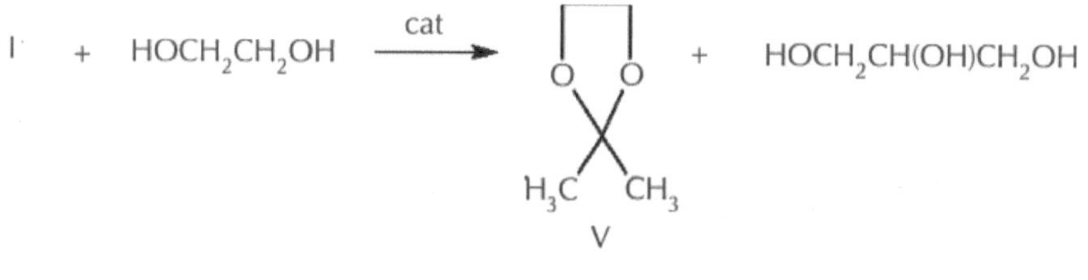

The shift of equilibrium toward the ketal (V) is due to its removal from the reaction system by distillation.

4.4 CONCLUSION

Cyclic ketals based on plant polyols such as glycerol and monosaccharide present a new type of fuel bioadditives with octane raising and stabilizing activities. The search of suitable catalytic system for ketal synthesis is of key importance for developing a technological scheme for their production. Acid-catalyzed glycerol–acetone condensation with solketal formation was studied as a model. The most attractive way to carry out the condensation with high yield of the product is based on the use of ethanol as drying agent and cocatalyst of the process. Another two versions of catalytic systems were found suitable for practical realization. Solketal obtained with high yield was used as starting compound in synthesis of a number of ketals by crossketalization with change of diol or carbonyl components.

KEYWORDS

- **Amberlyst**
- **Cyclohexanone**
- **Ethanol-gasoline blends**
- **Ketalization**
- **Sulfated zirconium oxide**

REFERENCES

1. Varfolomeev, S. D., Nikiforov, G. A., Volieva, V. B., Makarov, G. G., and Trusov, L. I. *The method for raising of octane number of gasoline fuel.* Patent RF № 2365617.
2. Centi, G., and van Santen, R. A. (Eds.). *Catalysis for Renewables.* Wiley–VCH (2007).
3. Corma, A., Huber, G. W., Sauvanaud, L., and O'Connor, P. Biomass to chemical: Conversion of glycerol/water mixtures into acrolein. *J. Catal.*, **257**, 163 (2008).
4. Ott, L., Bicker, M., and Vogel, H. Catalytic dehydration of glycerol in sub and supercritical water: A new chemical process for acrolein production. *Green Chem.*, **8**, 214 (2006).
5. Atia, H., Armbruster, U., and Martin, A. Dehydratation of glycerol in gas phase using heteropolyacid catalyst as active compounds. *J. Catal.*, **258**, 71 (2008).
6. Villa, A., Campione, C., and Prati, L. Bimetallic gold/platinum catalyst for the selective liquid phase oxidation of glycerol. *Catal. Lett.*, **115**, 133 (2007).
7. Feng, J., Fu, H., Wang, J., Li, R., Chen, H., and Li, X. Hydrogenolysis of glycols over ruthenium catalysis. *Catal. Commun.*, **9**, 1458 (2008).
8. Alhanash, A., Kozhevnikova, E. F., and Kozhevnikov, I. V. Hydrogenolysis of Hydrogenolysis of glycols over to propanediole over Ru. *Catal. Lett.*, **120**, 307 (2008).
9. Marczewski, M., Jakubiak, A., and Marczewska, H. Acidity of sulfated oxides *Chem. Phys.*, **6**, 2513 (2004).
10. Zubakova, L. B., Tevlina, A. S., and Davankov, A. B. *Sinteticheskie ionoobmennye materialy* (Synthetic Ion Exchange Materials), Khimiya, Moscow (1978).

5 Interrelation of Viscoelastic and Electromagnetic Properties of Densely Cross-linked Polymers: Part I

T. R. Deberdeev, N. V. Ulitin, R. Ya. Deberdeev, and G. E. Zaikov

CONTENTS

5.1 Introduction 35
5.2 Topological Structure of Experimental Objects 36
5.3 Constants of Mathematical Formalism 37
5.3.1 Some Thermal Constants 37
5.3.2 Equilibrium Constatnts 39
5.3.3 Weighting Coefficients 44
5.4 Parameters of A-Relaxation Times' Spectrum 47
5.5 Modeling of Thermomechanical and Thermo Optical Experiments' Course... 47
5.6 Conclusion 47
Keywords 48
References 48

5.1 INTRODUCTION

In the network of urgent problem of optimum densely cross-linked polymer matrix's determination for high strength radioparent fiberglass product [1] are presented and confirmed on epoxy amine polymers new mathematical descriptions of cross-linked polymers' with high density of cross-links viscoelastic pliability and strain electromagnetic susceptibility. Though constants of introduced model for many tightly cross-linked polymers were partially published or can be determined on the base of these data, in the course of time are synthesize new cross-linked polymers, before creation of which it is nessesary to present prospects of their application. There upon in this chpater on the same experimental objects is demonstrated theoretical determination of mathematical formalism's constants and made rating of thermomechanical and thermooptical curves' trend.

5.2 TOPOLOGICAL STRUCTURE OF EXPERIMENTAL OBJECTS

Repeated fragment of polymer meshes will produce in general form for all series of epoxy amine polymers and will determine it, consisting of following elementary links (Figure 1): mesh pointsa nitrogen atom surrounded by three methylene groups; tetramethylene links; fragments of DGEBA[1]; links, which play the role of chain's extenders and consist of AH fragments, and two methylene groups near the nitrogen atom. Introduce designations: N_{2f} = quantity of links-extenders, mole; N_{3f} = quantity of primary amides, which transform into elastic effective points, mole; N_{σ} = quantity of tetramethylene links, mole; N_{π} = quantity of DGEBA fragments, mole.

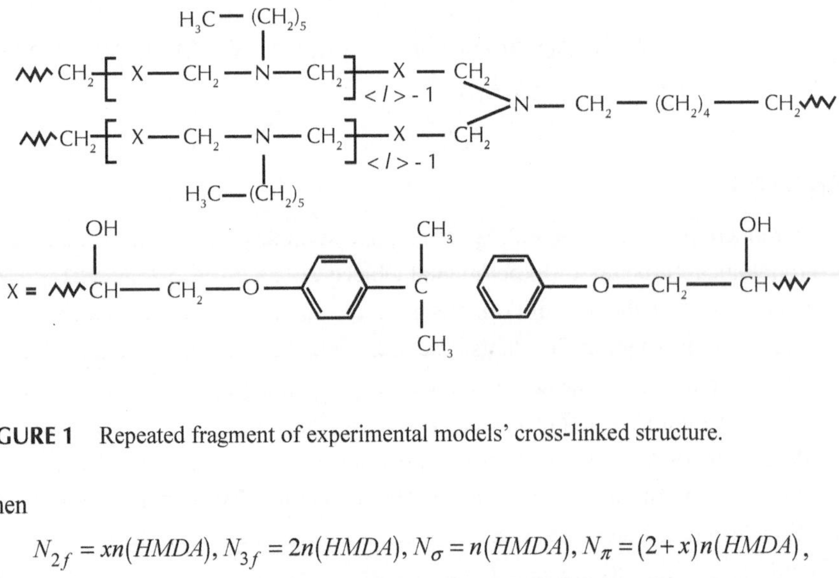

FIGURE 1 Repeated fragment of experimental models' cross-linked structure.

Then

$$N_{2f} = xn(HMDA), N_{3f} = 2n(HMDA), N_{\sigma} = n(HMDA), N_{\pi} = (2+x)n(HMDA),$$

$$N_{tot} = \sum_i N_i = (5+2x)n(HMDA)$$

Quantities of elastic effective points and fragments of DGEBA in meshes' structure later will define by statistical parameters

$$n_{3f} = N_{3f} / N_{tot} = 2/(5+2x), \quad n_{\pi} = N_{\pi} / N_{tot} = (2+x)/(5+2x) \cdot$$

[1]Designations and molar ratios are the same that in [1].

Main structural property, which determines structure of repeated meshes' fragment, is cross-site chain's average-numerical degree of polymerization $<l> = n_\pi / n_{3f} = 1 + 0.5x$.

5.3 CONSTANTS OF MATHEMATICAL FORMALISM

5.3.1 Some Thermal Constants

Further in many calculating equations, there are glass transition temperature and thermal expansion's coefficient. For cross-linked polymers glass transition temperature (T_g, K) [2]

$$T_g = \left(\sum_i \Delta V_i\right)_{r.f.} \Big/ \left(\left(\sum_i a_i \Delta V_i + \sum_j b_j\right)_l + \left(\sum_i K_i \Delta V_i\right)_y\right) \qquad (1)$$

where $\left(\sum_i \Delta V_i\right)_{r.f.}$ = Van der Waals's volume of repeated fragment, Å^3; a_i, b_j = increments, which characterize energy of poor dispersion and strong specific intermolecular interactions, respectively, K^{-1} и Å^3/K; ΔV_i = Van der Waals's volume of *i*th atom, Å^3; $\left(\sum_i a_i \Delta V_i + \sum_j b_j\right)_l$ –increments set for linear chains, which are part of the repeated fragment, Å^3/K; K_i = parameter, which is inputed for atoms of mesh point and is depended on energy of chemical bonds, K^{-1}; $\left(\sum_i K_i \Delta V_i\right)_y$ = increments set for mesh point, Å^3/K.

Van der Waals's volume of repeated fragment of topological structure [2]

$$\left(\sum_i \Delta V_i\right)_{r.f.} = (<l> - 1)\left(\sum_i \Delta V_i\right)_{l,1} + \left(\sum_i \Delta V_i\right)^*_{l,1} + \frac{1}{2}\left(\sum_i \Delta V_i\right)_{l,2} + \left(\sum_i \Delta V_i\right)_y,$$

where $\left(\sum_i \Delta V_i\right)_{l,1}, \left(\sum_i \Delta V_i\right)^*_{l,1}, \left(\sum_i \Delta V_i\right)_{l,2}, \left(\sum_i \Delta V_i\right)_y$ = Van der Waals's volumes of repeated link of sewn together chains' linear fragments, link of sewn together chains' linear fragments, artlessly adjoining to point, repeated link of sewn together bridges' linear fragment and mesh's point, respectively, Å^3.

$$\left(\sum_i \Delta V_i\right)_{r.f.} = 458.9 <l> - 57.4 \text{ Å}^3.$$

In calculation $\left(\sum_i a_i \Delta V_i + \sum_j b_j\right)_l$ there are peculiarities, due to following considerations. First, close to oxygen atom in -O-C_6H_4- there are four p-electrons, causing evenness's breach of distribution of electron density in aromatic ring. This atom may be as negative part of dipole in intermolecular interaction, therefore intermolecular dipoledipole interaction with it is participation take into account with multiplier 0.5. Secondly, there is hydrogen bond inside fragment >CH-OH, therefore it is ruled out from intermolecular interaction. Thirdly, since fragment -$(CH_2)_4$ does not take part in any of intermolecular interaction's forms, increments set defined energy of intermolecular interaction for it is equal to zero. Therefore,

$$\left(\sum_i a_i \Delta V_i + \sum_j b_j\right)_l = (1437.54 \langle l \rangle - 478.76)10^{-3}\ \text{Å}^3/\text{K}$$

For point received $\left(\sum_i K_i \Delta V_i\right)_y = 82.179 \cdot 10^{-3}\ \text{Å}^3/\text{K}$.

As provided by Equation (1), glass transition temperature (values look in Table 1)

$$T_g = \left((458.9 \langle l \rangle - 57.4)/(1437.54 \langle l \rangle - 396.581)\right) \cdot 10^3 - 273.15,\ {}^\circ\text{C}.$$

For cross-linked polymers in glassy state thermal expansion's coefficient (α_g, K^{-1}, or degree^{-1}) [2]

$$\alpha_g = \left(\left(\sum_i \alpha_i \Delta V_i + \sum_j \beta_j\right)_l + \left(\sum_i K_i \Delta V_i\right)_y\right) \Big/ \left(\sum_i \Delta V_i\right)_{r.f.} \qquad (2)$$

where α_i = partial volumetric coefficients of thermal expansion, which is caused by weak dispersion interaction of *i*th atom with the next, K^{-1}; β_j = parameter, which is defining contribution of intermolecular interaction's each type to thermal expansion's coefficient, Å^3/K.

Following the same considerations, that in calculation $\left(\sum_i a_i \Delta V_i + \sum_j b_j\right)_l$, was determined $\left(\sum_i \alpha_i \Delta V_i + \sum_j \beta_j\right)_l = (137.615 \langle l \rangle - 45.855)10^{-3}\ \text{Å}^3/\text{K}$.

Then in suitable for calculation form Equation (2) will be so:

$$\alpha_g = \left((137.615 \langle l \rangle + 36.324)/(458.9 \langle l \rangle - 57.4)\right) \cdot 10^{-3} \qquad (3)$$

TABLE 1 Glass transition temperature.

$x = n(AH)/n(HMDA)$	$\langle l \rangle$	T_g, °C calc.	T_g, °C exper. [1]	ε, %*
0.0	1.00	113	109	4
0.5	1.25	95	99	4
1.0	1.50	85	88	3
1.5	1.75	79	77	3
2.0	2.00	74	71	4

* Hereinafter relative discrepancy is calculated by equation.

$$\varepsilon = \left|(\text{experimental data} - \text{calculated value})/\text{emperimental data}\right| \cdot 100,\ \%$$

For cross-linked polymers in hyperelastic state thermal expansion's coefficient (α_{∞}, K^{-1}, or degree^{-1}) is expressed from Simkhi Boyer's equation:

$$\alpha_{\infty} = \left(0.106 + \alpha_g T_g\right)/T_g = \left(0.106/T_g\right) + \alpha_g \tag{4}$$

Derived by Equations (3) and (4) values of thermal expansion's coefficients were averaged off for each physical state of all experimental models: in glassy state –3.9 × 10^{-4} degree^{-1}, in hyperelastic state 6.8 × 10^{-4} degree^{-1}. For comparison will cite experimental values: in glassy state –4.3 × 10^{-4} degree^{-1} (ε = 9%), in hyperelastic state 7.0 × 10^{-4} degree^{-1} (ε = 3%).

5.3.2 Equilibrium Constatnts

For concerned experimental models hyperelastic state's constant (A_{∞}, K/MPa) can be written in the next expression [3]:

$$A_{\infty} = 1/\left(\frac{f}{2} C_{tot} n_{3f} RF\right) \tag{5}$$

where f = maximum functionality of points (for our models 3); C_{tot} = concentration of elementary links at structure glass transition temperature of cross-linked polymer, mole/cm^3;
R = universal gas constant, J/(mole×K); F = front-factor.

Unknown in Equation (5) are C_{tot} and F

As provided by expression

$$C_{tot} = N_{tot}/V_{tot}\left(T_g\right), \tag{6}$$

where $V_{tot}\left(T_g\right)$ = suumary volume of elementary links at glass transition temperature, cm^3.

Will assign the last quantity

$$V_{tot}\left(T_g\right) = N_{\sigma}V_{\sigma}\left(T_g\right) + N_{3f}V_{3f}\left(T_g\right) + N_{2f}V_{2f}\left(T_g\right) + N_{\pi}V_{\pi}\left(T_g\right),$$

where $V_{\sigma}(T_g), V_{3f}(T_g), V_{2f}(T_g), V_{\pi}(T_g)$ = molar volumes of appropriate elementary links at glass-transition temperature, cm^3/mol.

Substituting molar volumes of elementary links for ratios of their molar masses to density of polymer and substituting final expression $V_{tot}\left(T_g\right)$ to Equation (6), get:

$$C_{tot} = \left((5 + 2x)\, d\left(T_g\right)\right)/\left(M_{\sigma} + 2M_{3f} + xM_{2f} + (2 + x)M_{\pi}\right), \tag{7}$$

where $d\left(T_g\right)$ = density of polymer at glass transition temperature, g/cm^3; M_{σ}, M_{3f}, M_{2f}, M_{π} = molar masses of appropriate elementary links, g/mol.

At first find incoming in dependence (7) expression of experimental models' density at their structure glass transition temperatures. Dependence of cross-linked polymers on temperature is expressed by equation [2]

$$d(T)=k_g M_{r.f.}\Big/\left(10^{-24}\left[1+\alpha\left(T-T_g\right)\right]N_A\left(\sum_i \Delta V_i\right)_{r.f.}\right),\quad \alpha=\begin{cases}\alpha_g, & T<T_g\\ \alpha_\infty, & T>T_g\end{cases},$$

where $d(T)$ = density of polymer at temperature T, g/cm³; $k_g \approx 0.681$ = universal for densely cross-linked polymers value of molecular package's coefficient at T_g; $M_{r.f.}$ = molar mass of repeated mesh's fragment, g/mole; 10^{-24} —coefficient of conversion Å³ to cm³; $N_A = 6.023\cdot 10^{23}\ mole^{-1}$ Avogadro's constant.

As $M_{r.f.} = 441<l>-43$ g / mole (by analogy with determination of Van der Waals's volume of repeated mesh's fragment), so:

$$d(T)=1.13(441<l>-43)\Big/\left(\left[1+\alpha\left(T-T_g\right)\right](458.9<l>-57.4)\right) \quad (8)$$

Equation (8) satisfactorily describes experimental data. So, for example, averaged by all experimental models values at 25°C: $d_{calc.}(25)=1.14\ \tau/cm^3$, $d_{exp.}(25)=1.11\ \tau/cm^3$ (ε = 3%).

Molar masses of elementary links are equal to:

$$M_\sigma = \underbrace{4M_C+8M_H}_{-(CH_2)_4-} = 56 \text{ g/mole};\quad M_{3f}=\underbrace{M_N+3M_C+6M_H}_{N(CH_2)_3-}=56 \text{ g / mole};$$

$$M_{2f} = \underbrace{M_N+8M_C+17M_H}_{-(CH_2)_2N-CH_2-(CH_2)_4-CH_3} = 127 \text{ g / mole};$$

$$M_\pi = 2\left(\underbrace{M_C+M_O+2M_H}_{-CH-OH}+\underbrace{M_C+2M_H}_{-CH_2-}+\underbrace{M_O}_{-O-}+\underbrace{6M_C+4M_H}_{-C_6H_4-}\right)+\underbrace{3M_C+6M_H}_{>C(CH_3)_2}=$$

$$=314 \text{ g / mole}.$$

Expression of density (8) and values of elementary links' molar masses substitute to Equation (7)

$$C_{tot}=1.13(5+2x)(441<l>-43)\Big/\left((796+441x)(458.9<l>-57.4)\right) \quad (9)$$

Pass to determination of front-factor. Since for densely cross-linked polymers it is values increase with growth of mesh's points' frequency by linear law in range 0.650.85 [3], so

$$F = 1.125 n_{3f} + 0.4. \quad (10)$$

C_{tot} and F in Equation (5) substitute to expressions (9) and (10), respectively:

$$A_{\infty} = (796 + 441x)(458.9 < l > -57.4) / (1.695(1.125n_{3f} + 0.4)R(5 + 2x)(441 < l > -43)n_{3f}).$$

Derived theoretical and experimental values of A_{∞} and F are presented in Table 2. It is shown that experimental data are satisfactorily described by theoretical values of A_{∞} and F (Figure 2).

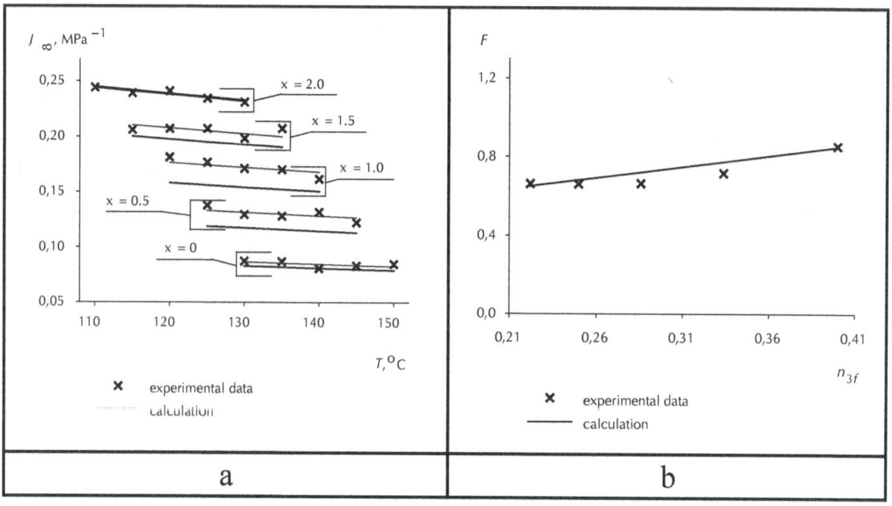

FIGURE 2 Dependence of equilibrium shear pliability on temperature (a); dependence of front-factor on n_{3f} (b).

Rating of electromagnetic susceptibility's equilibrium coefficient for densely cross-linked polymers will realize according to the equation [4]:

$$\xi_{\infty} = a_1 / \left(2\sqrt{\varepsilon_0}\right) \tag{11}$$

where a_1 = one of coefficients, introducing when record inductive capacity's tensor [5]; ε_0 = inductive capacity's of strainless polymer at structure glass transition temperature.

Expression, determined for a_1 (hyperelastic state), has next form:

$$a_1 = -(6K/\alpha_{\infty})(\partial\delta\varepsilon/\partial T),$$

where K = parameter, defined by topological structure of polymer (particularly, for linear polymers it is value is equal to 1); $\partial\delta\varepsilon/\partial T$ = summand of derivative of temperature from inductive capacity, reflecting appearance of electromagnetic anisotropy under the influence of mechanical field to polymer, K^{-1}, or degree^{-1}.

After substitution of expression for a_1 to Equation (11) get

$$\xi_\infty = -\left(3K/\left(\alpha_\infty\sqrt{\varepsilon_0}\right)\right)\left(\partial\delta\varepsilon/\partial T\right) \tag{12}$$

Unknown quantities in Equation (12) are K, ε_0, and $\partial\delta\varepsilon/\partial T$. Consider in detail determination each of them. Parameter K for experimental models, being in hyperelastic state, will calculate by equation [4]

$$K_{cr.}/K_{lin.} = u_{0,lin.}^{1.0}/u_{0,cr.}^{1.0} \tag{13}$$

where $K_{cr.}$ = parameter of cross-linked polymer; $K_{lin.}$ = parameter of abstract linear polymer, consisting of cross-linked polymer's repeated fragments (for epoxy amine polymers this property can be accepted as inverse to average worth of region by strains of epoxy oligomers); $u_0^{1.0}$ = region's worth of appropriate material by strains. Average value of worth of region by strains of epoxy oligomers, being in hyperelastic state is $u_{0,lin.}^{1.0} \approx 4.7\cdot10^{-2}$, and for densely cross-linked polymer meshes based on them $u_{0,cr.}^{1.0} \approx 3.0\cdot10^{-3}$ [4]. Then $K_{cr.} = 15.7$.

Express from dependence [2]

$$\left((\varepsilon_0-1)/(\varepsilon_0+2)\right)\left(M_{r.f.}/d(T)\right) = \left(\sum_i (R_D)_i + \sum_j (\Delta R_D)_j\right)_{r.f.}$$

(here: $(R_D)_i$ = atomic refractions of atoms by Eisenlor, cm³/mole; $(\Delta R_D)_j$ = adjustment for orientation of dipoles) with account of temperature function of density inductive capacity

$$\varepsilon_0 = \frac{10^{-24}\left[1+\alpha(T-T_g)\right]N_A\left(\sum_i \Delta V_i\right)_{r.f.} + 2k_g\left(\sum_i (R_D)_i + \sum_j (\Delta R_D)_j\right)_{r.f.}}{10^{-24}\left[1+\alpha(T-T_g)\right]N_A\left(\sum_i \Delta V_i\right)_{r.f.} - k_g\left(\sum_i (R_D)_i + \sum_j (\Delta R_D)_j\right)_{r.f.}}.$$

By the same principle of that determination of Van der Waals's volume of repeated mesh's fragment

$$\left(\sum_i (R_D)_i + \sum_j (\Delta R_D)_j\right)_{r.f.} = 131.476 < l > -14.954 \text{ sm}^3/\text{mole}.$$

Then

$$\varepsilon_0 = \frac{10^{-24}\left[1+\alpha(T-T_g)\right]N_A(458.9<l>-57.4) + 2k_g(131.476<l>-14.954)}{10^{-24}\left[1+\alpha(T-T_g)\right]N_A(458.9<l>-57.4) - k_g(131.476<l>-14.954)} \tag{14}$$

According to Equation (14), inductive capacity for strainless experimental models at their structure glass transition temperatures is equal to:

$$\varepsilon_0 = (455.465782<l>-54.939368)/(186.860314<l>-24.388346).$$

In [2] for calculation $\partial\delta\varepsilon/\partial T$ was suggested equation

$$\partial\delta\varepsilon/\partial T = \left(\sum_i \delta C_i \Delta V_i + \sum_i \delta C_{i,imi}\right)_{r.f.} \Big/ \left(\sum_i \Delta V_i\right)_{r.f.} \qquad (15)$$

As electromagnetic anisotropy, appearing in strain of polymers in hyperelastic state, is made for deformative and orientation effects, by calculation $\left(\sum_i \delta C_i \Delta V_i + \sum_i \delta C_{i,imi}\right)_{r.f.}$ increment δC_h for molecular fragment >CH-OH was taken into account in, though increments b_h and β_h for that fragment were eliminated from intermolecular interaction during calculation of glass transition temperature and thermal expansion's coefficients.

$$\left(\sum_i \delta C_i \Delta V_i + \sum_i \delta C_{i,imi}\right)_{r.f.} = (-212.7362 < l > -5.309)10^{-6}\ \mathring{A}^3/\mathrm{K}.$$

Final equation for determination of electromagnetic susceptibility's equilibrium elastic coefficient after substitution of the last expression and expression for calculation of inductive capacity taking into account coefficient $K_{cr.}$ and before determined thermal expansion's coefficient in hyperelastic state and Van der Waals's volume of repeated mesh's fragment to the Equation (12) is look as follows:

$$\xi_\infty = 0.07\left[\frac{212.7362 < l > +5.309}{458.9 < l > -57.4}\right]\sqrt{\frac{186.860314 < l > -24.388346}{455.465782 < l > -54.939368}} \qquad (16)$$

TABLE 2 Equilibrium constants.

x	$< l >$	n_{3f}	A_∞, K/MPa			F			ξ_∞		
			calc.	exper. [1]	ε, %	calc.	exper.	ε, %	calc.	exper. [1]	ε, %
0.0	1.00	0.4000	33.5	35.0	4	0.8500	0.8540	5	0.0240	0.0263	9
0.5	1.25	0.3333	47.3	53.0	11	0.7750	0.7164	8	0.0230	0.0235	2
1.0	1.50	0.2857	62.1	69.3	10	0.7214	0.6636	9	0.0220	0.0224	2
1.5	1.75	0.2500	77.7	81.6	5	0.6813	0.6608	3	0.0220	0.0207	6
2.0	2.00	0.2222	93.9	93.5	4	0.6500	0.6613	2	0.0220	0.0192	15

Determined by Equation (16) values are presented in Table 2. Testimonial, first of all about correctness of the introduced methodology and advanced on it is base computations is a little discrepancy between calculated and experimental values 215% and is clearly presented in Figure 3.

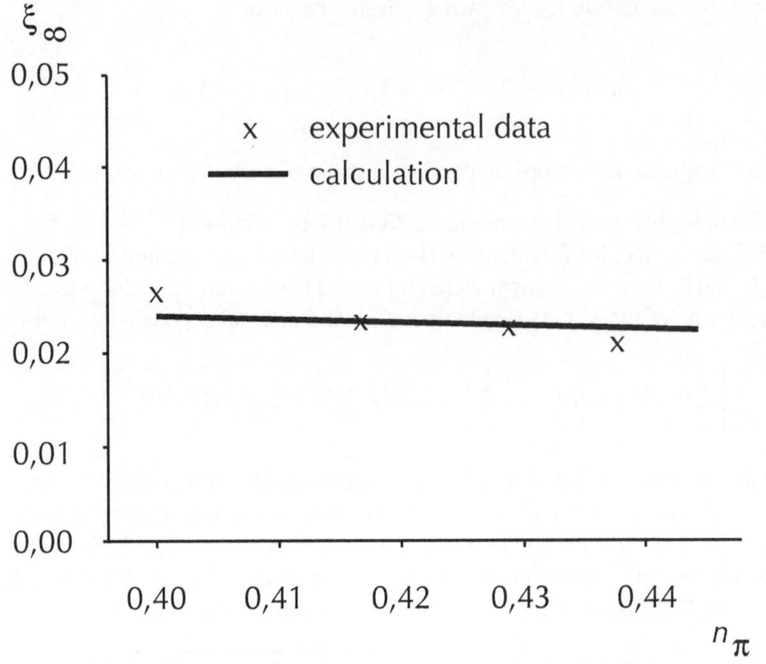

FIGURE 3 Dependence of electromagnetic susceptibility's equilibrium coefficient on statistical parameter, which defines amount of fragments DGEBA in experimental models' topological structure.

5.3.3 Weighting Coefficients

It is proposed following conception of weighting coefficient $w_{J,\beta}$ [1]:

$$w_{J,\beta} = \left(J_{\beta,\infty} T\right) / A_{\infty} \tag{17}$$

where $J_{\beta,\infty}$ = shear pliability of glassy state ($T < T_g - 15\ ^{\circ}\mathrm{C}$), MPa^{-1}. In experiment $J_{\beta,\infty}$ was determined at 25°C.

Equation for calculation of $J_{\beta,\infty}$ will deduce from expression for difference of main values of inductive capacity's tensor [3, 5]

$$\varepsilon_1 - \varepsilon_2 = (1/2) J a_1 \left(\tau_1 - \tau_2\right) \tag{18}$$

where J = shear pliability's relaxation operator, MPa^{-1}; τ_1, τ_2 = main values of shear tension, MPa.

Since inductive capacity is practically identical with square of refractive index $\varepsilon \approx n^2$ [5], so

$$\varepsilon_1 - \varepsilon_2 = n_1^2 - n_2^2 = \left(n_1 - n_2\right)\left(n_1 + n_2\right) = \left\{\begin{matrix} n_1 - n_2 = \Delta n \\ \left(n_1 + n_2\right)/2 \approx n_0 \quad [4] \end{matrix}\right\} = 2 n_0 \Delta n \tag{19}$$

where Δn = strain electromagnetic anisotropy; n_0 = refractive index of strainless polymer dielectric in glassy state.

After substitution of expression for $\varepsilon_1 - \varepsilon_2$ from Equation (19) to Equation (18)

$$\Delta n = \left(Ja_1/\left(4n_0\right)\right)\Delta\tau \tag{20}$$

Formula (20) is one of record forms of BrusterVertgame's law [4], [5], therefore

$$J_{\beta,\infty} = 4n_0 C_{\beta,\infty}/a_1 \tag{21}$$

where $C_{\beta,\infty}$ = strain electromagnetic susceptibility of glassy state, MPa^{-1}.
Will make a rating of all parameters, including in Equation (21)

As a result of that $\varepsilon \approx n^2$, for determination of refractive index at 25°C will use derived above expression of inductive capacity (14)

$$n_0 = \sqrt{\frac{0.6023\left[1+3.9\cdot10^{-4}\left(298-T_g\right)\right]\left(458.9<l>-57.4\right)+1.362\left(131.476<l>-14.954\right)}{0.6023\left[1+3.9\cdot10^{-4}\left(298-T_g\right)\right]\left(458.9<l>-57.4\right)-0.681\left(131.476<l>-14.954\right)}}.$$

Strain electromagnetic susceptibility of cross-linked polymers at 25°C will determine by introduced in [2] equation

$$C_{\beta,\infty} = \left(\left(\sum_i C_i\right)_{r.f.}\Big/\left(10^{-24}N_A\left(\sum_i V_i\right)_{r.f.}\right)\right) + \mathrm{P} \tag{22}$$

where $\left(\sum_i C_i\right)_{r.\delta.}$ = increments set of repeated mesh's fragment, which defines contribution of each atom and type of intermolecular interaction to $C_{\beta,\infty}$, cm^3/(MPa·mole); $\mathrm{P} = 3.61387\cdot10^{-4}$ MPa^{-1}—universal parameter.

Since, as electromagnetic anisotropy, appearing during strain of polymers in glassy state is connected with strain of valence angles and change of interatomic distances, so in calculation $\left(\sum_i C_i\right)_{r.f.}$ increment C_h for molecular fragment >CH-OH was follows to take into account with introduction of multiplier 0.5.

$$\left(\sum_i C_i\right)_{r.f.} = \left(11.4065-86.8863<l>\right)10^{-3}\ \mathrm{sm}^3/\left(\mathrm{MPa}\cdot\mathrm{mole}\right).$$

Final calculation equation gets after substitution of expression $\left(\sum_i C_i\right)_{r.f.}$ to formula (23):

$$C_{\beta,\infty} = \left(\left(\left(11.4065-86.8863<l>\right)\cdot10^{-3}\right)\Big/\left(0.6023\left(458.9<l>-57.4\right)\right)\right)+3.61387\cdot10^{-4},\ \mathrm{MPa}^{-1}.$$

Theoretical and experimental values of $C_{\beta,\infty} = 4.9\cdot10^{-5}$ MPa^{-1} and $5.3\cdot10^{-5}$ MPa^{-1}, respectively (ε = 7%) were averaged by all experimental models.

Expression a_1 for densely cross-linked polymers in glassy state is analogos to the same, derived for densely cross-linked polymers in hyperelastic state, but in the formula is used thermal expansion's coefficient of glassy state

$$a_1 = -\left(6K/\alpha_g\right)\left(\partial\delta\varepsilon/\partial T\right).$$

Coefficient K will calculate in the network of above described approach. For vitrificated epoxy oligomer average value $u_{0,lin.}^{1.0} \approx 15.4 \cdot 10^{-3}$, and for vitrificated epoxy amine polymers based on their $u_{0,cr.}^{1.0} \approx 7.0 \cdot 10^{-4}$ [4]. In concordance with Equation (13), $K_{cr} = 22 \cdot$

Derivative $\partial\delta\varepsilon/\partial T$ for experimental models in glassy state will determine by Equation (16). Peculiarity of increments set's calculation $\left(\sum_i \delta C_i \Delta V_i + \sum_i \delta C_{i,imi}\right)_{r.f.}$ for glassy state of experimental models consists in the fact that as in the case with calculation $\left(\sum_i C_i\right)_{r.f.}$, increment δC_h for molecular fragment >CH-OH follows to take into account with introduction of multiplier 0.5.

$$\left(\sum_i \delta C_i \Delta V_i + \sum_i \delta C_{i,imi}\right)_{r.f.} = \left(-146.6962 < l > -5.309\right)10^{-6}\ \mathring{A}^3/\mathrm{K}$$

TABLE 3 Weighting coefficients.

x	$<l>$	$w_{J,\beta}$			$w_{C,\beta}$		
		theor.	**exper. [1]**	**ε, %**	**theor.**	**exper. [1]**	**ε, %**
0.0	1.00	0.0220	0.0180	22	0.0360	0.0280	29
0.5	1.25	0.0157	0.0150	5	0.0270	0.0260	4
1.0	1.50	0.0120	0.0140	14	0.0210	0.0230	9
1.5	1.75	0.0096	0.0120	20	0.0170	0.0200	15
2.0	2.00	0.0080	0.0080	0	0.0140	0.0180	22

Thereby, substitution of n_0 and a_1 in Equation (21) for their expressions, and of $C_{\beta,\infty}$ for it is value supports at the average of identical for all experimental models quantity $J_{\beta,\infty}$ at 25°C2.5·× 10^{-3} MPa^{-1}. Average experimental value $J_{\beta,\infty}$ at 25°C2.6·× 10^{-3} MPa^{-1} (ε = 4%). Theoretical values of weighting coefficient $w_{J,\beta}$ (Table 3) are calculated from values $J_{\beta,\infty}$ and A_∞ by Equation (17).

Theoretical determination of weighting coefficient $w_{C,\beta}$ will bring to equation [1]

$$w_{C,\beta} = C_{\beta,\infty} T / \left(0.5\xi_\infty A_\infty\right) \tag{23}$$

Rating of all, taking part in Equation (23) parameters is demonstrated above. Experimental and derived by Equation (23) values of $w_{C,\beta}$ are presented in Table 3.

5.4 PARAMETERS OF α-RELAXATION TIMES' SPECTRUM

For theoretical rating of average α-relaxation times at one or another temperature will use derived by analytical way in [1] temperature function of α-relaxation times. In consequence of narrow molecular mass distribution of cross-site chains in experimental models will take $\Xi_{J,\alpha} = 0.5$.

5.5 MODELING OF THERMOMECHANICAL AND THERMO OPTICAL EXPERIMENTS' COURSE

Knowing A_∞, ξ_∞, $w_{J,\beta}$, $w_{C,\beta}$, $\Theta_{J,\alpha}(T)$, $\Xi_{J,\alpha}$, it is possible to model different temperature time realizations of cross-linked polymers' viscoelastic and electromagnetic properties, progressing without destruction of structure the latter (Figure 4).

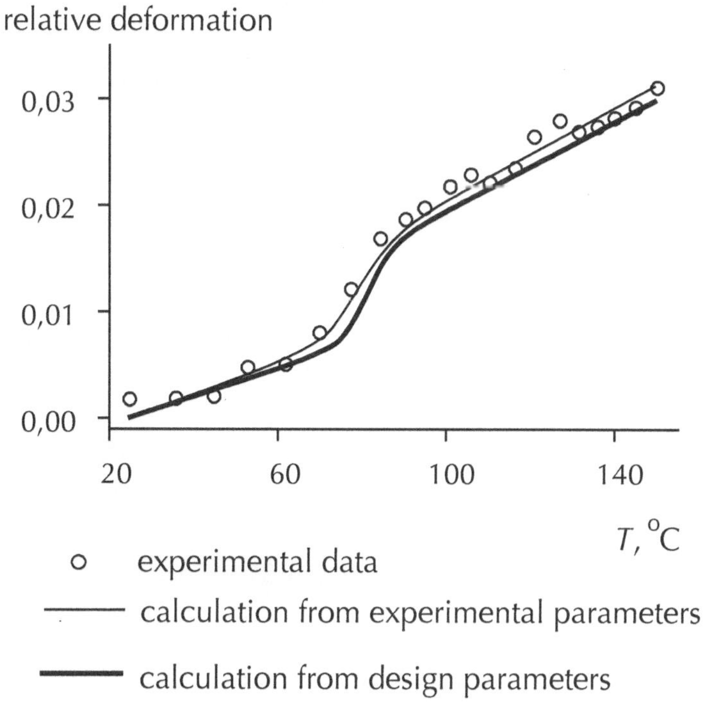

FIGURE 4 Thermomechanical dependence for model with structure x = 2 (heating ratio 0.9°C/min, load 720 g).

5.6 CONCLUSION

Thereby demonstrated in the chapter methodology of theoretical determination of developed in [1] mathematical model's constants will be helpful for apriori ratng of interconnected densely cross-linked polymers' viscoelastic and electromagnetic properties in all their physical states.

KEYWORDS

- **Cross-linked polymer**
- **Electromagnetic**
- **Glass transition temperature**
- **Linear fragments**
- **Van der Waals's volume**

REFERENCES

1. Ulitin, N. V., Deberdeev, T. R., and Deberdeev, R. Ya. *The new hereditary operators of interrelated viscoelastic and electromagnetic properties of densely cross-linked polymers* (2012).
2. Askadskii, A. A. and Kondrashenko, V. I. Computer science of materials of polymers. Vol.1. Atomic-molecular level, M. *Scientific world*, p. 544 (1999).
3. Bajenov, S. L., Berlin, Al. Al., Kulkov, A. A., and Oshmyan, V. G. Polymer composite materials: *Solidity and technology*. Dolgoprudnii, Intellect, p. 352 (2010).
4. In *The photoelasticity's method* vol 3. G. L. Khesin (Ed.). M. Gostechizdat, (1975).
5. Blythe, T. and Bloor, D. *Electrical properties of polymers*. Cambridge University Press, Cambridge, p. 492 (2005).

6 Interrelation of Viscoelastic and Electromagnetic Properties of Densely Cross-linked Polymers: Part II

N. V. Ulitin, T. R. Deberdeev, R. Ya. Deberdeev, and G. E. Zaikov

CONTENTS

6.1 Introduction 49
6.2 Theory 50
6.3 Synthesis and Methods of Study of Experimental Objects 52
6.3.1 Synthesis 53
6.3.2 Methods of Study 53
6.4 Discussing the Results 53
6.4.1 Thermal Properties of Experimental Models' Physical States 53
6.4.2 Hyperelastic State's Constant and Electromagnetic Susceptibility's Equilibrium Coefficient 53
6.4.3 Weighting Coefficients 54
6.4.4 Parameters of A -Relaxation Times' Spectrum 54
6.4.5 Prediction of Thermomechanical and Thermooptical Experiments Course 56
6.5 Conclusion 57
Keywords 59
References 59

6.1 INTRODUCTION

During projection of fabrication high strength radioparent fiberglass product's technological process appears a task of choice of optimum cross-linked polymer matrix, which has combination of as much as possible isotropic and stable radioparence coefficient and good viscoelastic propertieshigher value of viscoelastic module, or lower it is inverse propertypliability in specified fields of temperatures and tensions (determined by conditions of product's operation). Rating of these rates and accordingly substantiation of polymer system's choice is realized in the network of various mathematical descriptions [1]. However, calculations of them is not realized forever, since by modeling of viscoelastic properties module's or pliability's momentary constituent inputs without affixment to defined cross-linked polymers' physical state, and appropriateness of temperature time realization of cross-linked polymers' strain electromagnetic anisotropyquantity, inversely proportional to isotropy of radioparence's coefficient, were derived only for their hyperelastic state. Therefore objective of this chapter is to develop conceptions about change of cross-linked polymers interconnected viscoelastic and electromagnetic properties in different physical states.

6.2 THEORY

In polymer deformation's tensor [2]:

$$u_{ik} = (1/2) J \tau_{ik} - (1/3) B_\infty p \delta_{ik} \tag{1}$$

(where J = shear pliability's relaxation operator, MPa^{-1}; τ_{ik} = transverse strain's tensor, MPa; B_∞ = equilibrium volumetric pliability, MPa^{-1}; p = pressure, MPa; δ_{ik} = Krokener's tensor) will consider only shear pliability's relaxation operator, since contribution of B_∞ to u_{ik} for cross-linked polymers is slight. This will discuss the case, when deformation of polymer system happens without destruction of mesh's chemical structure. According to legacy conceptions [3]: (1) for densely cross-linked polymers with homogeneous topological structure (hereinafter all discussions are concerned only to such cross-linked polymers, since forming of said structure is guarantee of matrix's properties in full [2]) is specified single mode distribution of α-relaxation times, and for densely cross-linked polymers with heterogeneous topological structuremulti-modal distribution; (2) mesh points in local movements do not take part. Then will determine J as:

$$\boldsymbol{J} = J_\infty \left[w_{J,\beta} + (1 - w_{J,\beta}) \int_{-\infty}^{\infty} \frac{[1 - \exp(-t/\theta)] \sin[\pi(1 - \Xi_{J,\alpha})]}{2\pi\left(\mathrm{ch}[(1 - \Xi_{J,\alpha}) \ln(\theta/\Theta_{J,\alpha})] + \cos[\pi(1 - \Xi_{J,\alpha})]\right)} \mathrm{d}\ln\theta \right] \tag{2}$$

where J_∞ = equilibrium shear pliability in specified temperature, MPa^{-1}; $w_{J,\beta}$ = weighting coefficient, not depending on temperature and is being contribution to equilibrium shear pliability of local conformational mobility of polymeric meshe's cross-site chains;

$\boldsymbol{J}_{N,\alpha} = \int_{-\infty}^{\infty} L_{J,\alpha}(\theta)[1 - \exp(-t/\theta)] \mathrm{d}\ln\theta$ = normalized to 1, fractional exponential operator;

$$L_{J,\alpha}(\theta) = \frac{\sin[\pi(1 - \Xi_{J,\alpha})]}{2\pi\left(\mathrm{ch}[(1 - \Xi_{J,\alpha}) \ln(\theta/\Theta_{J,\alpha})] + \cos[\pi(1 - \Xi_{J,\alpha})]\right)}$$

Normalized to 1, distribution of α-relaxation times ($\Theta_{J,\alpha}$ = depending on temperature, average relaxation time of α-mode); $\Xi_{J,\alpha}$ = not depending on temperature, width of α = mode ($0 \le \Xi_{J,\alpha} \le 1$)) [4]; θ = α-relaxation time, t = current time.

Operator $J_{N,\alpha}$ in glassy and hyperelastic states of cross-linked polymers is equal to from 0 to 1, respectively, and in transition region between these conditions from 0 to 1. Therefore Equations (1) and (2) reproduce change of concerned cross-linked polymers hyperelastic properties in all their physical states: in hyperelastic, where is being momentary α-process, shear pliability's relaxation operator is equal to equilibrium shear pliability; in glassy, where is only local conformational mobility of polymeric mesh's cross-site chains, shear pliability's relaxation operator is equal to shear pliability of glassy state $J_{\beta,\infty} = w_{J,\beta} J_{\infty}$; in transition region between these states, where both types of relaxation processes progress together, shear pliability's relaxation operator realizes in full mathematical record.

Input analog conception for associated with viscoelastic pliability of electromagnetic properties.

Origin of strain electromagnetic anisotropy in deformated polymer dielectric is described by the second term of expression for inductive capacity's tensor [5]

$$\varepsilon_{ik} = \varepsilon_0 \delta_{ik} + a_1 \gamma_{ik} + (1/3)(a_1 + 3a_2) u_{ll} \delta_{ik} \tag{3}$$

where ε_0 = inductive capacity of strainless polymer object; γ_{ik} = shear deformation's tensor; a_1, a_2 = some coefficients; u_{ll} = compression deformation's tensor.

In other words, at the expense of shear deformation appears reducing to electromagnetic anisotropy of environment orientation of segments in every point of polymer dielectric, in this time is being it is polarization.

In that case, if through it passes electromagnetic radiation it becomes source of electromagnetic oscillations. From expression (3) may be derived Bruster Vertgame's equation

$$\Delta n = \xi_{\infty} \Delta \gamma = C_{\infty} \Delta \tau \tag{4}$$

where Δn = strain electromagnetic anisotropy; ξ_{∞} = electromagnetic susceptibility's equilibrium elastic coefficient; $\Delta \gamma$ and $\Delta \tau$ (MPa)differences of main shear deformations and tensions, respectively; C_{∞} = equilibrium strain electromagnetic susceptibility (MPa^{-1}), which is linked with J_{∞} for cross-linked polymers by appropriateness [5]

$$C_{\infty} = 0.5 \xi_{\infty} J_{\infty} \tag{5}$$

It was proved experimentally, that expressions (4) and (5) are effective only for hyperelastic state of cross-linked polymers. To describe change of strain electromagnetic anisotropy in all cross-linked polymers' physical states will consider strain electromagnetic susceptibility's C (MPa^{-1}) and electromagnetic susceptibility elastic coefficient's ξ relaxation operators. Suppose that they are submitted by such appropriateness, by which equilibrium properties in equation (5)

$$C = 0.5\xi J \tag{6}$$

According to Equation (6), electromagnetic susceptibility elastic coefficient's relaxation spectrum coincides with shearing modulus' spectrum. Then operator ξ will be written as

$$\xi = \xi_{\beta,\infty} + \xi_{\alpha,\infty} G_{N,\alpha},\ \xi_{\beta,\infty} = w_{\xi,\beta}\xi_{\infty},\ \xi_{\alpha,\infty} = (1 - w_{\xi,\beta})\xi_{\infty},$$

where $\xi_{\beta,\infty}$ = constituent of electromagnetic susceptibility elastic coefficient, due to local conformational mobility of polymeric mesh's cross-site chains, in fact it is electromagnetic susceptibility's elastic coefficient of glassy state; $\xi_{\alpha,\infty}$ = constituent of electromagnetic susceptibility's elastic coefficient, due to co-operative mobility of mesh's points; $G_{N,\alpha}$ = fractional exponential operator, which inverse $J_{N,\alpha}$; $w_{\xi,\beta}$ = weighting coefficient, not depending on temperature and describing part, which is inserted into value of electromagnetic susceptibility's elastic coefficient ξ_{∞} by local conformational mobility of mesh's cross-site chains.

Using multiplicative rule of normalized fractional exponential operators [4], from Equation (6) was derived expression of strain electromagnetic susceptibility's relaxation operator

$$C = C_{\beta,\infty} + C_{\alpha,\infty} J_{N,\alpha},\ C_{\beta,\infty} = w_{C,\beta} C_{\infty},\ C_{\alpha,\infty} = (1 - w_{C,\beta}) C_{\infty},\ w_{C,\beta} = w_{\xi,\beta} w_{J,\beta}, \tag{7}$$

where $C_{\beta,\infty}$ = constituent of strain electromagnetic susceptibility, due to local conformational mobility of mesh's cross-site chains in other words, it is strain electromagnetic susceptibility of glassy state, MPa^{-1}; $C_{\alpha,\infty}$ = constituent of strain electromagnetic susceptibility, due to co-operative mobility of mesh's points, MPa^{-1}; $w_{C,\beta}$ = weighting coefficient, not depending on temperature and describing part, which is inserted into value of equilibrium strain electromagnetic susceptibility C_{∞} by local conformational mobility of mesh's cross-site chains.

From Equation (7) follows conclusion about the same relaxation spectrum for strain electromagnetic susceptibility and shear pliability. This means that defined cooperative mobility of polymer meshe's points' α-relaxation times' spectrum reproduces both strain and electromagnetic response of cross-linked polymer system to external mechanical force. It should be noted that as well as shear pliability's relaxation operator, strain electromagnetic susceptibility's relaxation operator generalizes in it is record all physical states of cross-linked polymers.

6.3 SYNTHESIS AND METHODS OF STUDY OF EXPERIMENTALOBJECTS

Experimental restoration of inserted relaxation operators and testing of their prediction ability were made for cross-linked epoxy amine polymers from diglycide ether of bisphenol-A (DGEBA), which was solidificated by stoichiometric amounts of 1-aminohexane (AH) and hexamethylenediamine (HMDA) in variation of their molar ratio from 0 to 2 with pitch 0.5.

6.3.1 Synthesis

easured off amounts of DGEBA and HMDA were heated to 42°C; then to DGEBA was added required amount of AH and melt HDMA; compositions were mixed and degassed in rate of reiteration freezing defrosting cycles; after that ampoules were being filled by argon and soldered. On the base of principles, produced in [2], for the purpose of receipt of spatially homogeneous topological cross-linked structure was determined common for all compositions temperature time rate of their hardening: 20°C 72 hr, 50°C 72 hr, 80°C 72 hr, 120°C 72 hr. Part of gel-fraction was being determined in acetone and for all compositions was at the average 98%.

6.3.2 Methods of Study

All experiments (solid expansion, mechanical, and polarization optical) were made on installation [6]. In every certain experiment models were subjected by several independent tests. In solid expansion experiment models were cooled with average ratio 0.4°C/min range of temperatures 15025°C. Experimental values of shear pliability's relaxation operator in different physical states of study's objects were determined by Equation (1) on the base of experimental values of strain and known value of average transverse strain for usable form of models. In study's experiments of electromagnetic anisotropy in the capacity of electromagnetic radiation was used monochromatic radiation of spectrum's visible partgreen light (λ = 546 HM). Optical double diffraction, which in the case of monochromatic radiation of spectrum's visible part serves as a property of electromagnetic anisotropy, was measured in the model's (in the form of disk) centre. Experimental values of strain electromagnetic susceptibility's relaxation operator in different physical states of study's objects were determined by summary to all cross-linked polymers' physical states equation of Bruster Vertgame, proceeding from experimental values of double diffraction and known difference of main shear tensions in the centre of disk. Properties of cross-linked polymers depend on temperature history [2], therefore before each test was made burning: models were heated to 150°C, kept 3 hr, and then cooled with average ratio 0.2°C/min to temperature 25°C.

6.4 DISCUSSING THE RESULTS

6.4.1 Thermal Properties of Experimental Models' Physical States

By the results of experimental models' solid expansion study were determined temperature fields of their physical states. Values of glass transition temperature T_g are presented in Table 1. Thermal expansion's coefficients at the average are identical for all experimental models: in glassy state 4.3×10^{-4} degree^{-1}, in hyperelastic 7.0×10^{-4} degree^{-1}.

6.4.2 Hyperelastic State's Constant and Electromagnetic Susceptibility's Equilibrium Coefficient

Equilibrium shear pliability of cross-linked polymers with spatially homogeneous topological structure is inversely proportional to absolute temperature T (K) [2]

$$J_\infty = A_\infty / T \tag{8}$$

where A_∞ = hyperelastic state's constant, which does not depend on temperature, K/MPa.

In experiment was shown good equivalence of equilibrium shear pliability's temperature dependences to Equation (8) that proves spatially homogeneity of experimental models' topological structure and rightness of synthetic methodology. It became clear, that hyperelastic state's constant does not depend on temperature and is determined by the length of meshe's cross-site chains (Table 1). As hyperelastic state's constant electromagnetic susceptibility's equilibrium elastic coefficient does not depend on temperature and is parameter, which is sensitive to the density of cross-link (Table 1). That is in growth of hardener amounts' molar ratio the density of cross-link decreases, cross-site chains' average numerical degree of polymerization increases, and values of electromagnetic susceptibility's equilibrium elastic coefficient fall that is explained by increasing level of intermolecular interaction with increase of cross site chains' average numerical degree of polymerization.

6.4.3 Weighting Coefficients

Weighting coefficients (Table 1) were estimated by equations

$$w_{J,\beta} = J_{\beta,\infty}^{T_0} T_0 / A_\infty, \quad w_{C,\beta} = C_{\beta,\infty}^{T_0} T_0 / (0.5\xi_\infty A_\infty)$$

where $J_{\beta,\infty}^{T_0}$, $C_{\beta,\infty}^{T_0}$ = shear pliability and strain electromagnetic susceptibility of glassy state ($T_0 = 25^{\circ}\mathrm{C}$), MPa^{-1}.

Decrease of weighting coefficients' values with increase of hardeners' molar ratio, probably, also is being explained by increasing level of intermolecular interaction with elongation of cross-site chains.

6.4.4 Parameters of α-Relaxation Times' Spectrum

Parameters of α-relaxation times' spectrum were determined using the least-squares method by isothermal creep's and photocreep curves that were obtained for several temperatures in glass transition region. Experimentally was proved, that relaxation spectrum for shear pliability and strain electromagnetic susceptibility is identical. More one evidence of shear pliability's and strain electromagnetic susceptibility's spectrums coincidence may be slight discrepancy not more than 10% of photocreep's values, converted to mechanical scale (with use of relaxation spectrum's parameters defined by creep's curves) by dependence from the real experimental values.

$$C_N = \left(1 - \left(\left(1 - w_{C,\beta}\right) / \left(1 - w_{J,\beta}\right)\right)\right) + \left(\left(1 - w_{C,\beta}\right) / \left(1 - w_{J,\beta}\right)\right) J_N,$$

$$C_N = w_{C,\beta} + (1 - w_{C,\beta}) J_{N,\alpha}, \quad J_N = w_{J,\beta} + (1 - w_{J,\beta}) J_{N,\alpha},$$

For temperature function of cross-linked polymers' α-relaxation times was suggested following equation:

$$\lg\left(\Theta_{J,\alpha}(T) / \Theta_{J,\alpha}(T_g)\right) = \begin{cases} 40\left(\left(f_g / \left(f_g + \alpha_g (1 - f_g)(T - T_g)\right)\right) - 1\right), & T < T_g \\ 40\left(\left(f_g / \left(f_g + \alpha_\infty (1 - f_g)(T - T_g)\right)\right) - 1\right), & T > T_g \end{cases} \quad (9)$$

where 40 = coefficient, value of which does not depend on cross-linked polymer's nature and is determined by the basis of a logarithmic scale; f_g = part of fluctuating unconfined space at glass transition temperature; α_g, α_∞ = thermal expansion's coefficients in glassy and hyperelastic states, respectively, degree^{-1}.

Equation (9) was derived from the formula, suggested in [7] for relaxation behavior's description of linear, branched and rarely cross-linked polymers at temperatures above their glass transition temperatures

$$\lg\left(\Theta_{J,\alpha}(T)/\Theta_{J,\alpha}(T_g)\right)=\left(f_g^{/}+\left(0.025\alpha_\infty(1-f_g)/f_g\right)(T-T_g)\right)^{-1}-\left(f_g^{/}\right)^{-1}, \quad (10)$$

with the account of known thermal dependence of fluctuating unconfined space's part and next experimental data [2]: (1) value of fluctuating unconfined space's part for cross-linked polymers at glass transition temperature approximately = 0.09; (2) vitrifying begins at $\lg\Theta_{J,\alpha}(T_g+15)=2$. Value of quantity $f_g^{/}$ in Equation (10) fluctuates near 0.025 and in [7] is identified with unconfined space's part at glass transition temperature of polymer. It should be observed that this value is universal also for densely cross-linked polymers.

Temperature function of α-relaxation times consists of two branches linking at glass transition temperature of polymer , experimental points of each with good accuracy describe Equation (9) (Figure 1). Value f_g was averaged off by all experimental models and was equal to 0.095, that is reconciled with universal quantity for densely cross-linked polymers 0.09 and once again confirms validity of introduced α-relaxation times' temperature function's specific type for cross-linked polymers.

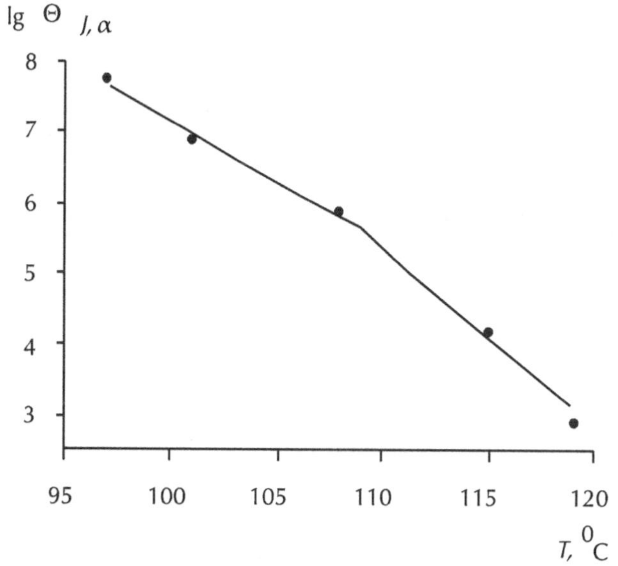

FIGURE 1 Connection of typical α-relaxation times with temperature for object with structure $x = 0$.

In experiment, it was showed, that quantity $\Xi_{J,\alpha}$, how it was supposed in mathematical substitution, does not depend on temperature and is structural sensitive parameter, that is to say it is values increase with growth of hardeners' molar ratio elongation of cross-site chains (Table 1). The latter regularity goes into well known conception about widening of relaxation spectrum with increase length of mesh's cross-site chains [3].

6.4.5. Prediction of Thermomechanical and Thermooptical Experiments Course

Thereby, by the results of different researches for the used experimental objects were determined introduced mathematical description of cross-linked polymers' viscoelastic and electromagnetic properties parameters. Verification of prediction abilities of such kind mathematical descriptions is realized by experiments, conditions of which are different from conditions of the other experiments, where unknown model's parameters are determined. In our case verification was prediction of Thermomechanical and Thermooptical curves' trend (Figure 2, 3)

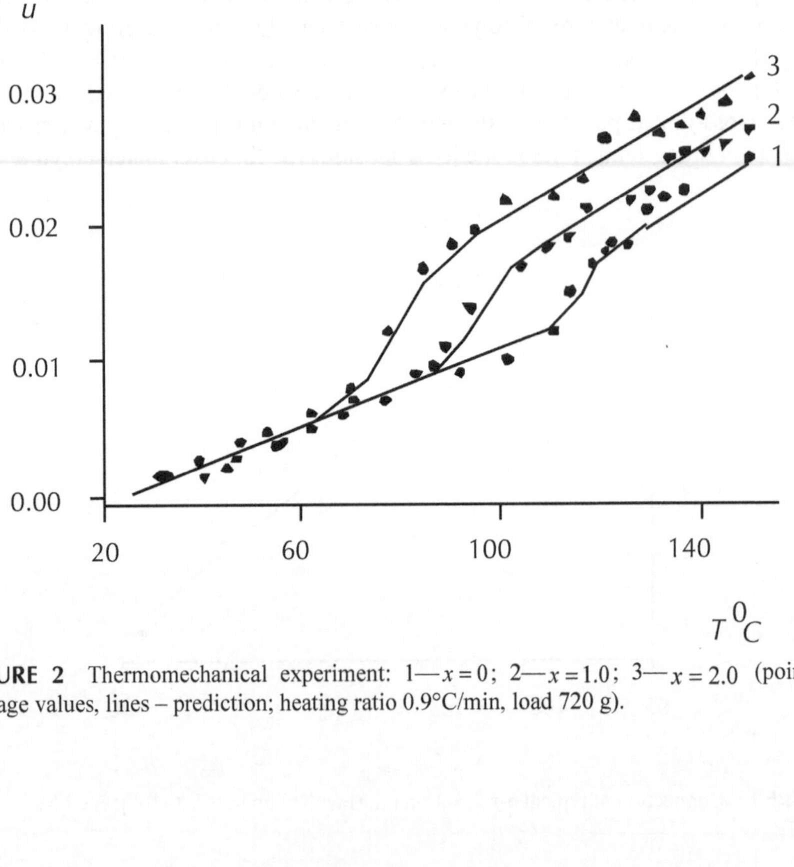

FIGURE 2 Thermomechanical experiment: 1—$x = 0$; 2—$x = 1.0$; 3—$x = 2.0$ (points – average values, lines – prediction; heating ratio 0.9°C/min, load 720 g).

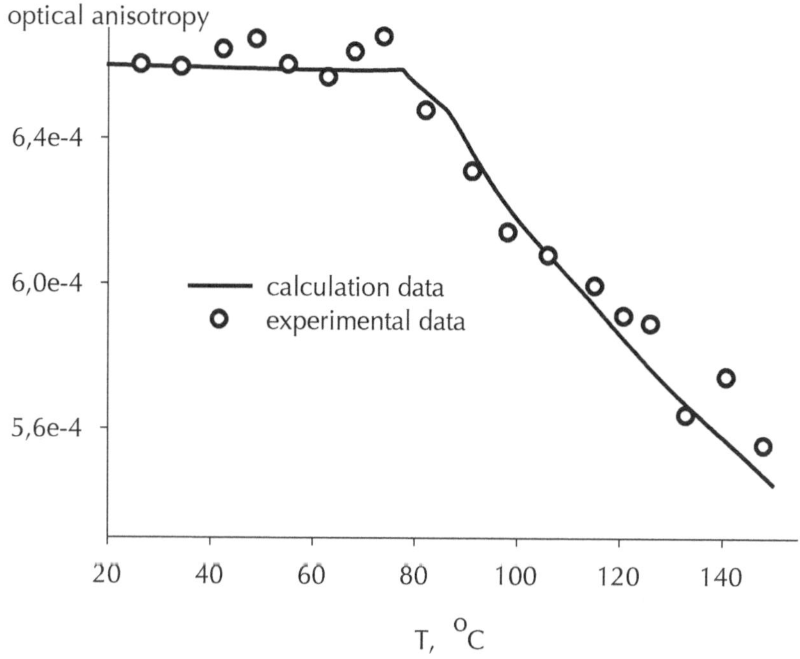

FIGURE 3 Thermooptical curve in cooling rate for object whith structure x = 1.5 (cooling ratio = 0.6°C/min).

TABLE 1 Values of glass transition temperature, hyperelastic state constant, electromagnetic susceptibility equilibrium elastic coefficient, weighting coefficients and width of α mode for the used experimental objects.

x	T_g, °C	A_∞, K/MPa	ξ_∞	$w_{J,\beta}$	$w_{C,\beta}$	$\Xi_{J,\alpha}$
0.0	109	35.0	0.0263	0.018	0.0280	0.374
0.5	99	53.0	0.0235	0.015	0.0260	0.440
1.0	88	69.3	0.0224	0.014	0.0230	0.497
1.5	77	81.6	0.0207	0.012	0.0200	0.557
2.0	71	93.5	0.0192	0.008	0.0180	0.636

6.5 CONCLUSION

Presented in this chapter, mathematical description of interconnected cross-linked polymers' viscoelastic and electromagnetic properties was developed on the base of common heredity's theory. In the main new and key moment was momentary components' of shear pliability and strain electromagnetic susceptibility modeling with the help of weighting coefficients, that allowed to formalize those properties in all their physical states of cross-linked polymers.

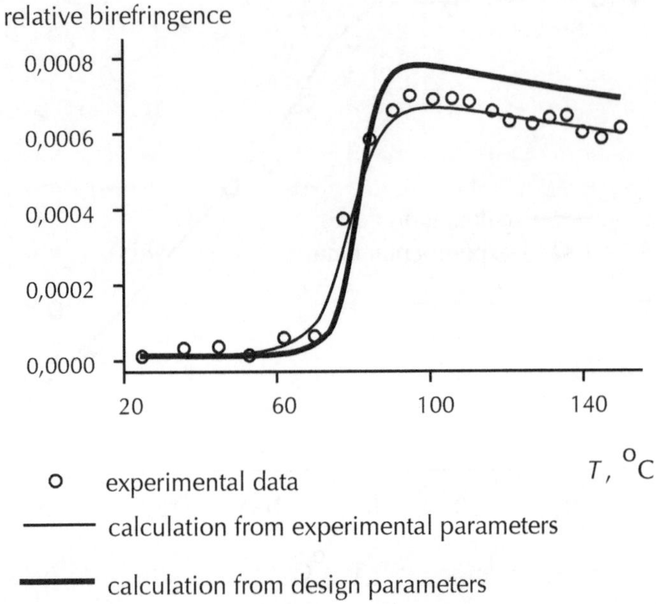

FIGURE 4 Thermooptical dependence for model with structure x = 2 (heaing ratio 0.9°C/min, load 720 g).

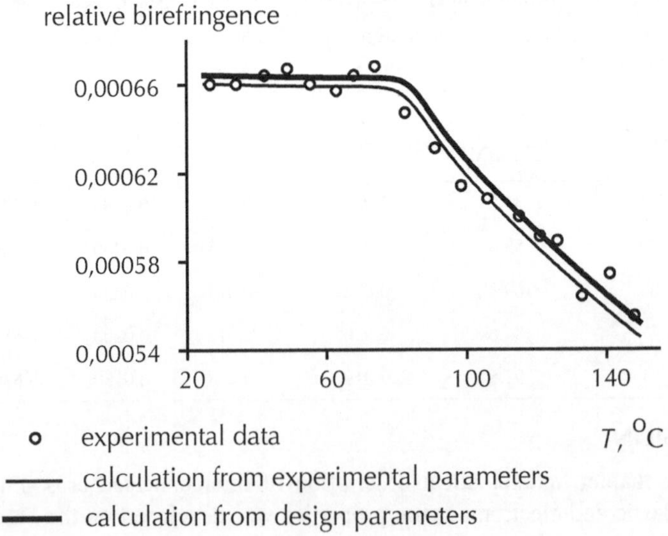

FIGURE 5 Freezing process of optical anisotropy for model with structure x = 1.5 (cooling ratio 0.6°C/min, load 720 g).

In the network of derived regularities can be realized substantiation of cross-linked polymer matrix with inevitable, but considerably reduced strain electromagnetic anisotropy's degree for high strength radioparent fiberglass product. The methodology consists in the following. Required values of viscoelastic properties, which must have cross-linked polymer matrix, on the assumption of known elastic properties of filler and viscoelastic properties of all fiberglass product in conditions of it is using, in present time it can be calculated most exactly by fractal models [8]. Then by equations, presented in this chapter, derived values of matrix's viscoelastic properties are recounted into the strain electromagnetic anisotropy. There is no doubt that indubitable positive side of developed mathematical formalism is operating of parameters, possessing physical sense, experimental values of which for many cross-linked polymers are given in literature and can be determined by results of short-term tests. Therefore, hardships with specific cross-linked matrix choice' substantiation should not appear.

In the end it should be noted that important application's field of presented mathematical models can be their use in calculations of fiberglass products' and constructions' with complicated configuration stress conditions.

KEYWORDS

- **Cross-linked polymers**
- **Fractional exponential operator**
- **Heterogeneous topological structure**
- **Homogeneous topological structure**
- **Mesh's cross-site**
- **Viscoelastic properties**

REFERENCES

1. Gurtovnik, I. G., Sokolov, V. I., Trofimov, N. N., and Shagunov, S. I. Radioparent fiberglass products. *M. Mir*, 368 (2003).
2. Bajenov, S. L., Berlin, A. A., Kulkov, A. A., and Oshmyan, V. G. Polymer composite materials: Solidity and technology. *Dolgoprudnii: Intellect*, 352 (2010).
3. Irzhak, V. I. Topological structure and relaxation properties of polymers. *Progress of Chemistry*, **74**(10), 10251056 (2005).
4. Rabotnov, U. N. Elements of solid matter hereditary mechanics. *Science*, 384 (1977).
5. Blythe, T. and Bloor, D. *Electrical properties of polymers*. Cambridge University Press, Cambridge, p. 492 (2005).
6. Zuev, B. M., Arkhireev, O. S., Filippova, A. P., and Gubanov, E. F. Opto-mechanical properties of tightly sewn polymers based on diallyl and divinyl monomers. *HMC*, **35A**(6), 669674 (1993).
7. Ferry, J. D. *Viscoelastic properties of polymers*. John Wiley & Sons, New York, p. 641 (1980).
8. Kulak, M. I. *Fractal mechanics of polymers*. Higher School, Minsk, p. 304 (2002).

7 Practical Hints on Recovery of Strain Electromagnetic Susceptibility Relaxation

T. R. Deberdeev, R. Ya. Deberdeev, N. V. Ulitin, and G. E. Zaikov

CONTENTS

7.1 Introduction 61
7.2 Experiment 62
7.3 Theoretical Determination of Densely Cross-linked Polymers' Strain Electromagnetic Anisotropy's Mathematical Model Constants 62
7.3.1 Electromagnetic Susceptibility's Equilibrium Elastic Coefficient 62
7.3.2 Weighting Coefficient 65
7.3.3 A-Relaxation Times' Spectrum and Parameters 65
7.4 Numerical Experiment, Modeling Change of Experimental Objects' Strain Electromagnetic Anisotropy under the Influence of Constant Load in Heating and Cooling Conditions 65
7.5 Conclusion 67
Keywords 67
References 67

7.1 INTRODUCTION

Densely cross-linked polymer, which is used as matrix in high strength radioparent fiberglass product with heightened isotropy of radioparence coefficient, must have extremely lower quantity of strain electromagnetic anisotropy in specified intensive polymer's state. For the purpose of creation fundamental basis for possibility of realization optimum densely cross-linked polymer matrixes' selection to these products in research [1] was suggested and tested in experiment by us mathematical model of cross-linked polymers' strain electromagnetic anisotropy, formalism of which included this property in all their relaxation states with clear physical meaning for the first time. This research is pointed at demonstration of theoretical determination constants of introduced model possibility and it is predictive abilities in rating of experimental objects' strain electromagnetic anisotropy' realization under combined action with them of mobile by some laws temperature and load.

7.2 EXPERIMENT

Description of experimental objects' synthesis, methods, and methodologies of their research is presented in [1, 2].

7.3 THEORETICAL DETERMINATION OF DENSELY CROSS-LINKED POLYMERS' STRAIN ELECTROMAGNETIC ANISOTROPY'S MATHEMATICAL MODEL CONSTANTS

Putting into operation of experimental objects' topological structure's statistical parameters and theoretical rating of hyperelastic state's constant are realized in [3].

7.3.1 Electromagnetic Susceptibility's Equilibrium Elastic Coefficient

Electromagnetic susceptibility's equilibrium elastic coefficient determine according to equation [4]:

$$\xi_\infty = a_1 / \left(2\sqrt{\varepsilon_0}\right) \tag{1}$$

where a_1 = one of coefficients, introducing in expression, which establishes link between inductive capacity tensor and deformation's tensor independent components [5]; ε_0 = inductive capacity of unstressed polymer at it is structure glass transition temperature.

For densely cross-linked polymers in glassy state theoretical rating a_1 was made according to equation, that was obtained by us in [3]. Equation of theoretical rating a_1 for densely cross-linked polymers in hyperelastic state is analog to the same derived in [3], but heat expansion's coefficient in glassy state substitutes for the same in hyperelastic:

$$a_1 = -\left(6\partial\delta\varepsilon/\partial T\right)\left(K/\alpha_\infty\right)$$

where $\partial\delta\varepsilon/\partial T$ = one of summand, on which is factorized inductive capacity's derivative by temperature and which is responsible for appearance of electromagnetic anisotropy in polymer during influence on it load, K^{-1}, or degree^{-1}; K = multiplier, value of which depends on polymer topology (for example, for linear polymers it is value is universal, does not determine by polymer nature and is equal to 1); α_∞ = heat expansion's coefficient in hyperelastic state, K^{-1}, or degree^{-1} (theoretical value 6.8×10^{-4} degree^{-1} is estimated in [3]).

Substitution of expression a_1 for densely cross-linked polymers in hyperelastic state to Equation (1) reduces following equation:

$$\xi_\infty = -\left(3K/\left(\alpha_\infty\sqrt{\varepsilon_0}\right)\right)\left(\partial\delta\varepsilon/\partial T\right) \tag{2}$$

Consider discovery of unknown quantities, taking part in formula (2).

Calculation K for densely cross-linked polymers in hyperelastic state will make in the network of the same approach, that in [3]. Average values of strain region's worth for hyperelastic state: epoxide resin $-u_{0,lin.}^{1.0} \approx 4.7 \cdot 10^{-2}$, densely cross-linked polymers based on them $u_{0,cr.}^{1.0} \approx 3.0 \cdot 10^{-3}$ [4]. In the total $K_{cr.} = 15.7$.

Inductive capacity of unstressed polymer at it is structure glass-transition temperature will determine from equation, derived in [3]

$$\varepsilon_0 = \frac{10^{-24}\left[1+\alpha_\infty\left(T-T_g\right)\right]N_A\left(458.9<l>-57.4\right)+2k_g\left(131.476<l>-14.954\right)}{10^{-24}\left[1+\alpha_\infty\left(T-T_g\right)\right]N_A\left(458.9<l>-57.4\right)-k_g\left(131.476<l>-14.954\right)} \quad (3)$$

where T = temperature, °C, or K; T_g = structure glass transition temperature, °C, or K; $N_A = 6.023 \cdot 10^{23}$ mole^{-1} = Avogadro's constant; $<l>$ = cross-site chains' average numerical degree of polymerization (calculating equation and values calculated by it look in [3]); $k_g \approx 0.681$ = universal for densely cross-linked polymers value of molecular package at glass transition temperature coefficient [6].

Put Equation (3) into the final calculating expression

$$\varepsilon_0 = \left(455.465782<l>-54.939368\right)/\left(186.860314<l>-24.388346\right).$$

For calculations of $\partial\delta\varepsilon/\partial T$ derivative will use increments' method [6]

$$\partial\delta\varepsilon/\partial T = \left(\sum_i \delta C_i \Delta V_i + \sum_i \delta C_{i,imi}\right)_{r.f.} \Big/ \left(\sum_i \Delta V_i\right)_{r.f.} \quad (4)$$

where ΔV_i = Van der Waals's volume of ith atom, taking part in repeated fragment of mesh, Å³; $\left(\sum_i \Delta V_i\right)_{r.f.}$ = Van der Waals's volume of repeated mesh's fragment, Å³ (look in [3]). Increments set $\left(\sum_i \delta C_i \Delta V_i + \sum_i \delta C_{i,imi}\right)_{r.f.}$ in Equation (4) for glassy state of experimental objects was calculated [3]. Appraising it in this case, will base on fact that during polymers deformation in hyperelastic state emergent electromagnetic anisotropy is be characterized by strain, and by orientation effects, therefore will consider increment δC_h for fragment >CH-OH in full.

$$\left(\sum_i \delta C_i \Delta V_i + \sum_i \delta C_{i,imi}\right)_{r.f.} = \left(-212.7362<l>-5.309\right)10^{-6}\ \text{Å}^3/\text{K}.$$

With an allowance for the latter expression, and also expressions $\left(\sum_i \Delta V_i\right)_{r.f.}$ [3] and ε_0 and coefficients $K_{cr.}$ and α_∞ values from Equation (2) get total equation for theoretical calculation of electromagnetic susceptibility's equilibrium elastic coefficient.

$$\xi_\infty = 0.07\left[\frac{212.7362<l>+5.309}{458.9<l>-57.4}\right]\sqrt{\frac{186.860314<l>-24.388346}{455.465782<l>-54.939368}} \quad (5)$$

In Table 1are Presented theoretical values, calculated according to Equation (5).

TABLE 1 Electromagnetic susceptibility's equilibrium elastic coefficient (theoretical and experimental values).

x	$<l>^{1}$	ξ_∞ calc.	ξ_∞ exper. [1]	ε, %
0.0	1.00	0.0240	0.0263	9
0.5	1.25	0.0230	0.0235	2
1.0	1.50	0.0220	0.0224	2
1.5	1.75	0.0220	0.0207	6
2.0	2.00	0.0220	0.0192	15

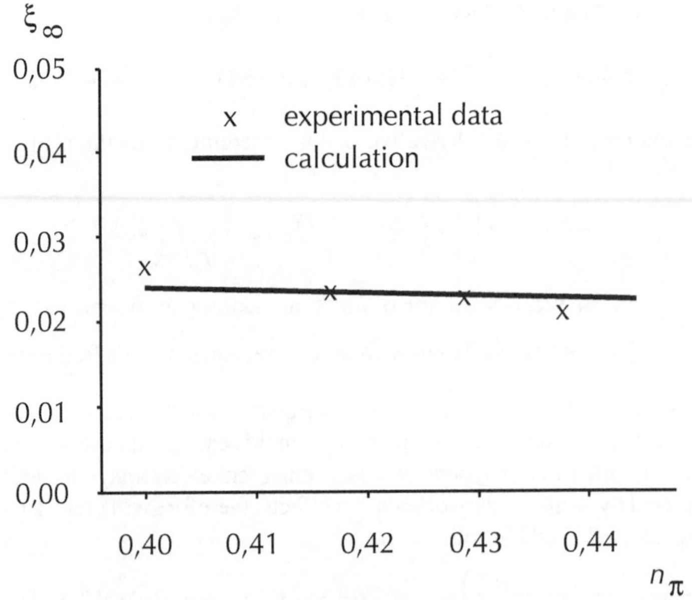

FIGURE 1 Dependence ξ_∞ on statistical parameter, which defines amount of fragments DGEBA1 in topological structure of experimental objects.

In Figure 1, theoretical and experimental values of ξ_∞ are compared graphically. Relatively little difference between theoretical and experimental values (215%) makes it possible to draw a conclusion about rightness of introduced algorithm.

Abbreviations of components and their ratio look in [2], topological structure's statistical parameters are included in [3].

7.3.2 Weighting Coefficient

Weighting coefficient $w_{C,\beta}$ is calculated according to the equation [1]

$$w_{C,\beta} = C_{\beta,\infty} T / (0.5 \xi_{\infty} A_{\infty}), \tag{6}$$

where $C_{\beta,\infty}$ = strain electromagnetic susceptibility of glassy state ($T < T_g - 15\ ^{\circ}C$). Discovered with involvement of increment's method quantity $C_{\beta,\infty}$ at T = 25°C was $4.9 \cdot 10^{-5}$ MPa^{-1}[3]. Theoretical values $w_{C,\beta}$, determined according to the Equation (6), look in Table 2.

TABLE 2 Weighting coefficient (theoretical and experimental values).

x	$< l >$	$w_{C,\beta}$		ε, %
		calc.	exper. [1]	
0.0	1.00	0.0360	0.0280	29
0.5	1.25	0.0270	0.0260	4
1.0	1.50	0.0210	0.0230	9
1.5	1.75	0.0170	0.0200	15
2.0	2.00	0.0140	0.0180	22

7.3.3 α-Relaxation Times' Spectrum and Parameters

So as strain electromagnetic susceptibility's and shear pliability's relaxation spectrums coincide [1, 2], theoretical account of **α**-relaxation times at specified temperature was made with the use of temperature function [2]. In the capacity of α-mode width's value analog [3] was taken 0.5.

7.4 NUMERICAL EXPERIMENT, MODELING CHANGE OF EXPERIMENTAL OBJECTS' STRAIN ELECTROMAGNETIC ANISOTROPY UNDER THE INFLUENCE OF CONSTANT LOAD IN HEATING AND COOLING CONDITIONS

In Figures 2 and 3 are presented graphic comparison of computer experimental objects' strain electromagnetic anisotropy's results, realized on the base of theoretical and empirical strain electromagnetic susceptibility's reconstruction, with experimental results.

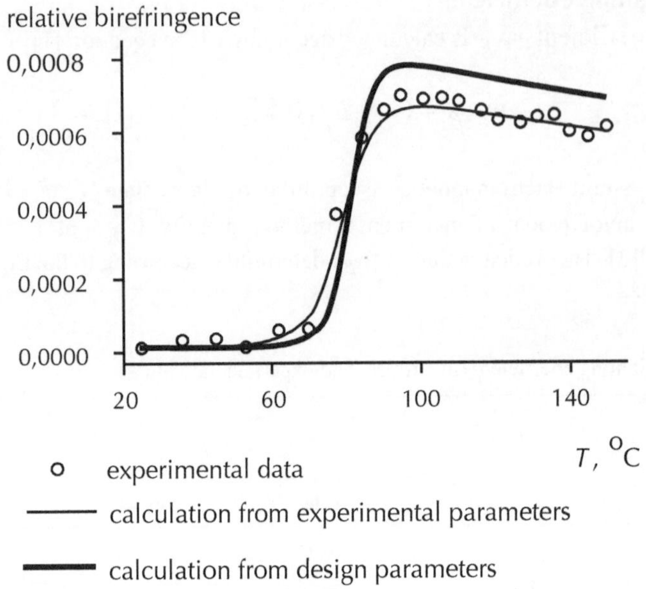

FIGURE 2 Thermooptic curve for experimental object with structure x = 2 (heat ratio 0.9°C/min, load 720 g).

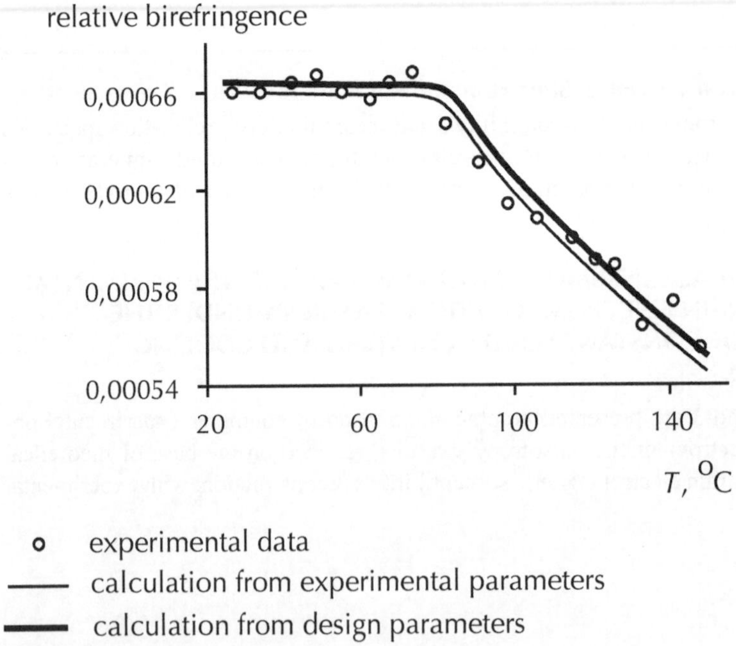

FIGURE 3 Freezing curve of optical anisotropy for experimental object with structure x = 1.5 (cooling ratio 0.6°C/min, load 720 g).

7.5 CONCLUSION

Developed mathematical model and demonstrated here by example of epoxy amine polymers approach to theoretical rating it is constant may be used for a *priori* choice of polymer binding agent, which generates during the forming of topological structure densely cross-linked matrix with minimum strain electromagnetic anisotropy's degree, of high strength radioparent fiberglass product with heightened isotropy of radioparence coefficient.

KEYWORDS

- **Cross-linked matrix**
- **Electromagnetic susceptibility**
- **Epoxy amine polymers**
- **Radioparence coefficient**
- **Radioparent fiberglass product**

REFERENCES

1. Deberdeev, T. R., Ulitin, N. V., and Deberdeev, R. Ya. *An operator form of strain electromagnetic susceptibility of cross-linked polymers with the new instant component submission* (2012).
2. Ulitin, N. V., Deberdeev, T. R., and Deberdeev, R. Ya. *Viscoelastic compliance of cross-linked polymers with nanoscale cross-site chains*. (2012).
3. Ulitin, N. V., Deberdeev, R. Ya., Samarin, E. V., and Deberdeev, T. R. *A computer simulation of the nanofragment polymer meshes' deformation behavior in all their physical states based on analytically founded parameters* (2012).
4. *The photoelasticity's method*, vol 3. G. L. Khesin (Ed.), Gostechizdat (1975).
5. Blythe, T. and Bloor, D. *Electrical properties of polymers*. Cambridge University Press, Cambridge, p. 492 (2005).
6. Askadskii, A. A. and Kondrashenko, V. I. Computer science of materials of polymers vol. 1. Atomic-molecular level. *Scientific world*, p. 544 (1999).

8 Susceptibility of Cross-linked Polymers: A Numerical Approach

T. R. Deberdeev, N. V. Ulitin, R. Ya. Deberdeev, and G. E. Zaikov

CONTENTS

8.1 Introduction 69
8.2 Mathematical Model 69
8.3 Experimental Part 71
8.4 Discussion and Results 71
8.5 Conclusion 73
Keywords 73
References 73

8.1 INTRODUCTION

Main claims, which are laid to densely cross-linked polymer matrixes used in radioparent high strength fiberglass products are besides high strength properties, maximum isotropy, and stability of radioparency coefficient [1]. Because of traditional mathematical formalism describes strain electromagnetic anisotropyproperty, which inversely as isotropy of radioparency coefficient only for polymers in hyperelastic state [2], when searching for binding agent, which satisfies all the designated requirements, appears a problem of quantity and stability of it is strain electromagnetic anisotropy at service of loaded products temperature ranges rating. Thereupon in this research was submitted mathematical model of nanofragment cross-linked polymers strain electromagnetic anisotropy, in which were covered all physical states of the latter.

8.2 MATHEMATICAL MODEL

Dependence between inductive capacity of deformated polymeric dielectric tensor and components of deformation's tensor can be produced by the following form [3]:

$$\varepsilon_{ik} = \varepsilon_0 \delta_{ik} + a_1 \gamma_{ik} + (1/3)(a_1 + 3a_2) u_{ll} \delta_{ik} \tag{1}$$

where ε_0 = inductive capacity of unstressed polymer model; δ_{ik} = Krokener's symbol;

γ_{ik} = shear strain tensor; a_1, a_2 = some coefficients; u_{ll} = deformation of compression tensor.

The second summand in right part of Equation (1) describes electromagnetic anisotropy, which appears in the issue of polymer products segments orientation during shear strain, in other words electromagnetic anisotropy is not linked to compression tensor [3]. According to BrusterVertgame's law, establishing dependence between deformative electromagnetic anisotropy of polymer in hyperelastic state and differences of it is main shear strains and tensions, deduces from formula (1)

$$\Delta n = \xi_\infty \Delta\gamma = C_\infty \Delta\tau \qquad (2)$$

where Δn = devormative electromagnetic anisotropy; ξ_∞ = equilibrium elastic coefficient of electromagnetic perception; $\Delta\gamma$ и $\Delta\tau$ (MPa) = differences of main shear strains and tensions, respectively; C_∞ = equilibrium strain electromagnetic perception, MPa^{-1}. Interconnection between the latter quantity and equilibrium shear pliability J_∞ (MPa^{-1}) of densely cross-linked polymer is subordinated to appropriateness [2]

$$C_\infty = 0.5\xi_\infty J_\infty \qquad (3)$$

For description strain electromagnetic anisotropy of densely polymer meshes in all their physical states include strain electromagnetic susceptibility relaxation operators C (MPa^{-1}), electromagnetic susceptibility elastic ξ and shear pliability J (MPa^{-1}) coefficients and consider that they are interconnected as appropriate equilibrium properties in Equation (3)

$$\boldsymbol{C} = 0.5\xi J \qquad (4)$$

For densely cross-linked polymers with spatial homogeneous topological structure mathematical formalization of relaxation operator J was identified by authors [4]. From Equation (4) follows that electromagnetic susceptibility elastic coefficient's relaxation spectrum is equal modulus of rigidity relaxation spectrum. Therefore operator ξ can be produced as:

$$\xi = \xi_{\beta,\infty} + \xi_{\alpha,\infty} G_{N,\alpha}, \quad \xi_{\beta,\infty} = w_{\xi,\beta}\xi_\infty, \quad \xi_{\alpha,\infty} = (1 - w_{\xi,\beta})\xi_\infty$$

where $\xi_{\beta,\infty}$ = part of local conformational mobility of polymeric meshe's cross-site chains in electromagnetic susceptibility elastic coefficient's quantity (in other words, it is electromagnetic susceptibility of glassy state elastic coefficient); $\xi_{\alpha,\infty}$ = part of mesh points' cooperative mobility in electromagnetic susceptibility elastic coefficient's quantity; $G_{N,\alpha}$ = fractional exponential operator inverse $J_{N,\alpha}$ [4]; $w_{\xi,\beta}$ =

weighting coefficient, which does not depends on temperature and is a part of local conformational mobility of meshe's cross-site chains in quantity ξ_∞.

Using Equation (4), multiplicative rule of normalized fractional exponential operators [5], receive final expression of strain electromagnetic susceptibility operational form:

$$\boldsymbol{C} = C_{\beta,\infty} + C_{\alpha,\infty}\boldsymbol{J}_{N,\alpha}$$
$$C_{\beta,\infty} = w_{C,\beta}C_\infty,\ \ C_{\alpha,\infty} = (1 - w_{C,\beta})C_\infty,\ \ w_{C,\beta} = w_{\xi,\beta}w_{J,\beta} \tag{5}$$

where $C_{\beta,\infty}$ = part of local conformational mobility of polymeric meshe's cross-site chains in strain electromagnetic susceptibility's quantity (in other words, it is strain electromagnetic susceptibility of glassy state), MPa^{-1}; $C_{\alpha,\infty}$ = part of cooperative meshe's points mobility in strain electromagnetic susceptibility's quantity, MPa^{-1}; $J_{N,\alpha}$ = normalized to 1 fractional exponential operator [4]; $w_{C,\beta}$ = weighting coefficient, which does not depends on temperature and is a part of local conformational mobility of meshe's cross-site chains in quantity C_∞.

Then, in concordance with Equation (5), shear pliability and strain electromagnetic susceptibility relaxational spectrums are identical. In conclusion of mathematical model's account it should be noted that operator *C* in it is record defines strain electromagnetic susceptibility in every physical state of densely cross-linked polymers, so as operator *J* [4].

8.3 EXPERIMENTAL PART

Experimental objects, detection operator *C* constants device and strain electromagnetic anisotropy of experimental objects researches are the same [4]. In the capacity of electromagnetic radiation was used green light (λ = 546 nm). Optical double refraction, that was electromagnetic anisotropy's property when using visible part of spectrum was being determined by ray's difference of move in centre of under consideration material's disc. Operator's *C* values in different physical states of experimental objects were calculated from BrusterVertgame's law coerced to operator's form on the base of experimental quantity of double diffraction and specified difference between main shear tensions in the disk's centre.

8.4 DISCUSSION AND RESULTS

It was determined, that electromagnetic susceptibility's equilibrium elastic coefficient does not depend on temperature (Table 1). It is values and values of weighting coefficient decrease slope oppositional to hardeners molar amounts ratio that can be explained by growth of intermolecular interaction's level during the increase of cross-site chains' average numerical degree of polymerization.

TABLE 1 Experimental values of electromagnetic susceptibility equilibrium elastic coefficient and weighting coefficient.

x^1	ξ_∞	wC,β
0.0	0.0263	0.0280
0.5	0.0235	0.0260
1.0	0.0224	0.0230
1.5	0.0207	0.0200
2.0	0.0192	0.0180

It was supposed that experimental restored relaxation spectrums of strain electromagnetic susceptibility and shear pliability operators are identical. Additional argument for equality of strain electromagnetic susceptibility and shear pliability relaxation spectrums was exact coincidence in practice of photocreep curves, converted according to the equation in mechanical scale and experimental received creep curves (difference between calculated and experimental points was not more than 10%).

$$C_N = (1-((1-w_{C,\beta})/(1-w_{J,\beta}))) + ((1-w_{C,\beta})/(1-w_{J,\beta}))J_N$$

$$J_N = J_\infty^{-1}J = w_{J,\beta} + (1-w_{J,\beta})J_{N,\alpha},\ C_N = C_\infty^{-1}C = w_{C,\beta} + (1-w_{C,\beta})J_{N,\alpha}$$

Demonstration of the fact that presented model describes strain electromagnetic anisotropy realization in all densely cross-linked physical states is showed in Figure1.

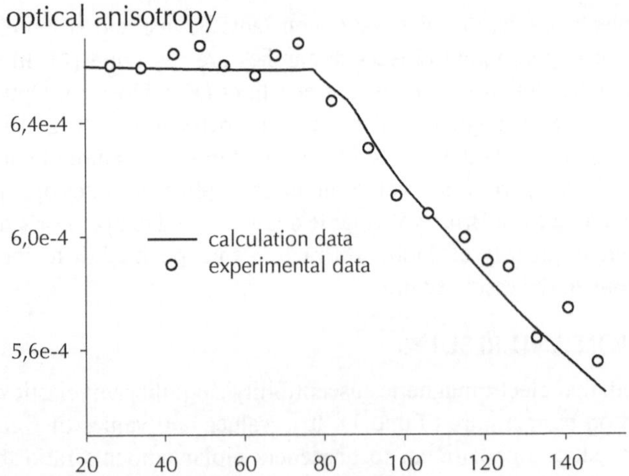

FIGURE 1 Freezing of optical anisotropy curve for system with structure x = 1.5 (cooling ratio 0.6°C/min, load 720 g).

[1]Equation for x look in [4].

8.5 CONCLUSION

Developed model will be useful in the time of choice substantiation for high strength radioparent fiberglass product of densely cross-linked polymer matrix with heightened isotropy of radioparence coefficient at specified temperature ranges and stress loading product's conditions. Important applied aspect of formalism is it is use for account of monolithic or collapsible fiberglass products with compound configuration stress using the polarization optical method.

KEYWORDS

- **Bruster Vertgame's law**
- **Electromagnetic susceptibility**
- **Polarization-optical method**
- **Photocreep curves**

REFERENCES

1. Gurtovik, I. G., Sokolov, V. I., Trofimov, N. N., and Shalgunov, S. I. Glass-fiber plastic radioparent products. *Mir*, 368 (2003).
2. Mott, P. H. and Roland, C. M. Mechanical and optical behavior of double network rubbers. *Macromolecules*, **33**(11), 41324137 (2000).
3. Blythe, T. and Bloor, D. *Electrical properties of polymers*. Cambridge University Press, Cambridge, p. 492 (2005).
4. Ulitin, N. V., Deberdeev, T. R., and Deberdeev, R. Ya. *Viscoelastic compliance of cross-linked polymers with nanoscale cross-site chains*. (2012).
5. Rabotnikov, U. N. Hereditary mechanics elements of solids. *Nauka*, 384 (1977).

9 Update on Cross-linked Polymers with Nanoscale Cross-site Chains

N. V. Ulitin, T. R. Deberdeev, R. Ya. Deberdeev, and G. E. Zaikov

CONTENTS

9.1 Introduction 75
9.2 Mathematical Model 76
9.3 Experimental Parts 77
9.4 Discussion of the Results 77
  9.4.1 Thermal Properties 77
  9.4.2 Hyperelastic State Constant and Weighting Coefficient 78
  9.4.3 A-Relaxation Times Spectrums Properties 78
  9.4.4 Prediction of Thermomechanical Experiment 80
9.5 Conclusion 81
Keywords 81
References 81

9.1 INTRODUCTION

The choice of nanofragment cross-linked matrix, which required value of viscoelastic module (or it is inverse property pliability) is one of the main problems in development of high strength fiberglass product technology [1]. The value of strain properties of matrix assessment with known conditions of product service (e.g., temperature, load's distribution, and period of time. and etc.) is only of formal nature, so long as the component of viscoelastic module or pliability momentary in legacy mathematical models such as spring-damper models, in equations of heredity's theory, in approximate functions (particularly, Kolraush's) is not attached to defined physical conditions of polymer mesh. Therefore, it was developed, new mathematical account of cross-linked polymers' pliability, in which it is momentary of component reproduces local conformational mobility of polymeric meshe's cross-site chains and can be determined experimentally in their glassy state.

9.2 MATHEMATICAL MODEL

As long as, contribution of equilibrium volumetric pliability strain tensor is negligibly small for cross-linked polymers consider shear pliability [1].

$$u_{ik} = (1/2) J\tau_{ik} - (1/3) Bp\delta_{ik} \quad (1)$$

In Equation (1): J = relaxation operator of shear pliability, MPa^{-1}; τ_{ik} = tensor of shear tension, MPa; B_∞ = equilibrium volumetric pliability, MPa^{-1}; p = pressure, MPa; δ_{ik} = Krokener's symbol.

It is believed that the deformation of polymeric mesh chemical bonds do not destroy then J as follows:

$$J = J_\infty J_N \quad (2)$$

where J_∞ = equilibrium shear pliability, MPa^{-1}; J_N = Volter's operator linearly associated with relaxation spectrum of shear pliability.

It is known that relaxational spectrum of shear pliability consists of α- and β - branches, which are combined with cooperative mobility of mesh's points and local conformational mobility of it is cross-site chains, respectively [2]. Suppose, that single-mode distribution of α-relaxations' times is typical for cross-linked polymers with spatially homogeneous topological structure, and multimodal distribution for cross-linked polymers with heterogeneous topological structure. All arguments keep only for the first case, so long as exactly in such structure properties of matrix fully realize [1]. Then, based on the fact that mesh's points do not assist in local movements [2], express operator J_N for cross-linked polymers with spatially homogeneous topological structure as:

$$J_N = w_{J,\beta} + (1 - w_{J,\beta}) J_{N,\alpha} \quad (3)$$

where $w_{J,\beta}$ = weighting coefficient, does not depending on temperature, which is contribution of local conformation mobility of cross-site mesh's chains to equillibrium of shear pliability; $J_{N,\alpha}$ = equipotential fractional operator normalized to 1, which is associated with distribution of α relaxations' times $L_{J,\alpha}(\theta)$ by next equation (θ = relaxational time, t = current time):

$$J_{N,\alpha} = \int_{-\infty}^{\infty} L_{J,\alpha}(\theta)[1 - \exp(-t/\theta)] \mathrm{d} \ln \theta \quad (4)$$

In the capacity of normalized to 1 distribution of α relaxations' times was chosen following distribution [3]:

$$L_{J,\alpha}(\theta) = \sin[\pi(1 - \Xi_{J,\alpha})]/2\pi(\mathrm{ch}[(1 - \Xi_{J,\alpha}) \ln(\theta/\Theta_{J,\alpha})] + \cos[\pi(1 - \Xi_{J,\alpha})])$$

where $\Theta_{J,\alpha}$ = average α-relaxations' time is temperature function; $\Xi_{J,\alpha}$ = width of α-mode, value of which does not depend on temperature and lies within the limits of 0–1.

Glassy and hyperelastic states of tightly sewn polymers are determined by values of operator $J_{N,\alpha}$ 0 and 1, respectively, and a transition region between these states by values in interval of 0–1. This make it possible to describe Equations (1)(4) tensor of deformation and pliability of dense polymer meshes in all physical states. So, in hyperelastic state cooperative α-transition is "momentary" and relaxation operator of shear pliability is equal to equilibrium shear pliability. In glassy state $J_{\beta,\infty} = w_{J,\beta} J_{\infty}$ due to "coldness" of α-transition it will be observed only local conformational mobility of cross-site mesh's chains and relaxation operator of shear pliability possesses the value which is equal to shear pliability. In transition region there are α- and β-mobilities, therefore, relaxation operator of shear pliability will be formalized in all it is recording.

9.3 EXPERIMENTAL PARTS

The determination of constants and testing of predictive capability of developed model were made for line of epoxy amine polymers based on diglycocide ether of bisphenol-A (DGEBA), which was solidificated by different molar ratios of 1-aminohexane (AH) and hexamethylenediamine (HMDA): $x = n(\text{AH})/n(\text{HMDA})$ was changing from 0 to 2 (Table 1).

The synthesis of experimental objects was made by next methodology. Required amounts of DGEBA and HMDA were heated to 42°C; after that melt of HMDA and required amount of AH were mixed with DGEBA; composition was being degassed in conditions of reiteration freezing defrosting cycles to invariable residual pressure; then argon was being pumped into an ampoule and then was being soldered. For the purpose of homogeneous topological structure receipt on the base of conceptions for all compositions was selected common hardening rate: 20°C–72 hr, 50°C–72 hr, 80°C–72 hr, and 120°C–72 hr [1]. Content of gel fraction for the limited hardened reactionary mixtures was determined in acetone and at the average for all compositions was 98%.

Model constants were found by the results of solid expansion and mechanical researches on installation [4] (measuring error was less than 3%). In first case models were chilled from 150 to 25°C with ratio 0.4°C/min. Relaxation operator of shear pliability was determined by Equation (1) on the base of experimental measured deformation of model and known value of applied mechanical tension. For exception of thermal prehistory influence on properties of experimental objects before each test models were heated to 150°C, were kept at that temperature 3 hr and chilled with ratio 0.2°C/min to temperature 25°C.

9.4 DISCUSSION OF THE RESULTS

9.4.1 Thermal Properties

In solid expansion research of experimental objects were determined structure glass transition temperature T_g (Table 1) and heat expansion coefficients. Values of the latter were averaged out for all models in every physical state. In the end: for glassy state $4.3\cdot10^{-4}$ degree^{-1}, for hyperelastic state $7.0\cdot10^{-4}$ degree^{-1}.

9.4.2 Hyperelastic State Constant and Weighting Coefficient

Equilibrium shear pliability of thick polymer meshes with spatially homogeneous topological structure is the inverse quantity of absolute temperature T [1]

$$J_{\infty} = A_{\infty} / T \tag{5}$$

where A_{∞} = hyperelastic state constant, which does not depend on temperature, K/MPa.

Researched experiment results showed sufficiently high accuracy of accounts with the use of Equation (5). It was shown that for all experimental objects hyperelastic state constant does not depend on temperature within error of experimental data (Table 1), though it is values rise with increase of hardeners molar ratio, in other words with growth of polymerization number average degree of cross-site chains and synchronous decrease of mesh points concentration.

Rating of weighting coefficient was made according to equation

$$w_{J,\beta} = J_{\beta,\infty}^{T_0} T_0 / A_{\infty}$$

where $J_{\beta,\infty}^{T_0}$ = shear pliability in glassy state at $T_0 = 25^{\circ}\mathrm{C}$, MPa^{-1}.

With increase of hardener molar ratio is observed decrease of weighting coefficient derived values (Table 1), that can be explained by increasing intermolecular interaction according to growth of cross-site chains length.

9.4.3 α-Relaxation Times Spectrums Properties

The α-relaxation times spectrums properties were considered with the help of least squares method by isothermal creep curves for several temperature values, which lie in transition region between glassy and hyperelastic states.

From introduced in [5] Equation

$$\lg\left(\Theta_{J,\alpha}(T)/\Theta_{J,\alpha}(T_g)\right) = \left(f_g^{/} + \left(0.025\alpha_{\infty}\left(1-f_g\right)/f_g\right)(T-T_g)\right)^{-1} - \left(f_g^{/}\right)^{-1} \tag{6}$$

described relaxation behavior of polymers with linear, branched and rarely cross-linked topology in hyperelastic state, taking into account that vitrifying for tightly sewn polymers begins at $\lg\Theta_{J,\alpha}(T_g+15)=2$ and quantity of their fluctuating unconfined space part at structure glass transition temperatures at the average = 0.09 [1], for α-relaxation times temperature dependence of dense polymer meshes approximation in all their physical states was deduced following dependence:

$$\lg\left(\Theta_{J,\alpha}(T)/\Theta_{J,\alpha}(T_g)\right) = 40\left(\left(f_g/\left(f_g+\alpha(1-f_g)(T-T_g)\right)\right)-1\right),\ \alpha = \begin{cases}\alpha_g,\ T<T_g\\ \alpha_{\infty},\ T>T_g\end{cases}, \tag{7}$$

where 40 = multiplier, which depends only on logarithmic scale base; f_g = fluctuating unconfined space part at structure glass transition temperature; α_g, α_∞ = heat expansion coefficients in glassy and hyperelastic states, respectively, degree^{-1}.

In [5] parameter $f_g^{/}$ correlates with unconfined space part at structure glass transition temperature of polymer, but how it was shown by unconnected with polymer relaxation properties studies of unconfined space, does not it. It was proved empirically that it is average value for linear, branched and rarely cross-linked polymers is universal and equal to 0.025. On the ground of made transformations it may be drawn a conclusion, that this value preserves for thick mesh polymers, too.

Derived experimental values are described by Equation (7) with good accuracyα-relaxation time's temperature function consists of two branches, linking up at glass transition temperature of system (Figure 1). Fairness of Equation (7) is confirmed by that averaged out for all experimental models quantity $f_g = 0.095$ conforms to literary data.

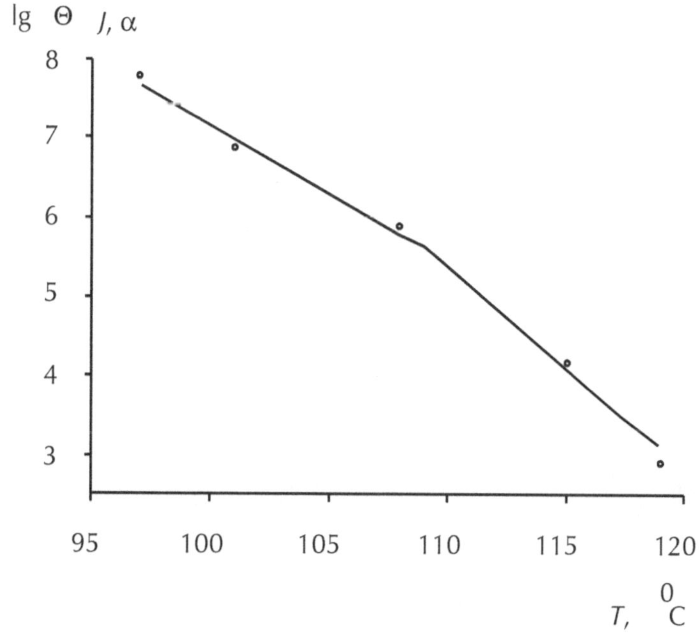

FIGURE 1 Dependence of α-relaxation average times decimal logarithm on temperature in region between glassy and hyperelastic states for model with structure $x = 0$.

With complication of molecular polymer structure relaxation spectrum expands [2]. This appropriateness is shown for our model, too values $\Xi_{J,\alpha}$ increase with growth of hardener molar ratioelongation cross-site chains. The $\Xi_{J,\alpha}$ does not depend on temperature (Table 1), and this confirms absence of α–transition decomposition once again, spatial homogeneity of models structure and rightness of synthetic methodology.

TABLE 1 Experimental values of glass transition temperature and hyperelastic state constant weighting coefficient and α-mode width.

x	T_g, °C	A_∞, K/MPa	$w_{J,\beta}$	$\Xi_{J,\alpha}$
0.0	109	35.0	0.018	0.374
0.5	99	53.0	0.015	0.440
1.0	88	69.3	0.014	0.497
1.5	77	81.6	0.012	0.557
2.0	71	93.5	0.008	0.636

9.4.4 Prediction of Thermomechanical Experiment

Thereby, for chosen experimental objects were determined unknown parameters of developed mathematical model: hyperelastic state constant, weighting coefficient and relaxation spectrum properties. Prediction results of thermomechanical curves trend successfully demonstrated prediction ability of introduced mathematical description of thick cross-linked polymers' viscoelastic pliability in all their physical states (Figure 2).

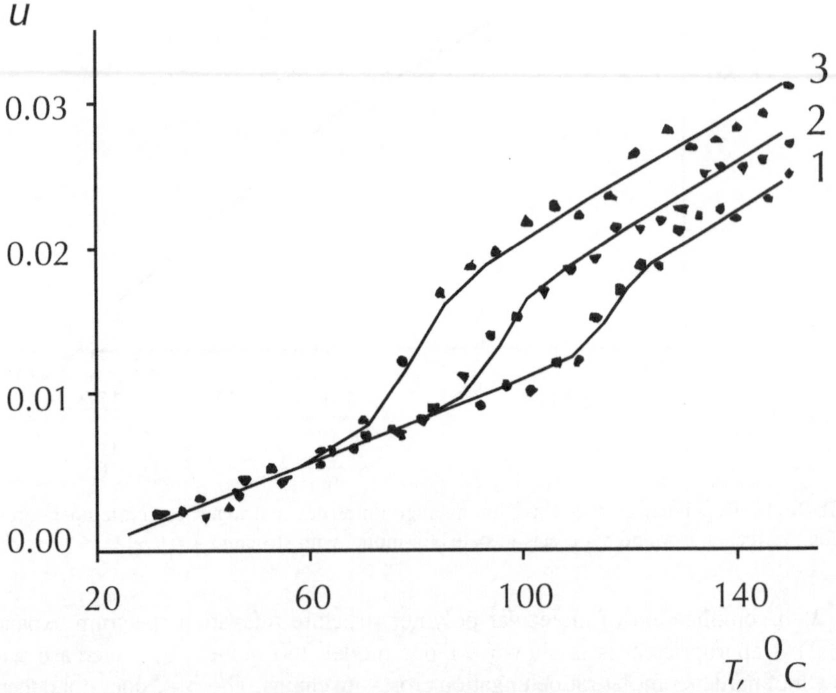

FIGURE 2 Thermomechanical curves: 1—$x = 0$; 2—$x = 1.0$; 3—$x = 2.0$ (points – average experiment, lines–prediction; heating rate 0.9°C/min, load 720 g).

9.5. CONCLUSION

The mathematical formalism thick cross-linked polymers' pliability at the expense of momentary constituent with use of weighting coefficient $w_{J,\beta}$ presentation was described in all it is physical states for the first time.

On conditions that on the base of fractal models [6] it is possible to calculate required value of binding agent's strain properties at the given elastic characteristics of filler and strain properties of all fiberglass product, so developed model makes it possible to substantiate choice of nanoscale cross-linked polymer matrix.

KEYWORDS

- **Cross-linked matrix**
- **Hyperelastic states**
- **Relaxation spectrum**
- **Strain tensor**
- **α-Relaxation**

REFERENCES

1. Bajenov, S. L., Berlin, A. A., Kulkov, A. A., and Oshmyan, V. G. *Polymer composite materials: Solidity and technology*. Dolgoprudnii, Intellect, p. 352 (2010).
2. Irzhak, V. I. Topological structure and relaxation properties of polymers *Russian Chemical Reviews*, **74**(10), 937967 (2005).
3. Rabotnov, U. N. Elements of solid matter hereditary mechanics. *M. Science*, p. 384 (1977).
4. Zuev, B. M., Arkhireev, O. S., Filippova, A. P., Gubanov, E. F. Opto-mechanical properties of tightly sewn polymers based on diallyl and divinyl monomers. *HMC*, **35**(A)(6), 669674 (1993).
5. Ferry, J. D. *Viscoelastic properties of polymers*. John Wiley & Sons, New York, p. 641 (1980).
6. Kulak, M. I. *Fractal mechanics of polymers*. Higher School Minsk, p. 304 (2002).

10 Role of Polymers in Technologies and Environment Protection

M. D. Goldfein and N. V. Kozhevnikov

CONTENTS

10.1 Introduction 83
10.2 Experimental 83
10.3 Discussion and Results 84
10.4 Conclusion 92
Keywords 92
References 92

10.1 INTRODUCTION

The chemical physics is the most important fundamental science in the modern natural sciences [1]. The concepts of it explain qualitatively and quantitatively, the mechanisms of various processes, such as the reactions of oxidation, burning, and explosion, obtaining food products and drugs, oil hydrocarbon cracking, polymeric material formation, biochemical reactions underlying metabolism, and genetic information transfer and so on. It comprises notions of the structure, properties, and reactivity of various substances, free radicals as active centers of chain processes proceeding in both mineral and organic nature, the scientific foundations of low-waste and resource saving technologies. The design and wide usage of synthetic polymeric materials is a lead in chemistry. This results in the appearance of new environmental problems due to pollution of the environment with these materials and the wastes of their production and monomer synthesis. The presented results of our studies of the kinetics and mechanism of vinyl monomer polymerization which can proceed in quite various conditions point to their usability in the scientific justification of technological regime optimization of monomer and polymer synthesis, and in solving both local and global environmental problems.

10.2 EXPERIMENTAL

To obtain reliable experimental data and to correctly interpret them, we used such physicochemical and analytical techniques as dilatometry, viscometry, UV and IR spectroscopy, electronic paramagnetic resonance, Raman light scattering spectroscopy, electron microscopy, and gas-liquid chromatography. To analyze the properties of polymeric dispersions, the turbidity spectrum method was used, and the efficiency of flocculants was estimated gravimetrically and by the sedimentation speed of special water-suspended imitators (e.g. copper oxide).

10.3 DISCUSSION AND RESULTS

Effect of salts of varivalent metals [2, 3] Additives of the stearates of iron (IS), copper (CpS), cobalt (CbS), zinc (ZS), and lead (LS) within a certain concentration range were found to increase the polymerization rate of styrene and methylmethacrylate (MMA) in comparison with thermal polymerization. By initiating activity, they can be arranged as LS < CbS < ZS < IS < CpS.

The effective activation energy decreases with the chain initiation, The activation energy of the initiating reaction depends on the kinetic reaction order with respect to the monomer's active participation.. The IR spectroscopy data shows that an intermediate monomer–stearate complex is formed and then decomposed into active radicals to initiate polymerization. The benzoyl peroxide (BP)–IS (or CpS) systems can be used for effective polymerization initiation. Concentration inversion of the catalytic properties of stearates has been found, which depends on salt concentration and conversion degree. The accelerating action efficiency decreases with increasing temperature, the BP–IS system possessing the highest initiating activity. The initiating mechanism for these systems principally differs from the redox one. It follows from our experimental data (color changes before and in the course of polymerization, the absorption and IR spectra of reactive mixtures, electron microscopy observations, etc.) that initiation occurs due to the stearate radicals formed at decomposition of the complex consisting of one BP molecule and two stearate ones.

Concentration and temperature inversion of the catalytic properties of gold, platinum, osmium, and palladium chlorides at thermal and initiated polymerization of styrene and MMA has been discovered. The mechanism of ambiguous action of noble metal salts is caused by the competition of the initiating influence of monomer complexes with colloidal metal particles and the inhibition reaction proceeding by ligand transfer.

Effect of some organic solvents [4-6]. The influence of acetonitrile (ACN) and dimethyl formamide (DMFA) on the radical polymerization of styrene and (meth) acrylic esters initiated with azoisobutyronitrile (AIBN) or BP was studied. Basic characteristics of the gross kinetics of polymerization in solution have been found. The reaction order with respect to initiator concentration is always 0.5, which points to bimolecular chain termination. The reaction order with respect to monomer varies within rather wide limits (above or below unity) depending on the chemical nature of the solvent, monomer, and initiator. The lowest (0.83) and highest (1.6) values were found for BP initiated polymerization of MMA in DMFA solution and for AIBN initiated polymerization of styrene in ACN solution, respectively. Such a high value of the order with respect to monomer is caused by the abnormally low rate of AIBN decomposition in ACN, which, in turn, is explained by the donor–acceptor interaction of the alkyl and nitrile groups of the initiator with the nitrile and alkyl groups of the same solvent molecule. The influence of DMFA and ACN on each elementary stage of polymerization has also been ascertained, which manifests itself in the dependences of the rate constants of the reactions of initiation, propagation, and termination on the monomer concentration. This is caused by such factors as changes in the initiator decomposition rate, macroradical solvation with the molecules of an electron–donor solvent, diffusional-controlled chain termination, and conformational changes of macromolecules in solution.

Homopolymerization and copolymerization of acrylonitrile in an aqueous solution of sodium sulfocyanide (ASSSC) [7–10]. When acrylonitrile (AN)-based fiber formation polymers are obtained, a spinning solution ready for fiber formation appears as a result of polymerization in some solvent. Organic solvents or solutions of inorganic salts are used as solvents in these cases. First, a comparative study was made of the kinetics and mechanism of polymerization of AN in DMFA and in an ASSSC initiated with AIBN and our newly synthesized azonitriles (azo*bis*cyanopentanol, azo*bis*cyanovalerian acid, azo*bis*dimethylethylamidoxime). The polymerization rate in ASSSC turns out to be significantly higher in comparison with the reaction in DMFA, in spite of the initiation rate in the presence of the said azonitriles being more than by 1.5 times lower than that in DMFA. The lower initiation rate in ASSSC is associated with stronger manifestation of the "cell effect" due to the higher viscosity of the water–salt solvent and the ability of water to form H bonds (which hinders initiation). The ratio of the rate constants of chain propagation and termination ($K_p/K_o^{0.5}$) was found to be ca. tenfold higher in ASSSC than in DMFA. The molecular mass of the polymer formed is correspondingly higher. These differences are caused by the medium viscosity influence on K_o and the formation of H bonds with the nitrile groups of the end chain of the macroradical and the monomer being added. But sodium sulfocyanide (which forms charge-transfer complexes with a molecule of AN or its radical to activate them) mainly contributes into the higher K_p.

The kinetics of AN copolymerization with methylacrylate (MA) or vinylacetate (VA) in ASSSC is qualitatively similar to AN homopolymerization in identical conditions. At the same time, the initial reaction rate, copolymer molecular mass, effective activation energy (E_{ef}), orders with respect to initiator and total monomer concentration differ. For example, the lower E_{ef} is due to the presence of a more reactive monomer (MA), and in the AN–VA system the non end monomer units in macroradicals influence the rate constant of cross chain termination. For these binary systems, the copolymerization constants have been estimated, whose values point to a certain mechanism of chain propagation, which leads to the MA concentration in the copolymer being significantly higher than in the source mixture. The same is observed in the case of AN with VA polymerization (it it naturally that the absolute amount of AN in the final product is much higher than that of MA or VA due to its higher initial concentration in the mixture).

Obtaining synthetic PAN fiber of a nitrone type is preceded by the preparation of a spinning solution by means of copolymerization of AN with MA (or VA) and itaconic acid (IA), or acrylic acid (AA), or methacrylic acid (MAA), or methallyl sulfonate (MAS). Usually, mixtures contain 15% of AN, 5–6% of the second monomer, 1–2% of the third one, and ca. 80 wt.% of 51.5% ASSSC. When a third monomer is introduced into the reaction mixture, the process rate and the molecular mass of the copolymer decrease, which allows treating these comonomers as peculiar low-effective inhibitors. In such a case, it becomes possible to estimate the inhibition (retardation) constant which is, in essence, the rate constant of one of the reactions of chain propagation. At copolymerization of ternary monomeric systems based on AN in ASSSC, the chain initiation rate increases with the total monomer concentration. The order of the initiation reaction with respect to the total monomer concentration varies from 0.5 to 0.8 depending on the degree of the retarding effect of the third monomer. Besides, AN in the three-component system is shown to participate less actively in the chain initiation reaction than at its homopolymerization and copolymerization with MA or VA.

Thus, the obtained results enable regulating the copolymerization kinetics and the structure of the copolymer formed, which, finally, is a way of chemical modification of synthetic fibers.

Stable radicals in the polymerization kinetics of vinyl monomers [11-18]. The laboratory of chemical physics (Saratov State University) is a leading research center in Russia to conduct studies of the influence of stable radicals on the kinetics and mechanism of polymerization of vinyl monomers. In general, as is known, free radicals are neutral or charged particles with one or several uncoupled electrons. Unlike usual (short-living) radicals, stable (long-living) ones are characteristic of paramagnetic substances whose chemical particles possess strongly delocalized uncoupled electrons and sterically screened reactive centers. This is the very cause of the high stability of many classes of nitroxyl radicals of aromatic, aliphatic-aromatic, and heterocyclic series, and radical ions and their complexes.

Peculiarities of thermal and initiated polymerization of vinyl monomers in the presence of tetracyanoquinodimethane (TCQM) radical anions were investigated. The $TCQM^{\cdot -}$ radical anions are shown to effectively inhibit both thermal and AIBN-initiated polymerization of styrene, MMA, and MA in ACN and dimethylformamide solutions. Inhibition is accomplished by the recombination mechanism and by electron transfer to the primary (relative to the initiator) or polymeric radical. The electron transfer reaction leads to the appearance of a neutral TCQM which regenerates the inhibitor in the medium of electron donor solvents. Our calculation of the corresponding radical chain scheme has allowed us to derive an equation to describe the dependence of the induction period duration on the initiator and inhibitor concentrations, and how the polymerization rate changes with time. The mechanism of the initiating effect of the peroxide—$TCQM^{-}$ system has been revealed, according to which a single-electron transfer reaction proceeds between an radical anion and a BP molecule, with subsequent reactions between the formed neutral TCQM and benzoate anion, and a benzoate radical and one more $TCQM^{-}$ radical anion. Free radicals initiating polymerization are formed at the redox interaction between the products of the aforesaid processes and peroxide molecules. The $TCQM^{-}$ radical anion interacts with peroxide only at a rather high affinity of this peroxide to electron (BP, lauryl peroxide); in the presence of cumyl peroxide, the radical anion inhibits polymerization only.

Iminoxyl radicals which are stable in air and are easily synthesized in chemically pure state (mainly, crystalline brightly-colored substances) present a principally new type of nitroxyl paramagnets. Organic paramagnets are used to intensify chemical processes, to increase the selectivity of catalytic systems, to improve the quality of production (anaerobic sealants, epoxy resins, and polyolefins). They have found application in biophysical and molecular biological studies as spin labels and probes, in forensic medicinal diagnostics, analytical chemistry, to improve the adhesion of polymeric coatings, at making cinema and photo materials, in device building, in oil extracting geophysics and flaw detection of solids, as effective inhibitors of polymerization, thermal and light oxidation of various materials, including polymers.

In this connection, systematic studies were made of the inhibiting effect of many stable mono and poly radicals on the kinetics and mechanism of vinyl monomer polymerization. The efficiency of nitroxyls as free radical acceptors has promoted their usage to explore the mechanism of polymerization by means of the inhibition technique. Usually, nitroxyls have time to only react with a part of the radicals formed at azonitrile decomposition, and they do not react at all with primary radicals at peroxide initiation. Iminoxyls have been found to terminate chains by both recombination and disproportionation in the presence of azonitriles. Inhibitor regeneration proceeds as a result of detachment of a hydrogen atom from an iminoxyl by an active radical to form the corresponding nitroso compound. The mechanism of inhibition by a nitroso compound is in addition of a propagating chain to a –N=O fragment to form a stable radical again. The interaction of iminoxyls with peroxides depends on the type of solvent. For example, in vinyl monomers, induced decomposition of BP occurs to form a heterocyclic oxide (nitrone) and benzoic acid. In contrast to iminoxyls, aromatic nitroxyls in a monomeric medium interact with peroxides to form non-radical products. Imidazoline-based nitroxyl radicals possess advantages over common azotoxides which are their stability in acidic media (owing to the presence of an imin or nitrone functional group) and the possibility of complex formation and cyclometalling with no radical center involved.

Monomer stabilization [19-25]. The practical importance of inhibitors is often associated with their usage for monomer stabilization and preventing various spontaneous and undesirable polymerization processes. In industrial conditions, polymerization may proceed in the presence of air oxygen and, hence, peroxide radicals MOO are active centers of this chain reaction. In such cases, compounds with mobile hydrogen atoms, for example phenols and aromatic amines, are used for monomer stabilization. They inhibit polymerization in the presence of oxygen only, that are antioxidants. As inhibitors of polymerization of (meth) acrylates proceeding in the atmosphere of air, some aromatic amines known as polymer stabilizers were studied, namely, dimethyldi-(*n*-phenyl-aminophenoxy)silane, dimethyldi-(*n*-β-naphthyl aminophenoxy)silane, 2-oxy-1,3-di-(*n*-phenylamino phenoxy)propane, 2-oxy-1,3-di-(*n*-β-naphthyl aminophenoxy) propane. These compounds have proven to be much more effective stabilizers in comparison with the widely used hydroquinone (HQ), which is evidenced by high values of the stoichiometric inhibition coefficients (by 3–5 times higher than that of HQ). It has been found that inhibition of thermal polymerization of the esters of acrylic and methacrylic acids at relatively high temperatures (100°C and higher) is characterized by a sharp increase of the induction periods when some critical concentration of the inhibitor $[X]_{cr}$ is exceeded. This is caused by that the formation of polymeric peroxides as a result of copolymerization of the monomers with oxygen should be taken into account when polymerization proceeds in air. Decomposition of polyperoxides occurs during the induction period as well and can be regarded as degenerated branching. The presence of critical phenomena is characteristic of chain branching. However, in early works describing inhibition of thermal oxidative polymerization, no degenerated chain branching on polymeric peroxides was taken into account. It follows from the results obtained that the value of critical concentration of inhibitor $[X]_{cr}$ can be one of the basic characteristics of its efficacy.

Inhibition of spontaneous polymerization of (meth) acrylates is necessary not only at their storage but also in the conditions of their synthesis proceeding in the presence of sulfuric acid. In this case, monomer stabilization is more urgent, since sulfuric acid not only deactivates many inhibitors but also is capable of intensifying polymer formation. The concentration dependence of induction periods in these conditions has a brightly expressed nonlinear character. And, unlike polymerization in bulk, decomposition of polymeric peroxides is observed at relatively low temperatures in the presence of sulfuric acid, and the values $[X]_{cr}$ of the amines studied are by ca. 10 times lower than $[HQ]_{cr}$.

Synthesis of MMA from acetone cyanohydrin is a widely spread technique of its industrial production. The process proceeds in the presence of sulfuric acid in several stages, when various monomers are formed and interconverted. Separate stages of this synthesis were modeled with reaction systems containing, along with MMA, methacrylamide and methacrylic acid, and water with sulfuric acid in various ratios. As heterogeneous as well as homogeneous systems appeared, inhibition was studied in both static and dynamic conditions. The aforesaid aromatic amines appear to effectively suppress polymerization at different stages of the synthesis and purification of MMA. Their advantages over HQ are strongly exhibited in the presence of sulfuric acid in homogeneous conditions, or under stirring in biphasic reaction systems. Besides, application of polymerization inhibitors is highly needed in dynamic conditions at the stage of esterification.

The usage of monomer stabilizers to prevent various spontaneous polymerization processes implies further release of the monomer from the inhibitor prior to its processing into a polymer. Usually, it is achieved by monomer rectification, often with preliminary extraction or chemical deactivation of the inhibitor, which requires high energy expenses and entails large monomer losses and extra pollution of the environment. It would be optimal to develop such a way of stabilization, at which the inhibitor would effectively suppress polymerization at monomer storage but would almost not affect it at polymer synthesis. The usage of inhibitors low-soluble in the monomer is one of possible variants. When the monomer is stored and the rate of polymerization initiation is low, the quantity of the inhibitor dissolved could be enough for stabilization. Besides, as the inhibitor is being spent, its permanent replenishment is possible due to additional dissolution of the earlier unsolved substance. The ammonium salt of N-nitroso-N-phenylhydroxylamine (cupferron) and some cupferronates were studied as such low-soluble inhibitors. The solubility of these compounds in acrylates, its dependence on the monomer moisture degree, the influence of the quantity of the inhibitor and the duration of its dissolution on subsequent polymerization was studied.

Differences in the action of cupferronates are due to their solubility in monomers, their various stability in solution, and the ability of deactivation. All this results in poorer influence of the inhibitor on monomer polymerization at producing polymer.

Some peculiarities of the kinetics and mechanism of emulsion polymerization [24-36]. Emulsion polymerization, being one of the methods of polymer synthesis, enables the process to proceed with a high rate to form a polymer with a high molecular weight, high concentrated latexes with a relatively low viscosity to be obtained, polymeric dispersions to be used at their processing without separation of the polymer from the reaction mixture, and the fire resistance of the product to be significantly raised. At the same time, the kinetics and mechanism of emulsion polymerization feature ambiguity caused by such specific factors as the multiphase nature of the reaction system and the variety of kinetic parameters, whose values depend not so much on the reagent reactivity as on the character of their distribution over phases, reaction topochemistry, the way and mechanism of nucleation and stabilization of particles. The obtained results pointing to the discrepancy with classical concepts can be characterized by the following effects:

(1) Recombination of radicals in an aqueous phase, leading to a reduction of the number of particles and to the formation of surfactant oligomers capable of acting as emulsifiers.
(2) The presence of several growing radicals in polymer–monomer particles, which causes the gel effect appearance and an increase in the polymerization rate at high conversion degrees.
(3) A decrease in the number of latex particles with the growing conversion degree, which is associated with their flocculation at various polymerization stages.
(4) An increase in the number of particles in the reaction course when using monomer-soluble emulsifiers, and also due to the formation of an own emulsifier (oligomers).

Surfactants (emulsifiers of various chemical nature) are usually applied as stabilizers of disperse systems, they are rather stable, poorly destructed under the influence of natural factors, and contaminate the environment. The principal possibility to synthesize emulsifier free latexes was shown. In the absence of emulsifier (but in emulsion polymerization conditions) with the usage of persulfate type initiators (e.g. ammonium persulfate), the particles of acrylate latexes can be stabilized by the ionized end groups of macromolecules. The $M_nSO_4^-$ radical ions appearing in the aqueous phase of the reaction medium, having reached some critical chain length, precipitate to form primary particles, which flocculate up to the formation of aggregates with a charge density providing their stability. Besides, due to radical recombination, oligomeric molecules are formed in the aqueous phase, which possess properties of surfactants and are able to form micelle-like structures. Then, the monomer and oligomeric radicals are absorbed by these "micelles", where chains can grow. In the absence of a specially introduced emulsifier, all basic kinetic regularities of emulsion polymerization are observed, and differences are only concerned with the stage of particle generation and the mechanism of their stabilization, which can be amplified at copolymerization of hydrophobic monomers with highly hydrophilic comonomers. Increasing temperature results in a higher polymerization rate and a growth of the number of latex particles in the dispersion formed, in decreasing their sizes and the quantity of the coagulum formed, and in improved stability of the dispersion. At emulsifier free polymerization of alkyl acrylates, the stability of emulsions and obtained dispersions rises in the monomer row: MA < ethylacrylate (EA) < butylacrylate (BA), that is the stability grows at lowering the polarity of the main monomer.

Our account of the aforesaid factors influencing the kinetics and mechanism of emulsion polymerization (in both presence and absence of an emulsifier) has enabled the influence of comonomers on the processes of formation of polymeric dispersions based on (meth)acrylates to be explained. Changes of some reaction conditions have turned out to affect the influence character of other ones. For example increasing the concentration of MAA at its copolymerization with MA at a relatively low initiation rate leads to a decrease in the rate and particle number and an increase in the coagulum amount. At high initiation rates, the number of particles in the dispersion in the presence of MAA rises and their stability improves. The same effects were revealed for emulsifier free polymerization of BA as well, when its partial replacement by MAA at high temperatures results in better stabilization of the dispersion, an increase in the reaction rate and the number of particles (whereas their decrease was observed in the presence of an emulsifier). Similar effects were found for AN as well, which worsens the stability of the dispersion at relatively low temperatures but improves it at high ones. Increasing the AN concentration in the ternary monomeric system with a high MAA content leads to a higher number of particles and better stability of the dispersion at relatively low temperatures as well.

Emulsion copolymerization of acrylic monomers and unconjugated dienes was studied with the aim to explore the possibility to synthesize dispersions whose particles would contain reactive polymeric molecules with free multiple C=C bonds. The usage of such latexes to finish fabrics and some other materials promotes getting strong indelible coatings. The kinetics and mechanism of emulsion copolymerization of EA and BA with allylacrylate (AlA) (with ammonium persulfate (APS) or the APS—sodium thiosulfate system as initiators) were studied. The found constants of copolymerization of AlA with EA (r_{AlA} = 1.05; r_{EA} = 0.8) and AlA with BA (r_{AlA} = 1.1; r_{BA} = 0.4) point to different degrees of the copolymer unsaturation with AlA units. Emulsion copolymerization of multicomponent monomeric BA based systems (with AN, MAA, and the unconjugated diene acryloxyethylmaleate (AOEM) as comonomers) was also studied. The effect of AN and MAA is described above. Copolymerization with AOEM depends on reaction conditions. For example AOEM reduces the polymerization rate in MAA free systems. In the presence of MAA, the rate of the process at high conversion degrees increases with the AOEM concentration, this monomer promoting the gel effect due to partial chain linking in polymer–monomer particles by the side groups with C=C bonds. The unsaturation degree of diene units in the copolymer is subject to the composition of monomers, AOEM concentration, and temperature. In the BA—AOEM system, an increase of the diene concentration results in an increase in the unsaturation degree. This means that the diene radical is added to its own monomer with a higher rate than to BA (r_{AOEM} = 7.7). The higher unsaturation degree of diene units in the MAA-containing systems points to that the AOEM radical interacts with MAA with a higher rate than with BA, and the probability of cyclization reduces, the degree of polymer unsaturation increases, and the diene's retarding effect upon polymerization weakens.

Compositions for production of rigid foamed polyurethane [25, 37, 38, 40]. In the field of polymer physicochemistry, studies were made according to the requirements of the Montreal Protocol on Substances That Deplete the Ozone Layer (1987), which demands to drastically reduce the production and consumption of chlorofluorocarbons (CFC, Freons) and even replace ozone dangerous substances by ozone safe ones. Our investigations dealt with the replacement of trichlorofluoromethane (Freon-11), which had been used as a foaming agent in the synthesis of rigid foam polyurethane (FPU) over a long period of time, which was thermal insulator in freezing chambers and building constructions. On the basis of our experimental dependences of the kinetic parameters of the foaming process (the instant of start, the time of structurization, and the instant of foam ending rising), the values of density and heat conductivity of pilot foam plastic samples on the concentration of the reaction mixture components and physicochemical conditions of FPU synthesis, optimal compositions (recipes) of mixtures with Freon-11 replaced by an ozone safe (with ozone destruction potentials by an order of magnitude lower in comparison with Freon-11) azeotropic mixture of dichlorotrifluoroethane and dichlorofluoroethane were found. Their practical implementation requires no principal changes of known technological procedures and no usage of new chemical reagents, which is an important merit of our achievements.

Scientific basics of the synthesis of a high-molecular-weight flocculent [24, 25, 39, 40]. There exists a problem of purification of natural water and industrial sewage from various pollutants, including suspended and colloid particles, associated with the growing consumption of water and its deteriorating quality (owing to anthropogenic influence). It is known that flocculants can be used for these purposes, which high-molecular-weight compounds capable of adsorption are on disperse particles to form quickly sedimenting aggregates. Polyacrylamide (PAA) is most active of them. As many countries (including Russian Federation) suffer from acrylamide (AA) deficit, we have developed modifications of the PAA flocculent synthesis by means of the usage of AN and sulfuric acid to bring about the reactions of hydrolysis and polymerization. The AN has turned out to participate in both processes simultaneously in the presence of a radical initiator of polymerization and sulfuric acid, and, as AA is being formed from AN, their joint polymerization begins. Desired polymer properties were achieved at AN polymerization in an aqueous solution of sulfuric acid up to a certain conversion degree with subsequent hydrolysis of the polymerizate (a two stage synthesis scheme) or at achieving an optimal ratio of the rates of these reactions proceeding in one stage. The influence of the nature and concentration of initiator, the sulfuric acid content, temperature and reaction duration on the quantity and molecular mass of the polymer contained in the final product, its solubility in water and flocculating properties were studied. The required conversion degree at the first stage of synthesis by the two-stage scheme is determined by the concentrations of AN and the aqueous solution of sulfuric acid. At one-stage synthesis, changes in temperature and monomer amount almost equally affect the hydrolysis and polymerization rates and do not strongly affect the copolymer composition. But changes in the concentration of either acid or initiator rather strongly influence the molecular mass and composition of macromolecules, which causes external dependences of the flocculating activity on these factors.

10.4 CONCLUSION

New kinetic regularities at polymerization of vinyl monomers in homophase and heterophase conditions in the presence of additives of transition metal salts, azonitriles, peroxides, stable nitroxyl radicals and radical anions (and their complexes), aromatic amines and their derivatives, emulsifiers and solvents of various nature were revealed. The mechanisms of the studied processes have been established in the whole and as elementary stages, their basic kinetic characteristics have been determined. Equations to describe the behavior of the studied chemical systems in polymerization reactions proceeding in various physicochemical conditions have been derived. Scientific principles of regulating polymer synthesis processes have been elaborated, which allows optimization of some industrial technologies and solving most important problems of environment protection.

KEYWORDS

- **Benzoyl peroxide**
- **Dimethyl formamide**
- **Methacrylic acid**
- **Methylacrylate**
- **Methylmethacrylate**
- **Tetracyanoquinodimethane**

REFERENCES

1. Semenov, N. N. Chemical Physics, *Chemical Physics on the Threshold of the 21 Century*. Moscow, Nauka, p. 5–9 (1996).
2. Goldfein, M. D. *Kinetics and mechanism of radical polymerization of vinyl monomers*. Saratov, Saratov Univ. Press, p.139 (1986).
3. Goldfein, M. D., Kozhevnikov, N. V., and Trubnikov, A. V. *Kinetics and regulation mechanism of polymer formation processes*. Saratov, Saratov Univ. Press, p. 178 (1989).
4. Kozhevnikov, N. V., Gayvoronskaya, S. I., and Leontieva, L. T. In *Kinetics and mechanism of radical and polymerization processes*. Saratov, Saratov Univ. Press, p. 85–93 (1973).
5. Stepukhovich, A. D., Kozhevnikov, N. V., and Leontieva, L. T. *Polymer Science, Series A.*, **16**(7), 1522–1529 (1974).
6. Kozhevnikov, N. V. and Stepukhovich, A. D. *Polymer Science, Series A.*, **21**(7), 1593–1599 (1979).
7. Goldfein, M. D., Kozhevnikov, N. V., Rafikov, E. A. et al. *Polymer Science, Series A.*, **17**(10), 2282–2287 (1975).
8. Goldfein, M. D., Rafikov, E. A., Kozhevnikov, N. V. et al. *Polymer Science, Series A.*, **19**(2), 275–280 (1977).
9. Goldfein, M. D., Rafikov, E. A., Kozhevnikov, N. V. et al. *Polymer Science, Series A.*, **19**(11), 2557–2562 (1977).

10. Goldfein, M. D. and Zyubin, B. A. *Polymer Science, Series A.*, **32**(11), 2243–2263 (1990).
11. Stepukhovich, A. D. and Kozhevnikov, N. V. *Polymer Science, Series A.*, **18**(4) 872–878 (1976).
12. Kozhevnikov, N. V. and Stepukhovich, A. D. *Polymer Science, Series A.*, **22**(5), 963–971 (1980).
13. Kozhevnikov, N. V. Proceedings of Russian Higher-Educational Establishments. *Chemistry and Chemical Technology*, **30**(4), 103–106 (1987).
14. Rozantsev, E. G., Goldfein, M. D., and Trubnikov, A. V. *Advances in Chemistry*, **55**(11), 1881–1897 (1986).
15. Rozantsev, E. G., Goldfein, M. D., and Pulin, V. F. *Organic Paramagnets*. Saratov, Saratov Univ. Press, p. 340 (2000).
16. Rozantsev, E. G., Goldfein, M. D. Oxidation Communications. *Sofia.*, **31**(2), 241–263 (2000).
17. Rozantsev, E. G. and Goldfein, M. D. *Polymers Research Journal*. Nova Science Publishers, New York, **2**(1), 5–28 (2000).
18. Rozantsev, E. G. and Goldfein, M. D. *Chemistry and Biochemistry. From Pure to Applied Science, New Horizons*, New York, **3**, 145–169 (2009).
19. Goldfein, M. D., Kozhevnikov, N. V., and Trubnikov, A. V. *Polymer Science, Series B.*, **25**(4), 268–271 (1983).
20. Goldfein, M. D. and Gladyshev, G. P. *Advances in Chemistry*, **57**(11), 1888–1912 (1988).
21. Goldfein, M. D., Kozhevnikov, N. V., and Trubnikov, A. V. *Russian Chemical Industry*, (1), 20–22 (1989).
22. Simontseva, N. S., Kashanova, T. T., and Goldfein, M. D. *Russian Plastics*, (12), 25–28 (1989).
23. Goldfein, M. D., Gladyshev, G. P., and Trubnikov, A. V. *Polymer Yearbook*, (13), 163–190 (1996).
24. Kozhevnikov, N. V., Goldfein, M. D., and Kozhevnikova, N. I. *Journal of the Balkan Tribological Association. Sofia*, **14**(4), 560–571 (2008).
25. Kozhevnikov, N. V., Kozhevnikova, N. I., and Goldfein, M. D. Proceedings of Saratov University. New Series. *Chemistry, Biology, Ecology.*, **10**(2), 34–42 (2010).
26. Kozhevnikov, N. V., Goldfein, M. D., Zyubin, B. A., and Trubnikov, A. V. *Polymer Science, Series A.*, **33**(6), 1272–1280 (1991).
27. Goldfein, M. D., Kozhevnikov, N. V., and Trubnikov, A. V. *Polymer Science, Series A.*, **33**(10), 2035–2049 (1991).
28. Goldfein, M. D., Kozhevnikov, N. V., Trubnikov, A. V. *Polymer Yearbook.*, **12**, 89–104 (1995).
29. Kozhevnikov, N. V., Zyubin, B. A., and Simontsev, D. V. *Polymer Science, Series A.*, **37**(5). 758–763 (1995).
30. Kozhevnikov, N. V., Goldfein, M. D., and Terekhina, N. V. *Russian Chemical Physics*, **16**(12), 97–102 (1997).
31. Kozhevnikov, N. V., Terekhina, N. V., and Goldfein, M. D. Proceedings of Russian Higher-Educational Establishments. *Chemistry and Chemical Technology*, **41**(4), 83–87 (1998).
32. Kozhevnikov, N. V., Goldfein, M. D., and Trubnikov, A. V. *Inter. Journal Polymer Mater.*, **46**, 95–105 (2000).
33. Kozhevnikov, N. V., Goldfein, M. D., and Trubnikov, A. V. *Preparation and Properties of Monomers, Polymers, and Composite Materials*. Nova Science Publishers, New York, p. 155–163 (2007).
34. Kozhevnikov, N. V., Goldfein, M. D., Trubnikov, A. V., Kozhevnikova, N. I. *Journal of the Balkan Tribological Association. Sofia.*, **13**(3), 379–386 (2007).
35. Kozhevnikov, N. V., Kozhevnikova, N. I., and Goldfein, M. D. Proceedings of Russian Higher-Educational Establishments. *Chemistry and Chemical Technology.*, **53**(2), 64–68 (2010).
36. Kozhevnikov, N. V., Goldfein, M. D., and Kozhevnikova, N. I. *Journal of Characterization and Development of Novel Materials.*, **2**(1), 53–62 (2011).
37. Goldfein M. D. and Kozhevnikov N. V. *Problems of Regional Ecology.*, (4), 92–95 (2005).
38. Goldfein, M. D. *Proceedings of Saratov University. New Series. Chemistry, Biology, Ecology.*, **9**(2), 79–83 (2009).

39. Kozhevnikov, N. V., Goldfein, M. D., and Kozhevnikova, N. I. *Modern Tendencies in Organic and Bioorganic Chemistry: Today and Tomorrow*. Nova Science Publishers, Inc., New York, p. 379–384 (2008).
40. Goldfein, M. D., Ivanov, A. V., Kozhevnikov, N. V. *Fundamentals of General Ecology, Life Safety and Environment Protection*. Nova Science Publishers, New York, p. 210 (2010).

11 Update on Quantum-chemical Calculation

V. A. Babkin, D. S. Zakharov, and G. E. Zaikov

CONTENTS

11.1 Introduction 95
11.2 Aims and Backgrounds 95
11.3 Methodical Part 95
Keywords 100
References 100

11.1 INTRODUCTION

Quantum-chemical calculation of molecules of α-cyclopropyl-p-izopropylstyrene, α-cyclopropyl-2,4-dimethylstyrene, α-cyclopropyl-p-ftorstyrene was done by method MNDO. Optimized by all parameters geometric and electronic structures of these compound was received. The universal factor of acidity was calculated (pKa = 32). Molecules of α-cyclopropyl-p-izopropylstyrene, α-cyclopropyl-2,4-dimethylstyrene, α-cyclopropyl-p-ftorstyrene pertain to class of very weak H-acids (pKa > 14).

11.2 AIMS AND BACKGROUNDS

The aim of this work is a study of electronic structure of molecules α-cyclopropyl-p-izopropylstyrene, α-cyclopropyl-p-izopropylstyrene, α-cyclopropyl-2,4-dimethyl styrene, α-cyclopropyl-p-ftorstyrene [1] and theoretical estimation its acid power by quantum-chemical method MNDO. The calculation was done with optimization of all parameters by standard gradient method built-in in PC GAMESS [2]. The calculation was executed in approach the insulated molecule in gas phase. Program MacMolPlt was used for visual presentation of the model of the molecule [3].

11.3 METHODICAL PART

Geometric and electronic structures, general and electronic energies of molecules α-cyclopropyl-p-izopropylstyrene, α-cyclopropyl-2,4-dimethylstyrene, α-cyclopropyl-p-ftorstyrene was received by method MNDO and are shown on Figure 1–3, and in Table 1–3, respectively. The universal factor of acidity was calculated by formula: $pKa = 49{,}4-134{,}61*q_{max}H^+$ [4], which used with success, for example in [5-35] (where, $q_{max}H^+$ – a maximum positive charge on atom of the hydrogen (by Milliken [1]) R =

0.97, R– a coefficient of correlations, $q_{max}H^+$ = +0.06, +0.06, +0.08, respectively. pKa = 30–33.

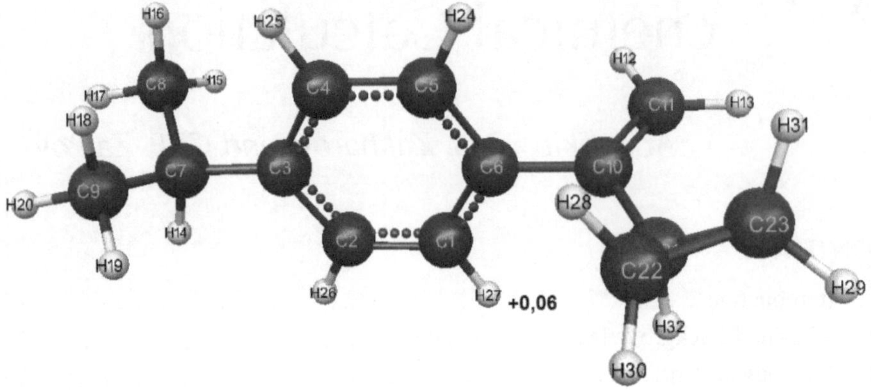

FIGURE 1 Geometric and electronic molecular structure of α-cyclopropyl-p-izopropylstyrene. (E0 = –196875 kDg/mol, Eel = –1215375 kDg/mol).

TABLE 1 Optimized bond lengths, valence corners and charges on atoms of the molecule of α-cyclopropyl-p-izopropylstyrene.

Bond lengths	R,A	Valence corners	Grad	Atom	Charge (by Milliken)
C(1)–C(6)	1.40	C(6)–C(1)–C(2)	121	C(1)	–0.04
C(2)–C(1)	1.39	C(1)–C(2)–C(3)	121	C(2)	–0.05
C(3)–C(2)	1.40	C(2)–C(3)–C(4)	119	C(3)	–0.07
C(3)–C(7)	1.50	C(3)–C(4)–C(5)	121	C(4)	–0.05
C(4)–C(3)	1.40	C(4)–C(5)–C(6)	121	C(5)	–0.05
C(5)–C(4)	1.39	C(2)–C(3)–C(7)	120	C(6)	–0.03
C(6)–C(5)	1.40	C(3)–C(7)–C(8)	111	C(7)	–0.02
C(6)–C(10)	1.47	C(3)–C(7)–C(9)	111	C(8)	+0.04
C(7)–C(8)	1.52	C(8)–C(7)–C(9)	110	C(9)	+0.04
C(7)–C(9)	1.52	C(1)–C(6)–C(10)	121	C(10)	–0.06
C(10)–C(11)	1.34	C(6)–C(10)–C(11)	122	C(11)	–0.03
C(10)–C(21)	1.47	C(10)–C(11)–H(12)	122	H(12)	+0.04
H(12)–C(11)	1.10	C(10)–C(11)–H(13)	122	H(13)	+0.04
H(13)–C(11)	1.10	C(3)–C(7)–H(14)	108	H(14)	+0.01
H(14)–C(7)	1.13	C(7)–C(8)–H(15)	110	H(15)	0.00
H(15)–C(8)	1.12	C(7)–C(8)–H(16)	111	H(16)	0.00

TABLE 1 *(Continued)*

Bond lengths	R,A	Valence corners	Grad	Atom	Charge (by Milliken)
H(16)–C(8)	1.12	C(7)–C(8)–H(17)	110	H(17)	–0.01
H(17)–C(8)	1.12	C(7)–C(9)–H(18)	111	H(18)	0.00
H(18)–C(9)	1.12	C(7)–C(9)–H(19)	110	H(19)	0.00
H(19)–C(9)	1.12	C(7)–C(9)–H(20)	110	H(20)	–0.01
H(20)–C(9)	1.12	C(6)–C(10)–C(21)	116	C(21)	–0.07
C(21)–C(22)	1.51	C(11)–C(10)–C(21)	123	C(22)	–0.05
C(22)–C(23)	1.50	C(22)–C(23)–C(21)	60	C(23)	–0.06
C(23)–C(21)	1.51	C(10)–C(21)–C(22)	121	H(24)	+0.06
H(24)–C(5)	1.10	C(21)–C(23)–C(22)	60	H(25)	+0.06
H(25)–C(4)	1.10	C(21)–C(22)–C(23)	60	H(26)	+0.06
H(26)–C(2)	1.10	C(22)–C(21)–C(23)	60	H(27)	+0.06
H(27)–C(1)	1.10	C(4)–C(5)–H(24)	120	H(28)	+0.04
H(28)–C(22)	1.10	C(3)–C(4)–H(25)	120	H(29)	+0.04
H(29)–C(23)	1.10	C(1)–C(2)–H(26)	120	H(30)	+0.04
H(30)–C(22)	1.10	C(2)–C(1)–H(27)	120	H(31)	+0.04
H(31)–C(23)	1.10	C(21)–C(22)–H(28)	119	H(32)	+0.05
H(32)–C(21)	1.11	C(21)–C(23)–H(29)	118		
		C(21)–C(22)–H(30)	119		
		C(21)–C(23)–H(31)	120		
		C(10)–C(21)–H(32)	111		

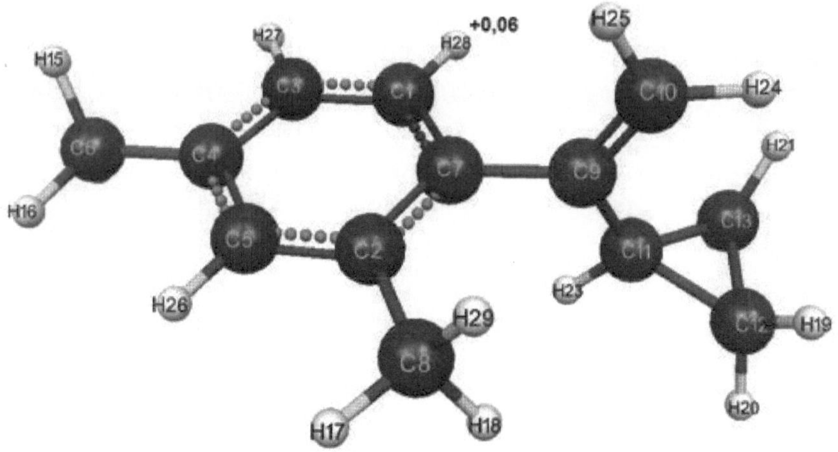

FIGURE 2 Geometric and electronic molecular structure of **α**-cyclopropyl-2,4-dimethylstyrene. (E_0 = –181125 kDg/mol, Eel= –1084125 kDg/mol).

TABLE 2 Optimized bond lengths, valence corners and charges on atoms of the molecule of α-cyclopropyl-p-izopropylstyrene.

Bond lengths	R,A	Valence corners	Grad	Atom	Charge (by Milliken)
C(1)–C(7)	1.42	C(1)–C(7)–C(2)	119	C(1)	–0.0508
C(2)–C(5)	1.42	C(7)–C(1)–C(3)	121	C(2)	–0.0835
C(3)–C(1)	1.40	C(1)–C(3)–C(4)	121	C(3)	–0.0412
C(4)–C(3)	1.41	C(2)–C(5)–C(4)	123	C(4)	–0.1009
C(5)–C(4)	1.41	C(3)–C(4)–C(5)	118	C(5)	–0.0280
C(6)–C(4)	1.51	C(3)–C(4)–C(6)	121	C(6)	0.0814
C(7)–C(2)	1.42	C(5)–C(2)–C(7)	119	C(7)	–0.0187
C(7)–C(9)	1.50	C(5)–C(2)–C(8)	119	C(8)	0.0807
C(10)–C(9)	1.35	C(1)–C(7)–C(9)	118	C(9)	–0.0545
C(11)–C(9)	1.50	C(7)–C(9)–C(10)	120	C(10)	–0.0416
C(12)–C(11)	1.54	C(7)–C(9)–C(11)	115	C(11)	–0.0617
C(13)–C(12)	1.52	C(9)–C(11)–C(12)	125	C(12)	–0.0561
C(13)–C(11)	1.54	C(9)–C(11)–C(13)	125	C(13)	–0.0568
H(14)–C(6)	1.11	C(4)–C(6)–H(14)	111	H(14)	–0.0028
H(15)–C(6)	1.11	C(4)–C(6)–H(15)	111	H(15)	–0.0027
H(16)–C(6)	1.11	C(4)–C(6)–H(16)	113	H(16)	–0.0050
H(17)–C(8)	1.11	C(2)–C(8)–H(17)	112	H(17)	–0.0072
H(18)–C(8)	1.11	C(2)–C(8)–H(18)	111	H(18)	–0.0002
H(19)–C(12)	1.10	C(11)–C(12)–H(19)	121	H(19)	0.0389
H(20)–C(12)	1.10	C(11)–C(12)–H(20)	118	H(20)	0.0368
H(21)–C(13)	1.10	C(11)–C(13)–H(21)	121	H(21)	0.0387
H(22)–C(13)	1.10	C(11)–C(13)–H(22)	118	H(22)	0.0370
H(23)–C(11)	1.10 1.09	C(9)–C(11)–H(23)	111	H(23)	0.0451
H(24)–C(10)	1.09	C(9)–C(10)–H(24)	124	H(24)	0.0394
H(25)–C(10)	1.09	C(9)–C(10)–H(25)	123	H(25)	0.0425
H(26)–C(5)	1.09	C(2)–C(5)–H(26)	119	H(26)	0.0550
H(27)–C(3)	1.09	C(1)–C(3)–H(27)	119	H(27)	0.0581
H(28)–C(1)	1.11	C(3)–C(1)–H(28)	119	H(28)	0.0600
H(29)–C(8)		C(2)–C(8)–H(29)	111	H(29)	–0.0019

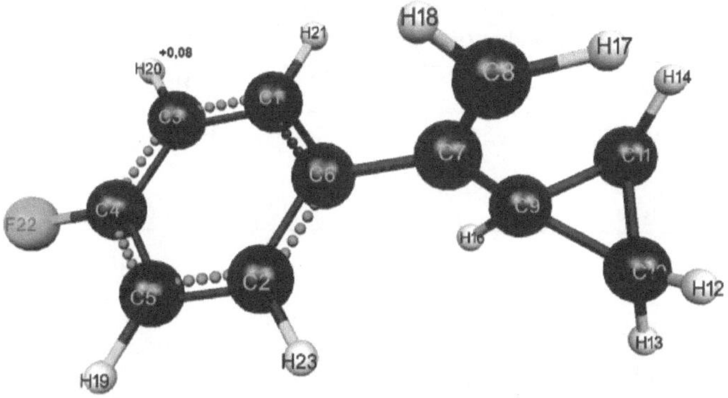

FIGURE 3 Geometric and electronic molecular structure of α-cyclopropyl-p-ftorstyrene. (E0 = –196875 kDg/mol, Eel = –1155000 kDg/mol).

TABLE 3 Optimized bond lengths, valence corners and charges on atoms of the molecule of α-cyclopropyl-p-izopropylstyrene.

Bond lengths	R,A	Valence corners	Grad	Atom	Charge (by Milliken)
C(1)–C(3)	1.40	C(1)–C(6)–C(2)	119	C(1)	–0.02
C(2)–C(6)	1.41	C(4)–C(5)–C(2)	120	C(2)	–0.02
C(3)–C(4)	1.42	C(5)–C(4)–C(3)	120	C(3)	–0.09
C(4)–C(5)	1.42	C(1)–C(3)–C(4)	120	C(4)	+0.15
C(5)–C(2)	1.40	C(6)–C(2)–C(5)	121	C(5)	–0.09
C(6)–C(7)	1.49	C(3)–C(1)–C(6)	121	C(6)	–0.06
C(7)–C(9)	1.50	C(1)–C(6)–C(7)	121	C(7)	–0.06
C(8)–C(7)	1.35	C(6)–C(7)–C(8)	120	C(8)	–0.04
C(9)–C(10)	1.54	C(6)–C(7)–C(9)	115	C(9)	–0.07
C(10)–C(11)	1.52	C(7)–C(9)–C(10)	124	C(10)	–0.06
C(11)–C(9)	1.54	C(7)–C(9)–C(11)	125	C(11)	–0.06
H(12)–C(10)	1.10	C(9)–C(10)–H(12)	121	H(12)	+0.04
H(13)–C(10)	1.10	C(9)–C(10)–H(13)	118	H(13)	+0.04
H(14)–C(11)	1.10	C(9)–C(11)–H(14)	121	H(14)	+0.04
H(15)–C(11)	1.10	C(9)–C(11)–H(15)	118	H(15)	+0.04
H(16)–C(9)	1.10	C(7)–C(9)–H(16)	111	H(16)	+0.04
H(17)–C(8)	1.09	C(7)–C(8)–H(17)	124	H(17)	+0.04
H(18)–C(8)	1.09	C(7)–C(8)–H(18)	123	H(18)	+0.04
H(19)–C(5)	1.09	C(2)–C(5)–H(19)	120	H(19)	+0.08
H(20)–C(3)	1.09	C(1)–C(3)–H(20)	120	H(20)	+0.08
H(21)–C(1)	1.09	C(3)–C(1)–H(21)	119	H(21)	+0.07
F(22)–C(4)	1.33	C(3)–C(4)–F(22)	120	F(22)	–0.18
H(23)–C(2)	1.09	C(5)–C(2)–H(23)	119	H(23)	+0.07

TABLE 4 General and energies (E0), maximum positive charge on atom of the hydrogen (qmaxH+), universal factor of acidity (pKa).

Molecules of aromatic olefins	E_0	$q_{max}H^+$	pKa
α-cyclopropyl-p-izopropylstyrene.	−196875	+0.06	33
α-cyclopropyl-2,4-dimethylstyrene	−181125	+0.06	33
α-cyclopropyl-p-ftorstyrene.	−196875	+0.08	30

KEYWORDS

- **Optimized bond**
- **Quantum-chemical calculation**
- **α-Cyclopropyl-2,4-dimethylstyrene**
- **α-Cyclopropyl-p-ftorstyrene**
- **α-Cyclopropyl-p-izopropylstyrene**

REFERENCES

1. Kennedy, J. *Cation polymerization of olefins*. World, Moscow, p. 430 (1978).
2. Shmidt, M. W. Baldrosge, K. K. Elbert, J. A. Gordon, M. S. Enseh, J. H. Koseki, S. Matsvnaga., N. Nguyen, K. A., Su, S. J., and anothers. *J. Comput. CHEM.*, **14**, 1347–1363, (1993).
3. Babkin V. A., Fedunov R. G., Minsker K. S., and anothers. *Oxidation communication*, **1**, 25, 21-47 (2002).
4. Bode, B. M. and Gordon, M. S. *J. Mol. Graphics Mod.*, **16**, 133–138 (1998).
5. Babkin, V. A., Dmitriev, V. Yu., and Zaikov, G. E. Quantum chemical calculation of molecule hexene-1 by method MNDO. In *Quantum chemical calculation of unique molecular system. Vol. I.* Publisher VolSU, c. Volgograd, pp. 93–95 (2010).
6. Babkin V. A., Dmitriev V. Yu., and Zaikov G. E. Quantum chemical calculation of molecule heptene-1 by method MNDO. In *Quantum chemical calculation of unique molecular system. Vol. I.* Publisher VolSU, c. Volgograd, pp. 95–97 (2010).
7. Babkin V. A., Dmitriev V. Yu., and Zaikov G. E. Quantum chemical calculation of molecule decene-1 by method MNDO. In *Quantum chemical calculation of unique molecular system. Vol. I.* Publisher VolSU, c. Volgograd, pp. 97–99 (2010).
8. Babkin V. A., Dmitriev V. Yu., and Zaikov G. E. Quantum chemical calculation of molecule nonene-1 by method MNDO. In *Quantum chemical calculation of unique molecular system. Vol. I.* Publisher VolSU, c. Volgograd, pp. 99–102 (2010).
9. Babkin V. A. and Andreev D. S. Quantum chemical calculation of molecule isobutylene by method MNDO. In *Quantum chemical calculation of unique molecular system. Vol. I.* Publisher VolSU, c. Volgograd, pp. 176–177 (2010).
10. Babkin V. A. and Andreev D. S. Quantum chemical calculation of molecule 2-methylbutene-1 by method MNDO. In *Quantum chemical calculation of unique molecular system. Vol. I.* Publisher VolSU, c. Volgograd, pp. 177–179 (2010).
11. Babkin V. A. and Andreev D. S. Quantum chemical calculation of molecule 2-methylbutene-2 by method MNDO. In *Quantum chemical calculation of unique molecular system. Vol. I.* Publisher VolSU, c. Volgograd, pp. 179–180 (2010).
12. Babkin V. A. and Andreev D. S. Quantum chemical calculation of molecule 2-methylpentene-1 by method MNDO. In *Quantum chemical calculation of unique molecular system. Vol. I.* Publisher VolSU, c. Volgograd, pp. 181–182 (2010).

13. Babkin V. A., Dmitriev V. Yu. and Zaikov G. E. Quantum chemical calculation of molecule butene-1 by method MNDO. In *Quantum chemical calculation of unique molecular system. Vol. I.* Publisher VolSU, c. Volgograd, pp. 89–90 (2010).
14. Babkin V. A., Dmitriev V. Yu., and Zaikov G. E. Quantum chemical calculation of molecule hexene-1 by method MNDO. In *Quantum chemical calculation of unique molecular system. Vol. I.* Publisher VolSU, c. Volgograd, pp. 93–95 (2010).
15. Babkin V. A., Dmitriev V. Yu., and Zaikov G. E. Quantum chemical calculation of molecule octene-1 by method MNDO. In *Quantum chemical calculation of unique molecular system. Vol. I.* Publisher VolSU, c. Volgograd, pp. 103–105 (2010).
16. Babkin V. A., Dmitriev V. Yu., and Zaikov G. E. Quantum chemical calculation of molecule pentene-1 by method MNDO. In *Quantum chemical calculation of unique molecular system. Vol. I.* Publisher VolSU, c. Volgograd, pp. 105–107 (2010).
17. Babkin V. A., Dmitriev V. Yu., and Zaikov G. E. Quantum chemical calculation of molecule propene-1 by method MNDO. In *Quantum chemical calculation of unique molecular system. Vol. I.* Publisher VolSU, c. Volgograd, pp. 107–108 (2010).
18. Babkin V. A., Dmitriev V. Yu., and Zaikov G. E. Quantum chemical calculation of molecule ethylene-1 by method MNDO. In *Quantum chemical calculation of unique molecular system. Vol. I.* Publisher VolSU, c. Volgograd, pp. 108–109 (2010).
19. Babkin V. A. and Andreev D. S. Quantum chemical calculation of molecule butadien-1,3 by method MNDO. In *Quantum chemical calculation of unique molecular system. Vol. I.* Publisher VolSU, c. Volgograd, pp. 235–236 (2010).
20. Babkin V. A., Andreev D. S. Quantum chemical calculation of molecule 2-methylbutadien-1,3 by method MNDO. In *Quantum chemical calculation of unique molecular system. Vol. I.* Publisher VolSU, c. Volgograd, pp. 236–238 (2010).
21. Babkin V. A. and Andreev D. S. Quantum chemical calculation of molecule 2,3-dimethylbutadien-1,3 by method MNDO. In *Quantum chemical calculation of unique molecular system. Vol. I.* Publisher VolSU, c. Volgograd, pp. 238–239 (2010).
22. Babkin V. A., Andreev D. S. Quantum chemical calculation of molecule pentadien-1,3 by method MNDO. In *Quantum chemical calculation of unique molecular system. Vol. I.* Publisher VolSU, c. Volgograd, pp. 240–241 (2010).
23. Babkin V. A., Andreev D. S. Quantum chemical calculation of molecule trans-trans-hexadien-2,4 by method MNDO. In *Quantum chemical calculation of unique molecular system. Vol. I.* Publisher VolSU, c. Volgograd, pp. 241–243 (2010).
24. Babkin V. A. and Andreev D. S. Quantum chemical calculation of molecule cis-trans-hexadien-2,4 by method MNDO. In *Quantum chemical calculation of unique molecular system. Vol. I.* Publisher VolSU, c. Volgograd, pp. 243–245 (2010).
25. Babkin V. A. and Andreev D. S. Quantum chemical calculation of molecule cis-cis-hexadien-2,4 by method MNDO. In *Quantum chemical calculation of unique molecular system. Vol. I.* Publisher VolSU, c. Volgograd, pp. 245–246 (2010).
26. Babkin V. A. and Andreev D. S. Quantum chemical calculation of molecule trans-2-methylpentadien-1,3 by method MNDO. In *Quantum chemical calculation of unique molecular system. Vol. I.* Publisher VolSU, c. Volgograd, pp. 247–248 (2010).
27. Babkin V. A. and Andreev D. S. Quantum chemical calculation of molecule trans-3-methylpentadien-1,3 by method MNDO. In *Quantum chemical calculation of unique molecular system. Vol. I.* Publisher VolSU, c. Volgograd, pp. 249–250 (2010).
28. Babkin V. A. and Andreev D. S. Quantum chemical calculation of molecule cis-3-methylpentadien-1,3 by method MNDO. In *Quantum chemical calculation of unique molecular system. Vol. I.* Publisher VolSU, c. Volgograd, pp. 251–252 (2010).
29. Babkin V. A. and Andreev D. S. Quantum chemical calculation of molecule 4-methylpentadien-1,3 by method MNDO. In *Quantum chemical calculation of unique molecular system. Vol. I.* Publisher VolSU, c. Volgograd, pp. 252–254 (2010).

30. Babkin V. A. and Andreev D. S. Quantum chemical calculation of molecule cis-3-methylpentadien-1,3 by method MNDO. In *Quantum chemical calculation of unique molecular system. Vol. I*. Publisher VolSU, c. Volgograd, pp. 254–256 (2010).
31. Babkin V. A. and Andreev D. S. Quantum chemical calculation of molecule 1,1,4,4-tetramethylbutadien-1,3 by method MNDO. In *Quantum chemical calculation of unique molecular system. Vol. I*. Publisher VolSU, c. Volgograd, pp. 256–258 (2010).
32. Babkin V. A., Andreev D. S. Quantum chemical calculation of molecule 2-phenylbutadien-1,3 by method MNDO. In *Quantum chemical calculation of unique molecular system. Vol. I*. Publisher VolSU, c. Volgograd, pp. 260–262 (2010).
33. Babkin V. A. and Andreev D. S. Quantum chemical calculation of molecule 1-phenyl-4-methylbutadien-1,3 by method MNDO. In *Quantum chemical calculation of unique molecular system. Vol. I*. Publisher VolSU, c. Volgograd, pp. 262–264 (2010).
34. Babkin V. A. and Andreev D. S. Quantum chemical calculation of molecule chloropren by method MNDO. In *Quantum chemical calculation of unique molecular system. Vol. I*. Publisher VolSU, c. Volgograd, pp. 264–265 (2010).
35. Babkin V. A. and Andreev D. S. Quantum chemical calculation of molecule trans-hexathrien-1,3,5 by method MNDO. In *Quantum chemical calculation of unique molecular system. Vol. I*. Publisher VolSU, c. Volgograd, pp. 266–267 (2010).

12 Some Aspects of Silver Nanoparticles

N. I. Naumkina, O. V. Mikhailov, and T. Z. Lygina

CONTENTS

12.1 Introduction 103
12.2 Experimental 103
12.3 Discussion and Results 105
Keywords 115
Acknowledgment 115
References 117

12.1 INTRODUCTION

Colloid elemental silver with small size of the particles formed in gelatin layers at development of silver halide photographic emulsions, was mentioned about more than 40 years ago in [1]. In a number of works, there are indications of existence of separate phase of the element silver consisting from nanoparticles and received as a result of photochemical reduction of Ag(I) salts, was appeared lately in the literature, in particular [2-11]. It has been noted that, before in [10, 11], at development of gelatin layers of silver halide photographic emulsions by alkaline water solutions containing tin(II) dichloride and some inorganic or organic substance form stable coordination compounds with Ag(I), formation of element silver occurs too. However, besides the gelatin layer is tinged brown or red but not black color as it takes place at standard development by using hydroquinone developers. It is significant that with increase of optical densities of the gelatin layer with an elemental silver, red tone in color of gelatin layer becomes more and more clearly expressed. The similar phenomenon takes place, when instead of silver halide AgHal in gelatin matrix there is such silver(I) compound as silver(I) hexacyanoferrate(II) $Ag_4[Fe(CN)_6]$. Whether this totality of particles is a novel phase of element silver? Or it is only a variety of known phases of the given simple substance? These questions remains till now, opened and deserves special consideration.

12.2 EXPERIMENTAL

As initial material to obtain silver-containing gelatin-immobilized matrix implants (GIM), X-ray film Structurix D-10 (Agfa-Gevaert, Belges) was samples of the given film (which actually is nothing but AgHal-GIM) having format 20 × 30 cm^2 were exposed to X-ray radiation with an irradiation dose at range 0.050.50 Röntgen. These exposed samples were further subjected to processing according to the following technology [10, 11]:

- Development in D-19 standard developer as it was indicated in [10, 11], for 6 min at 20–25°C;
- Washing with running water for 2 min at 20–25°C;
- Fixing in 25% water solution of sodium trioxosulphidosulphate(VI) ($Na_2S_2O_3$) for 10 min at 20–25°C;
- Washing with running water for 15 min at 18–25°C.

First three stages of standard processing (development, washing, and fixing) were carried out at non-actinic green-yellow light, and final washing at natural light. The samples of GIM containing elemental silver (Ag-GIM), were processed according to next technology:

(1) Oxidation in water solution containing (gl^{-1})

Potassium hexacyanoferrate(III)	50.0
Potassium hexacyanoferrate(II)	20.0
Potassium hydroxide	10.0
Sodium trioxocarbonate(IV) (Na_2CO_3)	5.0
Water	up to 1000 ml

for 6 min at 20–25°C;

(2) Washing with running water for 2 min at 20–25°C;

(3) Reduction in water solution containing (gl^{-1})

Tin(II) chloride	50.0
Sodium N,N¢-ethylene diaminetetra acetate	35.0
Potassium hydroxide	50.0
Reagent formed water soluble complex with Ag(I)	1.0–100.0
Water	up to 1000 ml

for 1 min at 20–25°C;

(4) Washing with running water for 15 min at 18–25°C;

(5) Drying for 23 hr at 20–25°C.

As complex forming reagents that form water soluble complexes with Ag(I), ammonia NH_3, potassium thiocyanate KSCN, sodium trioxosulphidosulphate(VI) $Na_2S_2O_3$, ethanediamine-1,2 $H_2N–CH_2–CH_2–NH_2$, 2-aminoethanol $H_2N–CH_2–CH_2–OH$, and 3-(2-hydroxyethyl)-3-azapenthanediol-1,5 $N(CH_2–CH_2–OH)_3$, were used. At the first stage of given processing of Ag-GIM obtained, conversion of Ag-GIM into $Ag_4[Fe(CN)_6]$-GIM occurred, on the second stage, reduction of $Ag_4[Fe(CN)_6]$-GIM with Sn(II) from elemental silver took place. And so, peculiar "re-precipitation" of elemental silver into gelatin matrix occurred incidentally.

An isolation substances from Ag-GIM was carried out by means of influence on them of water solutions of some proteolytic enzymes (for example, trypsin or Bacillus mesentericus) destroying the polymeric carrier of GIM (gelatin) and the subsequent separation of a solid phase from mother solution according to a technique described in [16]. The substances isolated thus from GIM further analyzed by X-ray diffraction method with using of spectrometer D8 Advance (Bruker, Germany). A scanning was carried out in interval from 3 to 65° × 2θ, a step was 0.05 × 2θ.

Calculation of intensities of reflexes (*I*) and inter-plane distances (*d*) carried out with application of standard software package EVA. Theoretical XRD spectra (X-ray patterns) were calculated under Powder Cell program described in [17, 18]. Optical density Ag-GIM was measured by means of Macbeth TD504 photometer (Kodak, USA) in range 0.15.0 units with accuracy of ± 2% (rel.).

12.3 DISCUSSION AND RESULTS

Already at visual observation over a course of transformation process of $Ag_4[Fe(CN)_6]$-GIM to Ag-GIM, following circumstance attracts its attention. The Ag-GIM received as a result of standard processing of exposed AgHal-GIM, at small optical density (D^{Ag}), have gray color, at big D^{Ag}, black color. The color Ag-GIM containing the "re-precipitated" element silver, varies from black-brown to red depending on the nature and quantity of complex forming reagent present in solution.

It is significant, however, that absorption spectra of both initial and the "re-precipitated" element silver in visible area, do not contain any accurately expressed maxima. Besides, optical density Ag-GIM with the "re-precipitated" silver (D^{Ag}), at the same volume concentration of element silver (C_{Ag}^{V}) in GIM, as a rule, is essentially more than D^{Ag} values and also depends on nature and quantity of complex forming reagent in solution contacting with GIM. Examples of $D = f(D^{Ag})$ и $D = f(C_{Ag}^{V})$ dependence for inorganic and organic reagents are presented in Figures 1–6. It is significant.

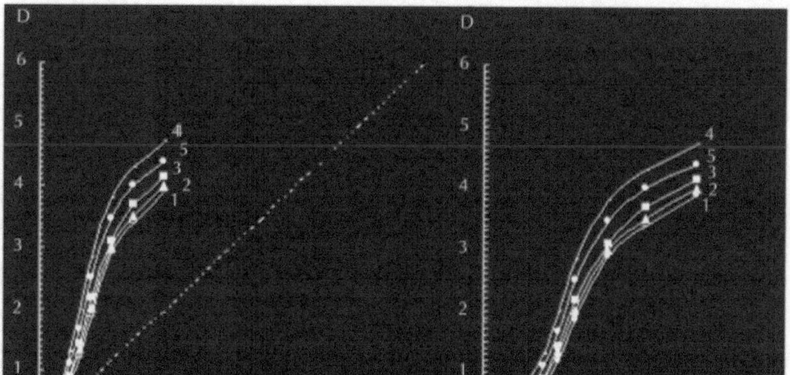

FIGURE 1 Dependence of $D = f(D^{Ag})$ and $D = f(C_{Ag}^{V})$ in reduction process of $Ag_4[Fe(CN)_6] \rightarrow Ag$ using NH_3 at concentration 1.5 gl^{-1} (curve 1), 3.0 gl^{-1} (2), 4.5 gl^{-1} (3), 6.0 gl^{-1} (4), and 7.5 gl^{-1} (5). Optical densities D^{Ag} and D were measured with blue light-filter with a transmission maximum at 450 nm.

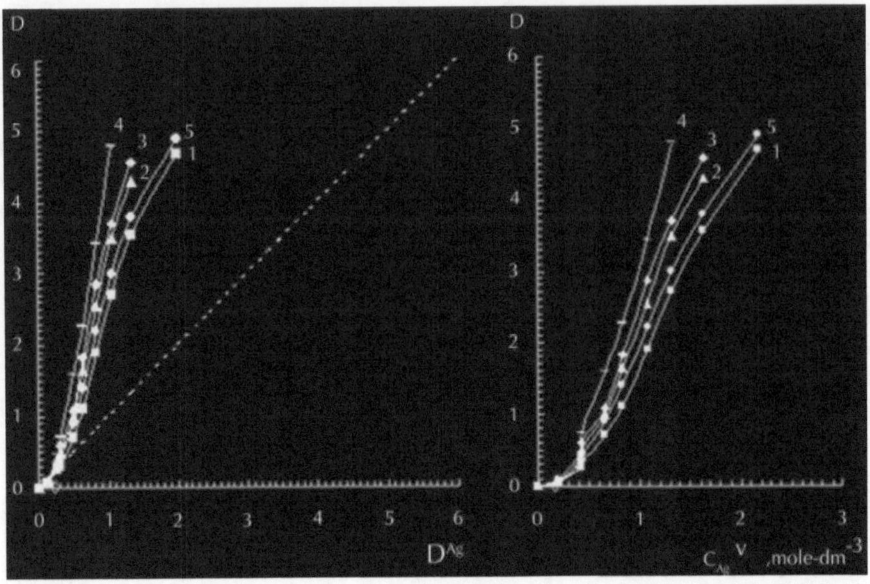

FIGURE 2 Dependence of $D = f(D^{Ag})$ and $D = f(C_{Ag}^{V})$ in reduction process of $Ag_4[Fe(CN)_6] \rightarrow Ag$ using $Na_2S_2O_3$ at concentration 2.0 gl^{-1} (curve 1), 4.0 gl^{-1} (2), 8.0 gl^{-1} (3), 24.0 gl^{-1} (4), and 40.0 gl^{-1} (5). Optical densities D^{Ag} and D were measured with blue light-filter with a transmission maximum at 450 nm.

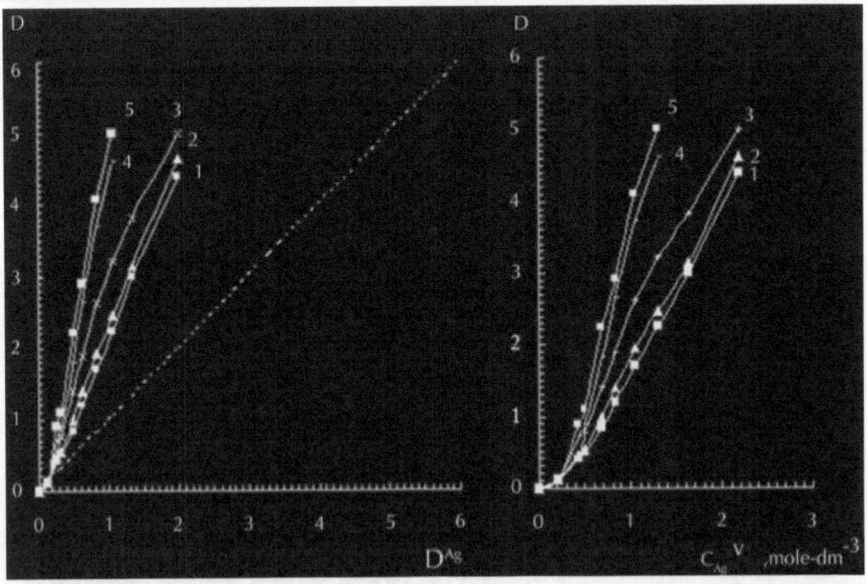

FIGURE 3 Dependence of $D = f(D^{Ag})$ and $D = f(C_{Ag}^{V})$ in reduction process $Ag_4[Fe(CN)_6] \rightarrow Ag$ using KSCN at concentration 2.0 gl^{-1} (curve 1), 4.0 gl^{-1} (2), 8.0 gl^{-1} (3), 24.0 gl^{-1} (4), and 60.0 gl^{-1} (5). Optical densities D^{Ag} and D were measured with blue light-filter with a transmission maximum at 450 nm.

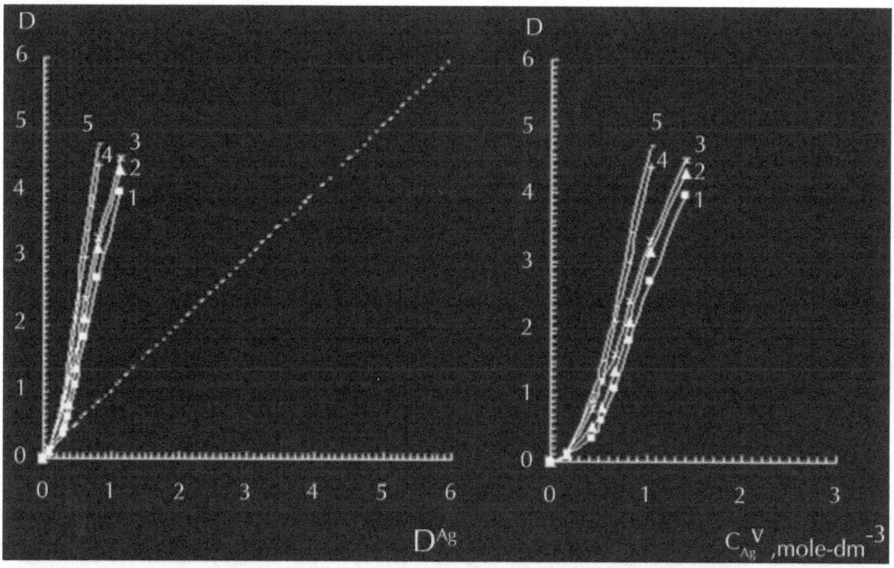

FIGURE 4 Dependence of $D = f(D^{Ag})$ and $D = f(C_{Ag}{}^{V})$ in reduction process of $Ag_4[Fe(CN)_6] \rightarrow Ag$ using of 2-aminoethanol $H_2N-(CH_2)_2-OH$ at concentration 7.5 gl^{-1} (curve 1), 15.0 gl^{-1} (2), 55.0 gl^{-1} (3), 110.0 gl^{-1} (4), and 150.0 gl^{-1} (5). Optical densities D^{Ag} and D were measured with blue light-filter with a transmission maximum at 450 nm.

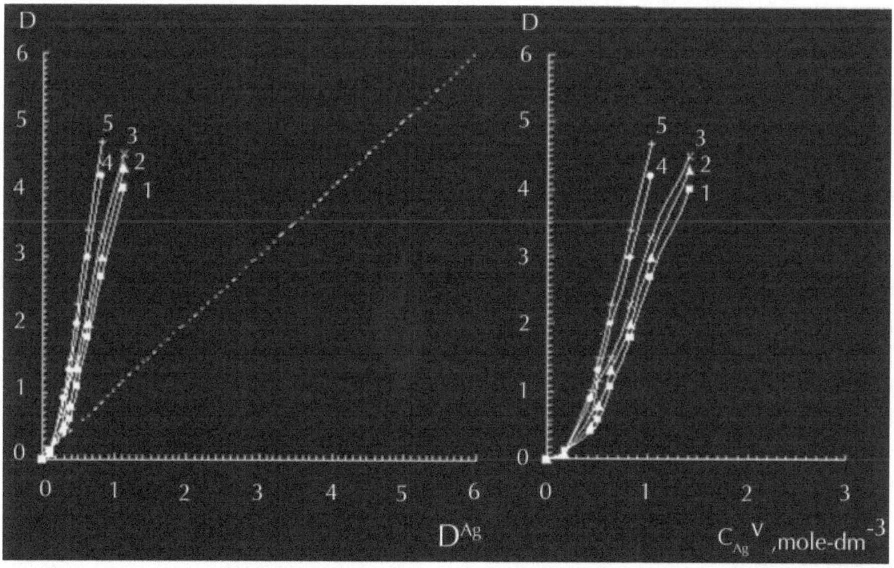

FIGURE 5 Dependence of $D = f(D^{Ag})$ and $D = f(C_{Ag}{}^{V})$ in reduction process $Ag_4[Fe(CN)_6] \rightarrow Ag$ using of ethanediamine-1,2 $H_2N-(CH_2)_2-NH_2$ at concentration 5.0 gl^{-1} (curve 1), 10.0 gl^{-1} (2), 20.0 gl^{-1} (3), 40.0 gl^{-1} (4), and 80.0 gl^{-1} (5). The optical densities D^{Ag} and D were measured with blue light-filter with a transmission maximum at 450 nm.

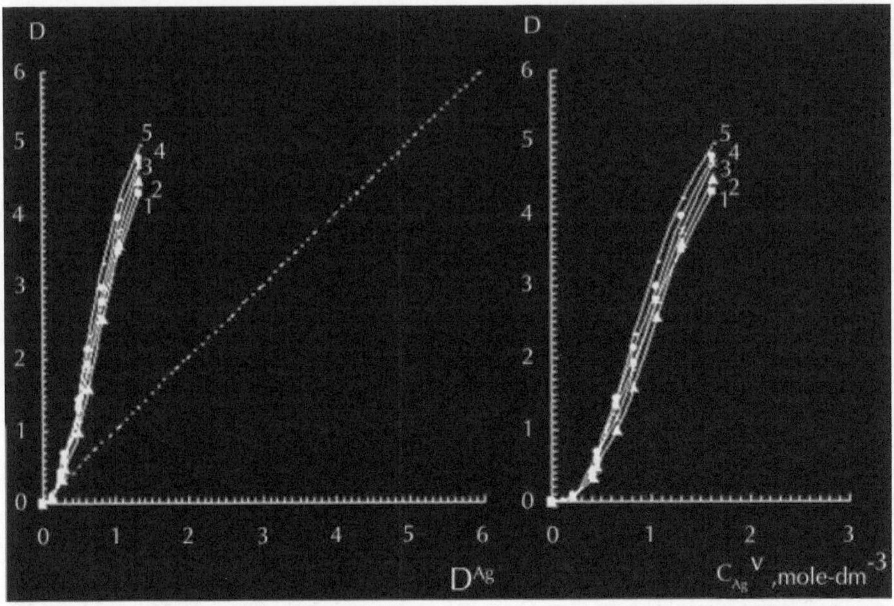

FIGURE 6 Dependence of $D=f(D^{Ag})$ and $D=f(C_{Ag}{}^{V})$ in reduction process of $Ag_4[Fe(CN)_6]\rightarrow Ag$ using of 3-(2-hydroxyethyl)-3-azapenthanediol-1, 5 $N(CH_2–CH_2–NH_2)_3$ at concentration 10.0 gl^{-1} (curve 1), 20.0 gl^{-1} (2), 35.0 gl^{-1} (3), 50.0 gl^{-1} (4), and 100.0 gl^{-1} (5). The optical densities D^{Ag} and D were measured with blue light-filter with a transmission maximum at 450 nm.

The D/D^{Ag} value, as a rule, greater than 1.0, and in some cases, it reaches very high values (as in the case of potassium thiocyanatenearly 5.0). Attention is drawn to the fact that the stronger the color of the gelatin layer with the "re-precipitated" elemental silver, different from the gray-black tones of the gelatinous layer and initially Ag-gelatin-immobilized matrix, the greater is the D/D^{Ag} value. The maximal degree of amplification $(D/D^{Ag})_{max}$ is also very much depends on the nature of the complex forming reagents agent (Table 1). The most profound effect on this parameter has etandiamin-1,2 $[(D/DAg)_{max} = 5.80]$, the least severe ammonia, the degree of possibility is also quite high $[(D/DAg)_{max} = 3.40]$. For the ammonia, the growth (D/D^{Ag}) value is typical with increasing concentration of NH_3 in reducing solution to relatively small value (~0.30 $moll^{-1}$), after which the optical density D begin to fall (Figure 1). It is noteworthy that red-brown color of gelatin layer attained at the indicated concentration, with further increase in the concentration of ammonia, does not change. Analogos situation occurs in the case of the other two we studied inorganic complex forming reagenttrioxosulfidosulfate(VI) and the thiocyanate anion (Figures 23), with the only difference being that in the case of $S_2O_3^{2-}$, maximum degree of amplification is achieved with less high in comparison with NH_3 concentration (0.15 $moll^{-1}$), in the case of SCN^-at a higher concentration (~0.70 $moll^{-1}$). In this regard, it was quite natural to try the difference marked with different stability of 1:2 complexes formed by Ag(I) with NH_3, $S_2O_3^{2-}$ and SCN^- (pK = 7.25, 13.32, and 8.39, respectively). However, in the presence of such correlation, the value of concentration indicated for SCN^- should be lower than for NH_3, which in fact is not observed. In reality, for the three organic complex forming reagent studied, molar concentrations at which the maximum value of D/D^{Ag} is reached, significantly greater than those for inorganic complex forming reagent (Table 1). Stability of the complexes of silver(I) with each of these ligands is lower than with NH_3, and the correlation function between these concentrations and the stability of coordination compounds of Ag(I) with given ligands in varying degrees, still visible. But any explosion term relationship, stability of the (D/D^{Ag}) values do not see: How it may be easily noticed when comparing the data of Table 1, the maximum degree of amplification decreases in the direction of ethanediamine-1,2 > 2-aminoethanol > SCN^- > $S_2O_3^{2-}$ > NH_3 > 3-(2-hydroxyethyl)-3-azapenthane-diol-1,5, while the resistance formed by these ligands complexes with silver(I)in the direction $S_2O_3^{2-}$ > SCN^- > ethanediamine-1,2 > NH_3 > 2-aminoethanol > 3-(2-hydroxyethyl)-3-azapenthanediol-1,5. Thus, the complex is though important but not the sole determinant of the degree of influence of complex forming agents on the redox process considered.

At the first stage of the given process, reaction described by general Equation (1), takes place (in the braces **{....}**, formulas of gelatin-immobilized chemical compounds have been indicated):

$$4\{Ag\} + 4[Fe(CN)_6]^{3-} \rightarrow \{Ag_4[Fe(CN)_6]\} + 3[Fe(CN)_6]^{4-} \quad (1)$$

TABLE 1 The maximal (D/D^{Ag}) and pK_s values of Ag(I) complexes for various complex forming reagents.

Complex forming reagent	$(D/D^{Ag})_{max}$	Concentration of complex forming reagent in solution at which reaches $(D/D^{Ag})_{max}$, gl^{-1} ($molel^{-1}$)	pK_s of Ag(I) complex having 1:2 composition
NH_3	3.40	4.5 (0.27)	7.25
$Na_2S_2O_3$	4.13	23.7 (0.15)	13.32
KSCN	4.92	69.7 (0.70)	8.39
$HO–(CH_2)_2–NH_2$	5.40	109.8 (1.80)	6.62
$H_2N–(CH_2)_2–NH_2$	5.80	77.0 (1.28)	7.84
$N(CH_2–CH_2–OH)_3$	3.93	99.8 (0.67)	3.64

Each of complex forming reagents under examination forms with Ag(I) soluble complex having a metal ion: ligand ratio of 1:2. That is why, formation of silver(I) complex with corresponding CR will occur to some extent when $Ag_4[Fe(CN)_6]$-GIM is at the contact with the solution containing any of complex forming reagent indicated. Gelatin-immobilized silver(I) hexacyanoferrate(II) as well as any of these soluble complexes, can participate in process of reduction with Sn(II). In this connection, proceeding two parallel processes Ag(I) → Ag(0) will take place at contact of $Ag_4[Fe(CN)_6]$-GIM with solution indicated above, containing Sn(II) and complex forming reagent:

- Gelatin-immobilized silver(I) hexacyanoferrate(II) reduction proceeding in a polymer layer,

- Ag(I) complex with complex forming reagent reduction proceeding on interface of phases GIM/solution.

In water solutions at pH = 1213, Sn(II) is mainly in a form of hydroxo complex $[Sn(OH)_3]^-$. In this connection, general Equation (2) may be offered for the first of these processes:

$$\{Ag_4[Fe(CN)_6]\} + 2[Sn(OH)_3]^- + 6OH^- \rightarrow 4\{Ag\} + 2[Sn(OH)_6]^{2-} + [Fe(CN)_6]^{4-} \quad (2)$$

For the second of these processes, general Equation (3)

$$2[AgL_2]^+ + [Sn(OH)_3]^- + 3OH^- \rightarrow 2Ag + 4L + [Sn(OH)_6]^{2-} \quad (3)$$

In the case of non-charged ligands and general Equation (4)

$$2[AgL_2]^{(2z-1)-} + [Sn(OH)_3]^- + 3OH^- \rightarrow 2Ag + 4L^{z-} + [Sn(OH)_6]^{2-} \quad (4)$$

In the case of negative charged "acid" ligands may be ascribed (Lsymbol of ligand and *z*its charge). The particles of element silver formed as a result of Equations (3) and (4), theoretically should have smaller sizes than the particles of element silver arising in polymer layer of GIM. To be a part of substance immobilized in GIM, these particles should place freely in intermolecular cavities of gelatin layer. Only in this case, they may diffuse in GIM and may be immobilized in gelatin mass.

Gelatin has an extremely high surface area and an extensive system of micropores. The fragment of its structure has been shown in Figure 7; as may be seen, it contains many intermolecular cavities. It may be valued the average size of intermolecular cavity in the gelatin structure [12-15].

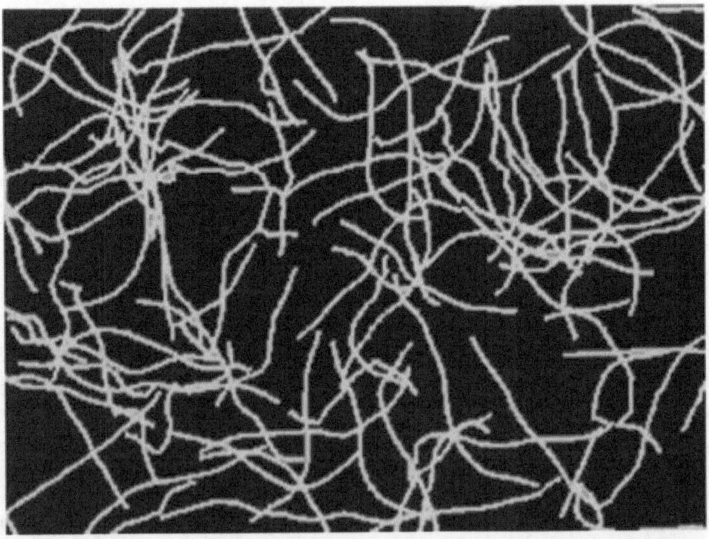

FIGURE 7 The fragment of gelatin structure containing intermolecular cavities.

For example, the volume of polymer layer of GIM (V_{gl}) having area 1 cm^2 and thickness 20 μm is $(1.0 \times 1.0 \times 20 \times 10^{-4})$ $cm^3 = 2.0 \times 10^{-3}$ cm^3, so that the mass of gelatin contained in such a layer, at average value of its density 0.5 gcm^{-3}, is $(0.5 \times 2.0 \times 10^{-3})$ g = 1.0×10^{-3} g. Molecular mass of gelatin (M_{Gel}) is known to be ~$(2.0–3.0)10^5$ c.u. [12, 13], the number of its molecules in given mass will be $(1.0 \cdot\times 10^{-3}/M_{Gel}) \cdot (6.02 \times 10^{23}) = (2.0–3.0)10^{15}$. As it was already mentioned above, gelatin molecule in average has length ~28,5000 pm and diameter ~1,400 pm, and if it is considered as narrow cylinder, total volume of gelatin molecules V_M will be equal to $(1/4)\pi D^2h = (1/4)$ 3.14 $(285{,}000 \times 10^{-10}$ cm$)(1{,}400 \times 10^{-10}$ cm$)^2 = 4.38 \cdot 10^{-19}$ cm^3. In the case of maximal compact arrangement, these molecules occupy total volume equal to 4.38×10^{-19} $cm^3 \times (2.03.0)10^{15} = (8.7613.15)10^{-4}$ cm^3. It may be postulated that the volume of cavities indicated, is equal to total volume of polymer massif minus the volume occupied by gelatin molecules, namely $(2.0 \times 10^{-3} - (8.7613.15)10^{-4})$ cm^3 that will be in the end $(0.691.12)10^{-3}$ cm^3. Then, the average volume of one intermolecular cavity may be found as a quotient from division of their total volume to number of gelatin molecules and, as it may be easily noted, will be $(3.45.6)10^{-19}$ $cm^3 = (3.45.6)10^{11}$ pm^3. The linear size of such an "average" cavity in the case when it has spherical form, will be equal to $d = (6V/\pi)^{1/3} = [6 \times (3.75.6)10^{11}$ $pm^3/3.14]^{1/3} = (89.1102.2)10^2$ pm; when it has cubic form, equal to $a = V^{1/3} = [(3.75.6)10^{11}$ $pm^3]^{1/3} = (71.882.4)10^2$ pm. As one can see from these values, these cavities are nanosized. Therefore, only nanoparticles of substance can entry into these cavities. By entering into such cavities, nanoparticles of element silver are isolated from each other. In consequence of thereof, their aggregation with each other becomes rather difficult.

With the concentration growth of any of the complex forming reagent mentioned above, concentration of coordination compounds formed given complex forming reagent with Ag(I) must increase. Correspondingly, the quantity of nanoparticles of the element silver formed as a result of reduction of these coordination compounds by $[Sn(OH)_3]^-$ complex, should also increase.. In this connection, it may be expected that when concentration of these complex forming reagent in solution increases, the share of nanoparticles contained in the "re-precipitated" elemental silver, should accrue gradually. Thus, at the same concentration nanoparticles of element silver owing to their higher dispersion degree in comparison with microparticles should provide higher degree of absorption of visible light (and, accordingly, higher optical density) in polymeric layer GIM. The experimental data presented in Figures 1–6, are in full conformity with the given prediction.

The particles of element silver formed as a result of Equations (3) and (4) are one or two nuclear. While it is not enough of them [it occurs, when concentration of Ag(I) complexes on interface of phases], these particles owing to their remoteness from each other have no time to be aggregated. It diffuses in polymeric layer of GIM, and immobilized without change of their sizes. With increase of complex forming reagent concentration [and, accordingly, of concentration of Ag(I) complex with given reagent], quantity of nanoparticles indicates on interface of phases the GIM/solution accrues. It leads to increase of a number of such particles, diffused into GIM. However, at some rather high concentration complex forming reagent in solution, the effect of aggregation of nanoparticles of element silver starts to affect. One and two nuclear particles of element silver formed at reduction of corresponding Ag(I) complex, to some extent begin to unite with each other in larger particles. Polynuclear particles of element silver resulting, such an association are not so mobile and consequently, will not be diffused into polymeric layer of GIM. They will be precipitated in it near to interface GIM/solution (or even to escape as solid phase in the solution contacting with GIM). As a result, rates of an increment of number of one and two nuclear particles of elemental silver with further growth of concentration complex forming reagent begin to be slowed down. Thus, inevitably there should come the moment when the number of similar particles will reach some limiting value. Certain "threshold" concentration complex forming reagent in solution, growth of D^{Ag} values must stop. Moreover, at excess of this "threshold" concentration, certain decrease D^{Ag} should begin. The point is that an alignment between number of the aggregated particles and number one and two nuclears with growth of concentration of Ag(I) complex continuously grows and has no restrictions. These polynuclear particles are precipitated in Frontier zone GIM on small depth and form, as a matter of fact, the microparticles of elemental silver formed as a result of reduction of gelatin-immobilized $Ag_4[Fe(CN)_6]$ according to Equation (2). Since, D^{Ag} values with increase of complex forming reagent concentration at first increase, reach a maximum and then decrease.

It may be assumed that "re-precipitated" gelatin-immobilized silver should contain, as a minimum, two phases of the silver particles, one of which is formed by nanoparticles, and another, by microparticles. In order to corroborate the given conclusion, we carried out the analysis of elemental silver isolated from initial Ag-GIM after end of "re-precipitation" process by X-ray powder diffraction method. X-ray powder diffraction patterns (XRD-patterns) of samples obtained are presented in Figures 8–10. As may be seen from them, XRD-pattern of initial elemental silver with gray-black color of gelatin layer (Figure 8) and XRD-patterns of "re-precipitated" elemental silver (Figures 9–10), rather essentially differ from each other. So, in XRD-patterns of "re-precipitated" elemental silver obtained at an availability of any of studied CR in solution contacting with GIM, there are accurate reflexes having d = 333.6, 288.5, 166.7, and 129.1 pm that are absent in XRD-pattern of initial elemental silver. At the same time, reflexes with d = 235.7, 204.1, 144.4, 123.1, and 117.9 pm are observed on them. These reflexes are characteristics for the known phase of elemental silver isolated from initial Ag-GIM. In this connection, there are all reasons to believe that the "re-precipitated" elemental silver obtained on using solution containing any of complex forming reagent indicated above, contains at least two structural modifications of elemental silver.

The next curious circumstance attracts its attention: reflexes with d = 333.6, 288.5, 204.2, 166.7, and 129.1 pm are rather close to d values of reflexes of silver(I) bromide AgBr (number of card PDF 06-0438, parameter of an elementary cell a_0 = 577.45 pm, face-centered lattice, cubic syngonia, *Fm3m* group of symmetry according to the international classification [14, 15]). In this connection, it may be assumed that the structure of the novel phase contained in "re-precipitated" elemental silver, at least in outline, resembles structure AgBr and its crystal lattice is similar to a lattice of silver(I) bromide where positions of atoms Br occupy atoms of silver.

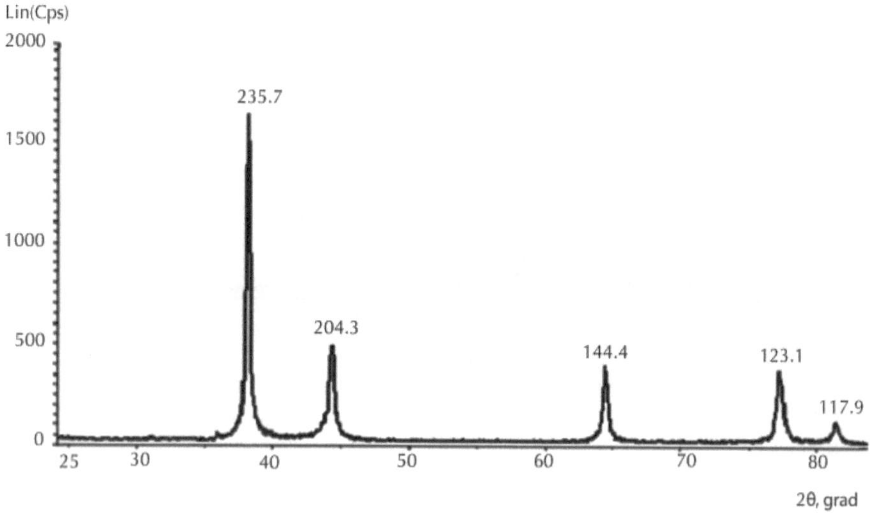

FIGURE 8 The XRD-pattern of elemental silver isolated from initial Ag-GIM.

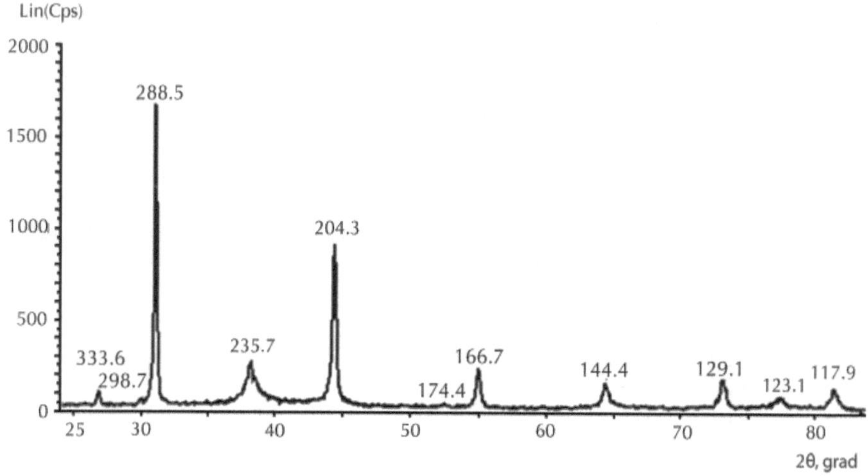

FIGURE 9 The XRD-pattern of substance isolated from Ag-GIM containing "re-precipitated" elemental silver and obtained with using of solution containing $Na_2S_2O_3$ in concentration 20.0 gl^{-1}.

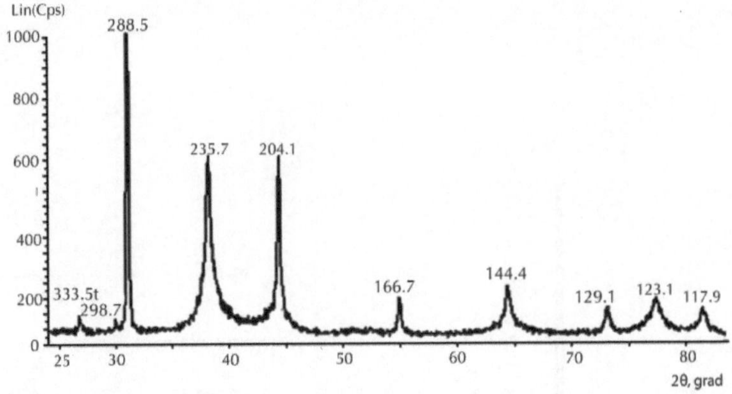

FIGURE 10 The XRD-pattern of substance isolated from Ag-GIM containing "re-precipitated" elemental silver and obtained on using solution containing ethanediamine-1,2 in concentration 20.0 gl–1.

To answer on the question, whether the reflexes indicated can belong to elemental silver with such a space structure in principle, theoretical XRD-patterns of assumed structure of element silver with use of program powder cell described in work [18], have been constructed by us. These XRD-patterns are presented in Figure 11. As may be seen from them, the theoretical *d* values calculated by us (333.6, 288.7, 204.2, 174.1, 166.7, 144.4, 132.5, 129.1, and 117.9 pm) for specified above structure.

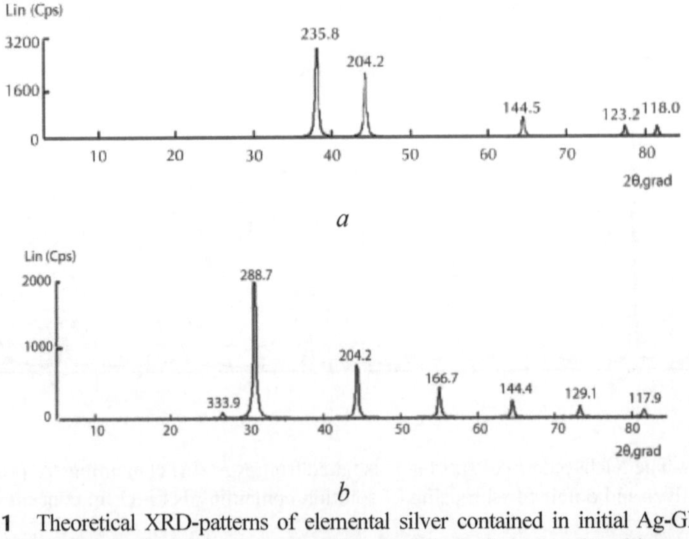

FIGURE 11 Theoretical XRD-patterns of elemental silver contained in initial Ag-GIM (a) and of elemental silver formed in GIM as a result of "re-precipitation" process (b) with an elementary cell parameter a = 288.72 pm and *Pm*3*m* symmetry group, practically coincide with *d* values experimentally observed in XRD-pattern of the "re-precipitated" elemental silver (d = 333.6, 288.5, 166.7, and 129.1 pm).

It should be noted in this connection that *d* values calculated theoretically for the elemental silver isolated from initial Ag-GIM (235.4, 204.3, 144.5, 123.2, and 118.0 pm), correspond to compact-packed crystal structure having an elementary cell parameter $a = 408.62$ pm and $Fm3m$ symmetry group, interplane distances in which are 235.7, 204.1, 144.4, 123.1, and 117.9 pm.

Thus, formation of a novel phase of elemental silver which, probably, was not described in the literature up to now, takes place here indeed.

KEYWORDS

- **Complex forming reagents**
- **Gelatin**
- **Gelatin-immobilized matrix**
- **Intermolecular cavity**
- **Light-filter**
- **Nanoparticles**

ACKNOWLEDGMENT

The Russian Foundation of Basic Researches (RFBR) is acknowledged for the financial support of given work (grant No. 09-03-97001).

REFERENCES

1. Skillman, D. G. and Berry, C. R. Effect of particle shape on the spectral absorption of colloid silver in gelatin. *J. Chem. Phys.*, **48**(7), 32973304 (1968).
2. Linnert, T., Mulvaney, P., Henglein, A., and Weller, H. Long-lived nonmetallic silver clusters in aqueous solution: Preparation and photolysis. *J. Am. Chem. Soc.*, **112**(12), 46574664 (1990).
3. Fedrigo, S., Harbich, W., Butter, J. Collective dipole oscillations in small silver clusters embedded in rare-gas matrices. *Phys. Rev.*, **47**(23), 1070610715 (1993).
4. Satoh, N., Hasegawa, H., Tsujii, K., and Kimura, K. Photoinduced coagulation of Ag nanocolloides. *J. Phys. Chem.*, **98**(7), 21432147 (1994).
5. Sato, T. Ishikawa, T., Ito, T., Yonazawa, Y., Kodono, K., Sakaguchi, T., and Miya, M. *Chem. Phys. Lett.*, **242**(3), 310314 (1995).
6. Al-Obaidi, A. H. R., Rigbi, S. J. McGarvey, J. J., Wamsley, D. G. Smith, K. W., Hellemans, I., and Snauwaert, J. Microstructural and spectroscopies studies of metal liquidlike films of silver and gold. *J. Phys. Chem.*, **98**(24), 1116311168 (1994).
7. Ershov, B. G. and Henglein, A. Reduction of Ag^+ on polyacrilate chains in aqueous solutions. *J. Phys. Chem., B*, **102**(24), 1066310666 (1998).
8. Kapoor, S. Surface modification of silver particles. *Langmuir*, **14** (5), 10211025 (1998).
9. Sergeev, B. M., Kiryukhin, M. V., and Prusov, A. N. Effect of light on the disperse composition of silver hydrosols stabilized by partially decarboxylated polyacrylate. *Mendeleev Commun.*, **11**(2), 6869 (2001).
10. Mikhailov, O. V., Guseva, M. V., and Krikunenko, R. I. An amplification of silver photographic images by using of processes changing disperse of image carrier. *Zh. Nauchn. Prikl. Foto-Kinematogr.*, **48**(4), 5257 (2003).
11. Mikhailov, O. V. Kondakov, A. V., and Krikunenko, R. I. Image intensification in silver halide photographic materials for detection of high-energy radiation by reprecipitation of elemental silver. *High Energy Chem.*, **39**(5), 324329 (2005).

12. Mikhailov, O. V. Reactions of nucleophilic, electrophilic substitution and template synthesis in the metalhexacyanoferrate(II) gelatin-immobilized matrix. *Rev. Inorg. Chem.*, **23**(1), 3174 (2003).
13. Mikhailov, O. V. Gelatin-Immobilized Metalcomplexes: Synthesis and Employment. *J. Coord. Chem.*, **61**(7), 13331384 (2008).
14. Mikhailov, O. V.. Self-Assembly of Molecules of Metal Macrocyclic Compounds in Nanoreactors on the Basis of Biopolymer-Immobilized Matrix Systems. *Nanotechnol. Russ.*, **5**(1–2), 1825 (2010).
15. Mikhailov, O. V. Soft template synthesis of Fe(II,III), Co(II,III), Ni(II) and Cu(II) metalmacrocyclic compounds into gelatin-immobilized matrix implants. *Rev. Inorg. Chem.*, **30**(4), 199273 (2010).
16. Mikhailov, O. V. Enzyme-assisted matrix isolation of novel dithiooxamide complexes of nickel(II). *Indian J. Chem.*, **30A**(2), 252254 (1991).
17. Kraus, W. and Nolze, G. Powder CellA program for the representation and manipulation of crystal structures and calculation of the resulting X-ray powder patterns. *J. Appl. Cryst.*, **29**(3), 301303 (1996).
18. Powder Diffract File. Search Manual Fink Method. Inorganic. USA, Pennsylvania: JCPDS – International Centre for Diffraction Data, 1995 (release 2000).

13 Mechanics of Textile Performance: Part I

A. K. Haghi and G. E. Zaikov

CONTENTS

13.1 Introduction 117
13.2 Model Aspect 119
13.2.1 Problem Plan 119
13.2.2 Mathematical Model 120
13.2.2.1 Weave Angle and Lateral Strain 120
13.2.2.2 Friction Coefficient 121
13.2.2.3 Lateral Forces and Normal Load 122
13.3 Experimental Studies 123
13.4 Discussion and Results 125
13.5 Conclusion 129
Keywords 129
References 129

13.1 INTRODUCTION

Fabrics are highly structured materials. Fine component of fabrics that is the yarns and the fibers make them flexible and capacitate to disclose diverse deformations. Variations in pattern, type of fibers, and the weave density make it possible to optimize fabrics performance and applications, but it may complicate studies on mechanical parameters and their behavior under loading. Two types of solutions have been developed, after facing this problem.

In complex deformations, involving either unsteady strain or out of plane loading, fabrics are treated as two-dimensional continuum sheets or membranes, considered as anisotropic materials. Such materials are usually composed of viscoelastic building blocks, often revealing large displacement instead of small displacement and therefore rarely completely conform to the rules of elasticity.

Another type of solution is based on the fact that the mechanical behavior of the fabrics is excessively affected by the physical properties of their constituents. In studies of simple deformations of fabrics, such as shear or extension, it would be practicable to model a fabric in detail as an assemblage of its component yarns or fibers. These techniques contain simplifying assumptions which result in inclusion of errors.

Several investigations reported in the scientific literatures, have dealt with the mechanical behavior of fabrics. All investigators established considerations on geometrical presumptions, describing fabric structure in a relaxed condition. The primal and the most detailed and elaborate model was introduced by Peirce [10]. It was a geometrical analysis of a woven fabric, characterizing some modes of fabric deformation under the tensile loading. Painter [9] and Love [6] followed his work with some modifications. Many authors such as De Jong and Postle [2], Hearle and Shanahan [3] and their coworkers employed energy methods to model various deformations of fabrics.

Pullout test is also a method, providing useful information about fabric tearing, its ability to absorb energy especially in ballistic applications, finishing efficiency, bending, and shearing hysteresis of the fabric and finally the frictional behavior of the fabric.

In this test a force is applied at one end of a single yarn of the weave to pull it out. This sort of loading is not a novel procedure and scholars have applied such tests to examine mechanical behavior of fabrics. The prior study belongs to Taylor [15] who studied fabric tear strength through a yarn pullout test, and represented a theory based on the yarn interactions. Upon his view the applied load overcomes the frictional force in crossovers and causes the thread to be pulled out of the fabric weave. When the applied load increases, it tears the fabric, because it becomes more than yarn breaking strength. Scelzo et al. [11] and his coworkers developed Taylor's work. They modeled fabric tear as a process composed of three stages, yarn extension, yarn pullout from the weave, and jamming. They emphasized that tear resistance of woven fabrics can be predicted solely based on yarn properties and fabric geometry.

In the field of fabric failure, Realff et al. [14] and Seo et al. [13] and their colleagues focused on the role of yarn-to-yarn friction and the slippage of yarns in crossovers when the woven fabric is applied by a force. They pointed that before fabric failure, the yarn pullout happens. They introduced interesting definitions on the relations between yarn and fabric properties and the interactions within the fabric.

Sebastian et al. [12], Motamedi et al. [7], and their colleagues developed physical models based on spring junction assumptions, to investigate frictional behavior of yarns within the fabrics. Pan and Young Youn [8] emphasized on geometrical and mechanical properties of fabric and the yarns. His model was able to predict the maximum pullout force, and found meaningful relations between the pullout force and some important properties of fabric such as the bending hysteresis, the shear hysteresis, and the tensile compliance.

Badrossamay et al. [1] introduced an oscillation model which was capable of anticipating yarn pullout force displacement profile. And finally Kirkwood et al. [4] and their coworkers studied yarn pullout process as an energy absorbent and characterized the yarn pullout force and energy as a function of pullout distance through a semi empirical model.

In this chapter an attempt is made to introduce a mechanistic model using the force balance analysis, and to calculate some fabric mechanical parameters that is yarn-to-yarn friction coefficient, normal load, lateral forces, weave angle variations, and the yarn pullout force. These parameters are useful in determination of the internal mechanical characteristics of the woven fabrics, and could be employed to simulate yarn pullout behavior of such fabrics.

To evaluate the competency of the theoretical controversies, suggested equations of the force balance model have been used to estimate the amount of the force required to pullout a single yarn from the weave in a yarn pullout test, the efficiency of the force balance model has been examined by calculation of the correlation coefficient between experimental data and that of the model results.

13.2 MODEL ASPECT

13.2.1 Problem Plan

In yarn pullout test a rectangular fabric sample (for simplicity a plain pattern is considered here) is clamped to the two opposite side of a U frame. The pullout force is applied to a single yarn (for instance at the middle) with a specific free end length as shown in Figure 1. The yarn is pulled out in three steps, the detail of which will be given in the discussion part of the chapter .

As is readily inferred, there are two kinds of stresses in such an experiment:

1. Shear stresses happening in the fabric plane parallel to the pulled yarn direction.
2. Tensile stresses occurring (a) in the pulled yarn and (b) in the longitudinal direction of the fabric, parallel to the opposed yarn direction.

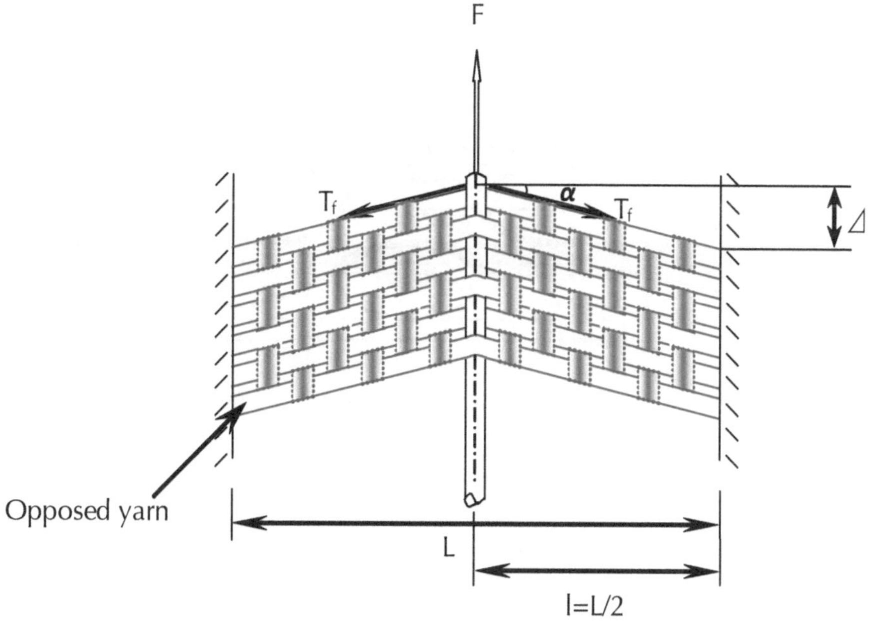

FIGURE 1 Schematic of fabric deformation in yarn pullout test.

After neglecting fabric thickness decrement due to uncrimping, the problem could be considered as an in plane situation and that the extensions are small in opposed directions. As it is seen in the experimental part of the study, some of the mechanical specifications and the geometrical details of the fabric are computed using the properties of the fabric constituents.

Fabrics are usually made of viscoelastic, nonlinear, and time dependent materials, but resolving of this problem in a simple linear elastic form may lead us to a sufficiently acceptable estimation under some circumstances. Besides it can offer a basic framework to give an account to some other complicate theories.

In this chapter the main attention is focused on internal deformation of the fabric, the normal load at each crossover and the lateral forces formed in the fabric.

13.2.2 Mathematical Model

13.2.2.1 Weave Angle and Lateral Strain

First of all the structure of the weave and its variations during the test has been considered. Figure 2 presents the cross section schematic of the plain woven fabric in the opposed direction.

In the test, the weave angle θ decreases to θ' due to the extension. If the height and the volume variations are considered to be negligible (this assumption could be reinforced by selecting a densely weaved fabric), θ' would be calculated as follows:

$$V = V' \Rightarrow t.h.L = t'.h'.L'$$
$$Or \quad t.h.p = t'.h'.p' = h(t-\Delta t)(p+\Delta p) \tag{1}$$
$$t' = t\frac{p}{p'} \Rightarrow \tan\theta' = \frac{t'}{2p'} = \frac{t.p}{2p'^2}$$

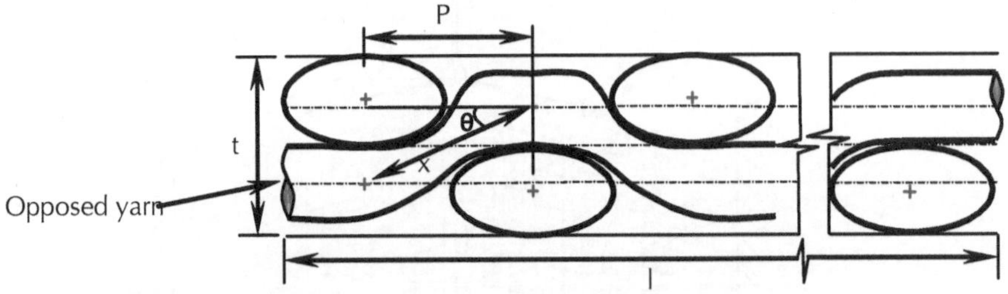

FIGURE 2 Fabric construction before yarn pulling (opposed direction).

$\tan\theta = \frac{t}{p}$, and on the other hand during pullout $p' = \frac{p}{\cos\alpha}$, consequently:

$$\tan\theta' = \frac{t\cos^2\alpha}{2p}$$

$$\text{or } \theta' = \text{Arc}\tan(\tan\theta.\cos^2\alpha) \tag{2}$$

Strain of the opposed yarns $\varepsilon_y = \frac{\Delta x}{x}$ are estimated as follows:

$$x = \frac{p}{\cos\theta}, \quad x' = \frac{p'}{\cos\theta'} = \frac{p}{\cos\alpha\cos\theta'}$$

and

$$\varepsilon_y = \frac{x' - x}{x} = \frac{\cos\theta - \cos\alpha\cos\theta'}{\cos\alpha\cos\theta'} \tag{3}$$

Note that when α is small θ' is close to θ whereupon ε is small too.

13.2.2.2 Friction Coefficient

By determination of the strain and the weave angle variations during the yarn pullout test, an account is given to the force balance. Figure 3 illustrates the schematic of a crossover during yarn pulling and its free body diagram. Effects of shear stresses have been implied in the friction force which is equal to the pullout force F according to Equation (4) (the Amounton's law):

$$F = N.\mu.F_N \tag{4}$$

The pullout force is normalized by the number of crossovers N in the direction of the pulled yarn, to represent the pullout force per each crossover. That is:

$$f = \frac{F}{N} - \mu.F_N \tag{5}$$

It is clear that during the process four main forces are formed: the normal load F_N, the normalized pullout force f, and two lateral forces T_f in the fabric plane. It should be noted that T_f is the projection of the lateral forces T_y, propagating in the opposed yarn direction, shown by the stipple line in Figures 3 and 4.

Referring to Figures 1 and 4, we have:

$$F_N = 2T_y \sin\theta' \tag{6}$$

$$f = 2T_f \sin\alpha \tag{7}$$

$$T_f = T_y \cos\theta' \tag{8}$$

With regard to Equations (5)-(8) a simple function for friction coefficient is obtained:

$$\mu = \frac{f}{F_N} = \frac{\sin\alpha}{\tan\theta'} \tag{9}$$

With regards to the fact that α and θ (or θ') are ever salient angles, so sinα and tanθ' and consequently yarn-to-yarn friction coefficient μ are ever positive too.

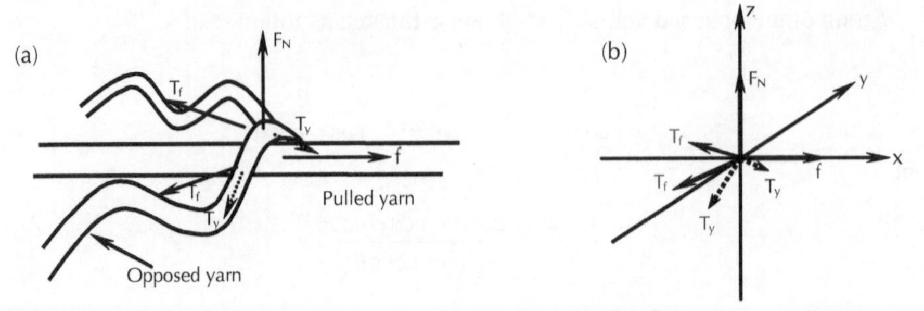

FIGURE 3 (a) Force propagation during pullout process (b) Free body diagram of forces.

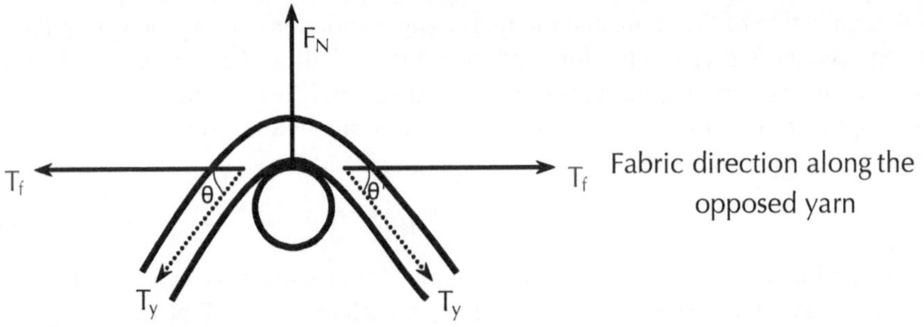

FIGURE 4 A crossover and its force component.

13.2.2.3 Lateral Forces and Normal Load

Finally, if E_y is considered as the opposed yarns modulus, then the lateral forces (T_y or T_f) and the normal load are obtained from Equations (6) and (8) as follows:

$$T_y = E_y.\varepsilon_y.\rho$$

Hence:

$$T_y = E_y.\rho\frac{\cos\theta - \cos\theta'\cos\alpha}{\cos\theta'\cos\alpha} \tag{10}$$

And

$$F_N = 2E_y.\rho\frac{\cos\theta - \cos\theta'\cos\alpha}{\cos\theta'\cos\alpha}.\sin\theta' \tag{11}$$

So that:

$$T_f = E_y.\rho(\frac{\cos\theta}{\cos\alpha} - \cos\theta') \tag{12}$$

In order to evaluate the efficiency of the force balance model, the yarn pullout force was simulated by Amonton's Law (Equation (6)). Therefore, the yarn-to-yarn friction coefficient and the normal load in crossovers were required to be known, these values were calculated through the force balance model. Therefore, the fabric modulus and its density along the warp direction, the number of crossovers and the weave angle of the fabric should be given. In this study these parameters were determined experimentally. The quantities of the fabric deformation angle α, and the variations of the weave angle θ' during a yarn pullout test at any instance should be known too. These factors were provided using the fabric deformation information, which were acquired by the image processing method.

13.3 EXPERIMENTAL STUDIES

This study was accomplished on five plain woven fabrics, three of which were 100% cotton fabrics, and others were 100% polyester fabrics, characteristics of which are given in Table 1. In order to measure the weave angle of the selected fabrics, resin hardener FK20 (from Esfahan Supply Co.) was used. Small samples of the fabrics were submerged in this resin harder for 48 hr. During this time resin stiffened. Cross section of the fabrics along the warp and the weft directions were prepared by a SLEE MAINZ microtome device "model cut 4055", and scanned by a motic microscope "model B3". Figure 5 shows the cross sections of all fabrics in the warp direction. Average weave angle of each fabric along the warp direction was calculated by the Photoshop software.

Table 1. Characteristics of samples.

	COT1	COT2	COT3	PET1	PET2
Fabric material	**Scoured cotton, weft and warp: ring yarns**	**Bleached cotton, weft and warp: ring yarns**	**Bleached cotton, weft and warp: open-end yarns**	**Polyester, weft and warp: twisted multi-filament**	**Polyester, weft textured multi-filament and warp: inter-mingled multi-filament**
Dimensions of rectangular sample (mm^3)	0.3×10×50	0.29×10×50	0.33×10×50	0.15×10×50	0.19×8.2×50
Ends/cm	25	25	28	45	34
Picks/cm	24	22	27	25	30
Linear density of warp yarns (tex)	25	22.5	30	8.9	17
Linear density of weft yarns (tex)	24	22	30	17	8.9
Number of crossovers (N)	24	24	28	44	28

The pullout test repeated five times for each fabric. In the execution of the yarn pullout test, a fabric sample was clamped to each side of a U form frame. The weft yarns of the fabric were free. The frame was connected to the movable lower head of a Zwick tensiometer “model 1440–60”. The middle weft yarn was tied to the stationary upper head while a free length at the end was set to be 1 cm. The length of the yarns between the upper head and the fabric was set to be 6 cm for all samples. The lower head moved down at a velocity of 10 mm/min. The pullout force was measured by a load cell with a maximum capacity of 500 N while the pulled yarn fully pulled out of the weave.

Pullout process was video recorded at the speed of 30 frames/s using a Sony video camera “model DCR-PC120E”. Video pictures were then digitalized with the aid of Fast-Dizzle software and the frames of each film were separated using Adobe Premier Software. These pictures contained fabric deformation data and were used to find the fabric deformation angle α. Some of the frames were selected proportional to tensiometer data recorder speed. *Via* a Matlab program, RGB images converted to gray scale ones. By specifying of the boundary of light and dark areas, the deformation angle was measured. Figure 6 reveals the result. Each frame consists of indicator marks make it possible to measure the fabric displacements correctly.

Fabrics tensile specifications have been measured using the above mentioned tensiometer

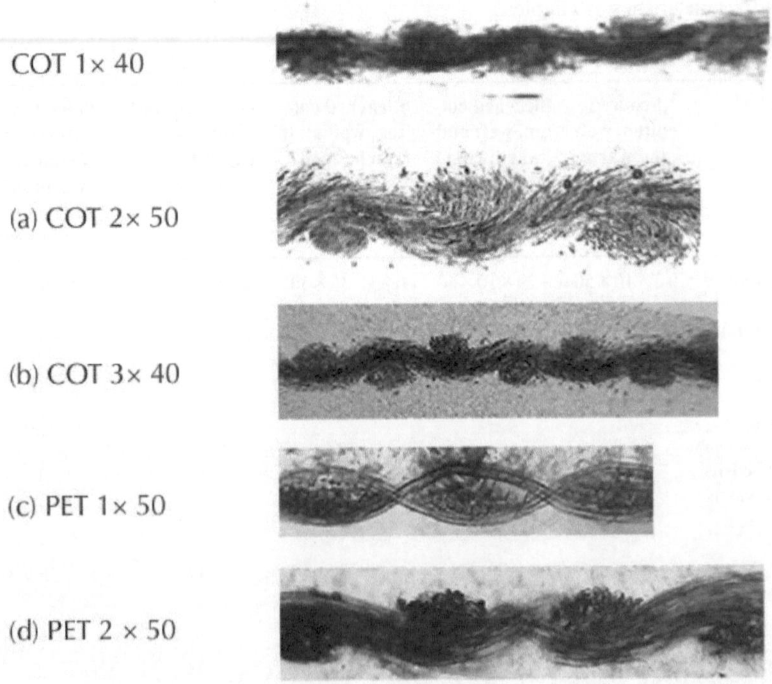

FIGURE 5 Cross sections of fabrics along warp direction.

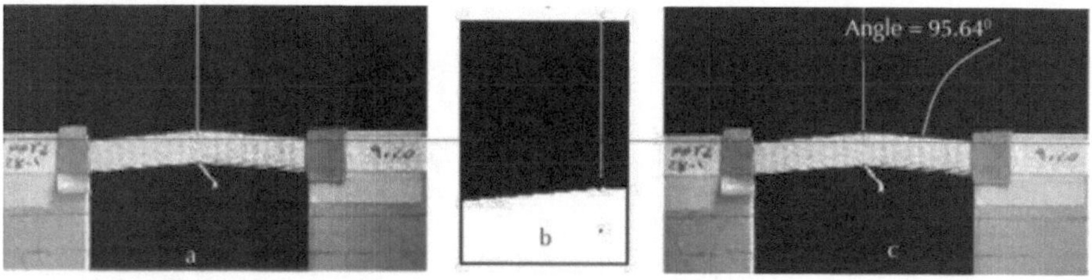

FIGURE 6 Image processing steps (a) main image (b) gray scale image (c) angle derivation.

13.4 DISCUSSION AND RESULTS

By applying a pulling load F on a yarn of a fabric, the crimps of the pulled yarn is opened and a deformation is created (Figure 1). After decrimping, which consume a little force compared with the fabric deformation, the force F is increased to a maximum value F_S, called the static friction force or the static maximum. At this step the yarn has no slippage, neither being pulled out through the opposed yarns. The F_S is equal to that of the required force to overcome the limiting static friction force. The strain and pullout force are functions of yarns interactions or the adhesive properties. The frictional behavior of yarns within the fabric is a complicated function of yarns surface roughness as well as the fabric structural characteristics. Hence Equation (9) which is a combined function of these two specifications would be a tolerable approximation of the yarns resistance to movement within the fabric. From the data of the maximum pullout point the static friction coefficient (μ_S) is calculated.

This is a new definition of the friction coefficient which in spite of other methods such as the point contact or the line contact, can estimate the real yarn-to-yarn resistant to movement within the fabrics eminently, as it depends on the yarns and the fabric specifications.

After the first junction rupture, the pulled yarn slips in the weave. Therefore, the pullout force and consequently the fabric displacements decrease to some extent. But it increases later due to new adhesion formation. This happens in the second step, called the stick-slip motion, and recurs until the free end of the pulled yarn leaves the first crossover of the weave. Thereafter the stick-slip motion adopts a reductive trend, creating the third step. Figure 7 presents the yarn pullout force displacement profile typically. It shows an oscillation behavior in the second and the third steps, due to the mentioned slips and adhesions.

In the two last steps, fabric displacements and its deformation angle named Δ_D and α_D respectively declaring their dynamic nature. Dynamic lateral forces T_{fD}, dynamic friction coefficient μ_D and dynamic normal load F_{ND} may be calculated using parameters such as α_D and θ'_D related to the fabric deformation, through Equations (9)-(11).

These parameters are usually less than the static ones, as crossover displacements and therefore α_D and θ'_D are smaller than α_S and θ'_S. This specifies that adhesion between the yarns decreases during the stick-slip motion. In some situations exceptional cases may occur. For instance when the end of the pulled yarn is tangled in the weave, the pullout force may exceed the static one.

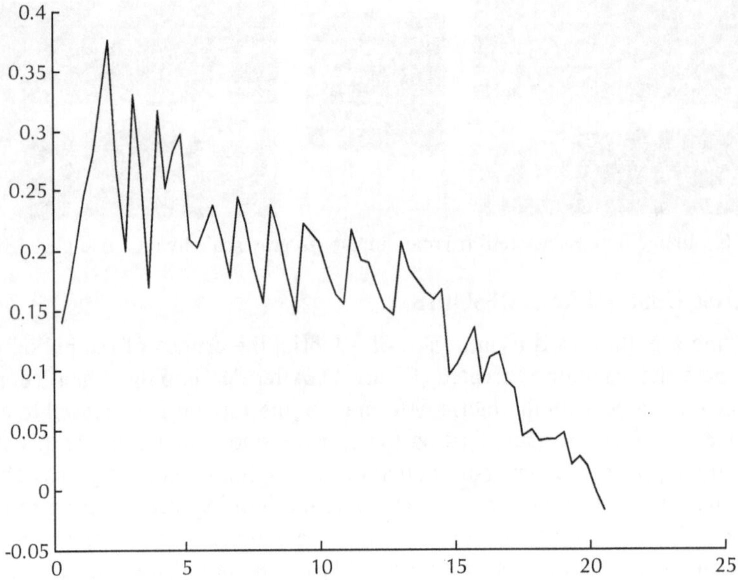

FIGURE 7 Schematic of yarn pullout force displacement profile.

The movement of the pulled yarn with in the fabric is accomplished by its bend and unbend phenomenon which consumes energy somehow. In the present model this energy is considered negligible as well as the decrimping ones. Bending and unbending of the pulled yarn could be considered as another reason for the oscillation behavior of the force displacement profile.

Another important factor that should be noted is the yarn modulus E_y. Fabrics are considered as composites of yarns. Thus due to the internal junctions, the yarns could not reveal their intrinsic properties. Therefore $E_{y,}$ employed in the presented equations, is not the same as the one measured by a tensiometer device. Measured values for E_y usually conduce to an overestimation of the parameters, contributing to calculation.

To resolve this problem, we accentuate the fabric modulus E_f, and introduce a relationship, transforming E_f to E_{yf} (modified yarn modulus) in a specific fabric direction (weft or warp):

$$E_{yf} = \frac{E_f \times sample\ \ thickness \times sample\,width}{\rho \times number\ \ of\ \ yarns\ \ in\ \ sample\,width} \tag{13}$$

As it can be concluded Equation (13) leads to a better approximation of E_y under such situation. The modified lateral forces T_{yf} can be calculated from Equation (10) by employing the modified yarn modulus. Table 2 represents the mean values of the weave angle in a relaxed situation, the fabric modulus along warp direction, and the modified warp yarns modulus.

TABLE 2 Structural and mechanical properties of fabrics.

	COT1	*COT2*	*COT3*	*PET1*	PET2
θ(degree)	19.5	20.5	17.4	14	15
$(E_f)_{warp}$ (N/mm²)	83.67	93.12	202.17	456.6	184.75
$(E_{yf})_{warp}$ (N/tex)	0.416	0.48	0.79	1.67	0.6

The mechanical properties of each fabric that is θ', ε_y, μ, F_N, T_{yf}, F could be calculated by reference to Equations (2–4), (9–11), and (13) of the force balance model respectively, after specifying the experimentally required factors. These parameters at their maximum static situation are reported in Table 3.

TABLE 3 The estimated mechanical parameters of fabrics in maximum static situation.

	COT1	*COT2*	*COT3*	*PET1*	PET2
θ'_s (degree)	18.72	19.76	16.99	13.89	14.77
α_s (degree)	11.93	11.36	10.11	5.1	7.31
ε_y	0.01734	0.015	0.013	0.0035	0.0071
μ_s	0.61	0.55	0.57	0.36	0.483
T_{ys} (N)	0.2014	0.1648	0.3085	0.0523	0.073
F_{NS} (N)	0.1157	0.1114	0.1803	0.0251	0.0371
F_s (N)	1.6995	1.475	2.92	0.4	0.472

Concerning to the relative equality of samples dimensions, the yarn-to-yarn friction coefficients of fabrics are compared. As it was expected the yarn-to-yarn friction coefficients of the cotton yarns were greater than that of the polyester filaments yarns. Besides smoother filaments of PET1 leads to lower friction coefficients towards PET2, textured and intermingled filament. The estimated Yarn-to-yarn friction coefficient of cotton fabrics are comparable to what measured by Briscoe and his coworker (1990).

The presented relationships explain the mechanical characteristics of yarns within the fabric. These characteristics are affected by the weave density, the yarns type, the weave pattern, and the finishing type of the fabric. Therefore, the suggested model is suitable for predicting of the mechanical properties of yarns and their behavior in the fabric construction. Hence, the parameters obtained by this method seem to be more realistic than the other methods, resulting more exact parameters.

To simulate the force displacement diagram using the force balance model, the force F was calculated in a regular time steps. The screening speed is definite. By setting a linear relation between this rate and the tensiometer speed, the force displacement profile was predicted for each sample by this method. In the oscillation method, we just need amounts of these parameters in the maximum static point and also one dynamic point. The dynamic point considered at the condition of the average deformation in the harmonic stick-slip motion.

Figure 8 illustrates a comparison between experimental profiles of the pullout test and the theoretical model. The diagrams are related to one selected sample of each fabric.

As it is seen, the results of force balance model are close to the experimental data. Validity of the suggested model has been assessed using the Pearson product correlation coefficient:

$$R = \frac{\sum (F_{es} - \overline{F}_{es})(F_{ex} - \overline{F}_{ex})}{(n-1)s_1 s_2} \tag{14}$$

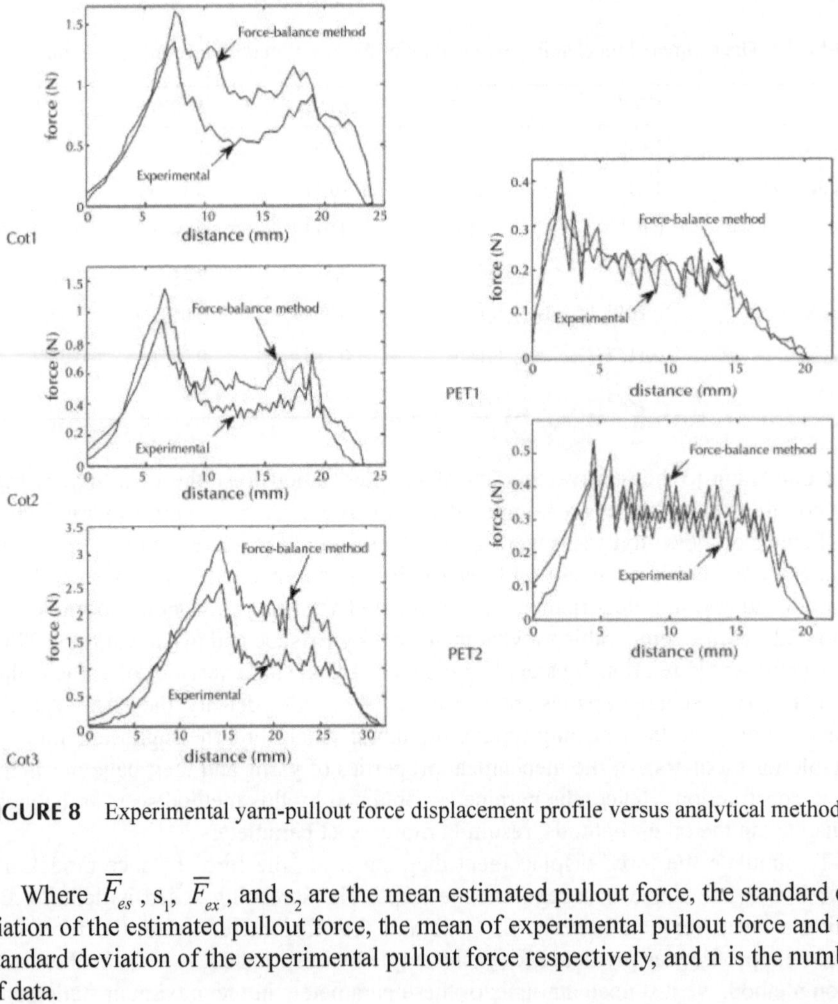

FIGURE 8 Experimental yarn-pullout force displacement profile versus analytical method.

Where $\overline{F}_{es}$, s_1, $\overline{F}_{ex}$, and s_2 are the mean estimated pullout force, the standard deviation of the estimated pullout force, the mean of experimental pullout force and the standard deviation of the experimental pullout force respectively, and n is the number of data.

The correlation coefficient was determined for each sample to present the amount of statistical conformity between the experimental results and the theoretical predictions. The mean values of the fabrics correlation coefficients are appeared in Table 4.

Table 4 Correlation coefficient of the model in compare with experimental results.

	COT1	COT2	COT3	PET1	PET2
Correlation coefficient of the force-balance method	0.86	0.92	0.93	0.86	0.79

13.5 CONCLUSION

In this study the yarn pullout test is applied to investigate internal mechanical properties of the plain woven fabrics. In the first step an analytical model was developed, inputs of which employs simple mechanical properties such as the fabric modulus, the weave angle, and the fabric deformation angles during the pullout test. This model predicts important mechanical parameters such as the weave angle variations, the yarn-to-yarn friction coefficient, the normal load in crossovers, the lateral forces, and the opposed yarn strain within the fabric. This approach may be extended to other types of the woven fabrics.

In the next step, the experimental analysis was employed to evaluate the validity of the predicted parameters of the force balance model. Here the force displacement profile of the yarn pullout test was simulated using the equations of the force balance model, and the characteristics of the plain woven fabric and its yarns.

Acceptable agreement between the experimental and the theoretical results reveals that the reported can predict the yarn pullout behavior to an admissible extent. Besides, it proves that the equations of the force balance method are suitable enough to be used for determining some important structural and mechanical properties of the woven fabrics. Simplicity of the model principals and the few numbers of the experimentally required factors are the main achievements of this model.

KEYWORDS

- **Force balance model**
- **Pullout test**
- **Static friction force**
- **Yarn-to-yarn friction coefficients**
- **Zwick tensiometer**

REFERENCES

1. Badrossamay, M. R., Hosseini Ravandi, S. A., and Morshed, M. Fundamental Parameters Affecting Yarn Pullout Behavior. *J. Tex. Inst.*, **92**, 280-287 (2001).

2. De Jong, S. and Postle, R. An Energy Analysis of Woven-Fabric Mechanics by Mean of Optimal-Control Theory Part I: Tensile Properties. *J. Text. Inst.*, **68**, 350-361 (1977).
3. Hearle, J. W. S. and Shanahan, W. J. An Energy Method for Calculations in fabric Mechanics Part I: Principle of The Method. *J. Text. Inst.*, **69**, 81-91 (1978).
4. Kirkwood, K. M., et al. Yarn Pullout As a Mechanism for Dissipating Ballistic Impact energy in Kevlar® KM-2 Fabric, Part I: Quasi Static Characterization of Yarn Pullout. *Tex. Res. J.*, **74**, 920-928 (2004).
5. Kirkwood, J. E., et al. Yarn Pullout as a Mechanism for Dissipating Ballistic Impact energy in Kevlar® KM-2 Fabric, Part II: Predicting Ballistic Performance. *Tex. Res. J.*, **74**, 939-948 (2004)
6. Love, L. Graphical Relationships in Cloth Geometry for Plain, twill and Sateen Weaves. *Tex. Res. J*, **24**, 1073-1083 (1954)
7. Motamedi, F., Baily, A. I., Briscoe, B. J., and Tabor, D. Theory and Practice of Localized Fabric Deformation. *Textile Res. J.*, **59**, 160-172 (1989).
8. Pan, N. and Young Youn, M. Behavior of Yarn Pullout from Woven Fabrics: Theoretical and Experimental. *Textile Res. J.*, **63**, 629-637 (1993)
9. Painter, E. V. Mechanics of Elastic Performance of Textile Materials, Part VIII: Graphical Analysis of Fabric Geometry. *Tex. Res. J.*, **22**, 153-169 (1952).
10. Peirce, F. T.. The Geometry of Cloth Structure. *J. Text. Inst.*, **28**, T45-69 (1937)
11. Scelzo, W. A., Backer, S., and Boyce, M. C. Mechanistic Role of Yarn and Fabric Structure in Determining Tear Resistance of Woven Cloth. Part II: Modeling Tongue Tear. *Tex. Res. J.*, **64**, 321-329 (1994),
12. Sebastian, S. A. R. D., Baily, A. I., Briscoe, B. J., and Tabor, D. Extension, Displacements and Forces Associated with Pulling a Single Yarn from a Fabric. *J. Phys. D. Appl. Phys.*, **20**, 130-139 (1987).
13. Seo, M. H., et al. Mechanical Properties of Fabric Woven from Yarns Produced by Different Spinning Technologies: Yarn Failure in Woven Fabric. *Textile Res. J.*, **63**, 123-134 (1993).
14. Realff, M. L., et al. Mechanical Properties of Fabric Woven from Yarns Produced by Different Spinning Technologies: Yarn Failure as a Function of Gauge Length. *Textile Res. J.*, **61**, 517-530 (1991).
15. Taylor, H. M. Tensile and Tearing Strength of Cotton Cloths, *J. Text. Inst.*, **50**, T161-188 (1959).

14 Mechanics of Textile Performance: Part II

A .K. Haghi and G. E. Zaikov

CONTENTS

14.1 Introduction 131
14.2 Definition of the Oscillation Model Parameters 132
14.3 Experimental Procedure 135
14.4 Discussion and Results 137
14.5 Conclusion 138
Keywords 139
References 139

14.1 INTRODUCTION

Frictional interactions between intersecting warp and weft yarns have a great influence on the behavior of different types of fabrics even for apparel application with deformations like yarn running, tearing and drape or industrial fabrics undergoing fragment impact, projectile penetration or other large deformations. Frictional forces, which are involved in load transfer and energy absorption, determine directly the magnitude of the interactions within the fabric. Yarn-to-yarn friction as an important mechanical characteristic can be measured by different methods, like twist friction (capstan) method, Howell hanging fiber method and inclined plane method [2]. Although all these methods are based on strong physical hypotheses, they can simulate the simple yarn-to-yarn friction and are incapable in remodeling the exact internal friction within the fabric, which is influenced by the interactions of neighbor yarns, weaving pattern and density.

As a physical study, yarn pullout test would be an appropriate approach to realize an estimation of the internal friction forces between the yarns within the fabric. Under the circumstances of this test, a pulled yarn slides along the intersecting perpendicular yarns during fabric deformation.

By three last decades, several models have been presented to simulate the behavior of woven fabrics in a yarn pullout test. The advantage of this test is the simulation of internal frictional interactions. Concerning to the principals of the method [38] developed several models to predict the mechanical behavior of textile during pullout procedure.

In this field y one of the latest studies have represented [1] . They reported an analytical model anticipating the force displacement profile of the yarn pullout test for plain woven fabrics. The model called "Oscillation Model" after is based on the vibration theory of the strings. It determines the pullout behavior apparently well but it was not investigated experimentally. In another recently reported study, advanced a new analytical model using force balance method, employing the deformational data of the plain woven fabric and its internal yarns in a yarn pullout test to predict the force displacement profile of the test [10]. This model is capable of calculating the internal yarn-to-yarn friction coefficient, normal load in interlacing points, and some other mechanical parameters as well.

In this study, force balance model has been used to calculate the internal yarn-to-yarn friction coefficient and the normal load of yarns in crossing points of plain woven fabrics, which are actually the required factors of oscillation model. Spring stiffness constant of the pulled yarn is another parameter, estimated using springs in series law. Finally, comparative experimental studies have been represented to investigate the validity of oscillation model.

14.2 DEFINITION OF THE OSCILLATION MODEL PARAMETERS

Yarn pullout test, can be implemented in different designs. The sample may be fixed from two or three sides but normally it is fixed from two opposite sides and is free from the other two sides. In pullout design of (Figure 1(a)), which has been used in this study, the middle yarn with a determined free end is being pulled. Force displacement profile of such a pullout test will be schematically like (Figure 1(b)), consisting of three different zones.

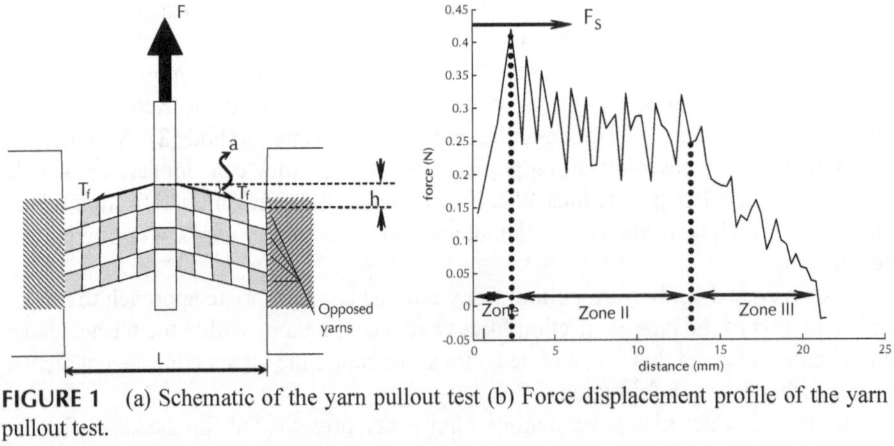

FIGURE 1 (a) Schematic of the yarn pullout test (b) Force displacement profile of the yarn pullout test.

In the first zone, the applied load overcomes the static friction resistance. It was assumed that prior to the first junction rupture (maximum static point F_S), the pullout force varies elastically and proportional to the yarn displacement. Therefore, in the oscillation model the force of this part was suggested according to Coulombs law:

$$F_S = N\ \mu_S\ F_{NS} \qquad \text{Zone I} \tag{1}$$

In the second and third zones, the pullout force overcomes the dynamic friction resistance F_D, in addition to the vibration motion of the opposed yarns. Therefore, the predictive equations of these sections (Equations (2) and (3)), composed of two forces: the frictional (F_D), and the oscillatory (F_{Os}) forces [1].

During these periods, the pulled yarn slips over the crossing points in its trace. This slippage causes the decrement of pullout force. Meanwhile the creation of new adhesive interactions increases the frictional resistance as well as pullout force. This part of yarn displacement is called stick-slip motion. In the third zone, the pulled yarn releases the weave gradually. Therefore, the total pullout force decreases gently according to the crossovers decrement.

$$F_{Cr}^{S} = N\ \mu_D\ F_{ND} + N\ K\ y(\frac{L}{2},t) \qquad \text{Zone II} \tag{2}$$

$$F_{Cr}^{S} = (N-i)\ \mu_D\ F_{ND} + (N-i)\ K\ y(\frac{L}{2},t) \qquad \text{Zone III} \tag{3}$$

Here i is the number of released crossovers and $y(L/2, t)$ is the displacement function of the vibration.

The vibration part of the oscillation model has been developed on the principals of the vibration of flexible strings in a continuous system. Here the material is assumed homogenous and isotropic.

If the string of (Figure 2(a)) is plucked at the midpoint and released with zero initial velocity, according to (Figure 2(b)) the equation of the vibration is:

$$\rho dx \frac{\partial^2 u}{\partial t^2} = T\left(\varphi + \frac{\partial \varphi}{\partial x}dx\right) - T\varphi \tag{4}$$

Solving this partial differential equation with the initial condition of Equation (5), the vibratory displacement $y(x, t)$ would be as Equation (6) [9].

$$y(x,0) = \begin{cases} \dfrac{2hx}{L} & 0 \le x \le \dfrac{L}{2} \\ 2h(1-\dfrac{x}{L}) & \dfrac{L}{2} \le x \le L \end{cases} \tag{5}$$

$$y(x,t) = \frac{8h}{\pi^2}\sum_{n=0}^{\infty}\frac{(-1)^n}{(2n+1)^2}\left[\sin\frac{(2n+1)\pi x}{l} + \cos\frac{(2n+1)\pi C t}{l}\right] \tag{6}$$

Where $C = \sqrt{\dfrac{T}{\rho}}$

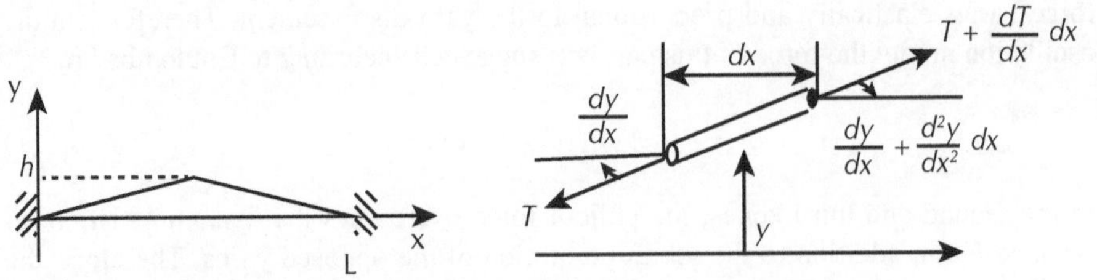

FIGURE 2 Vibration of a flexible string (a) Initial vibration (b) Element dx.

To evaluate the oscillation model, the components of its equations ought to be measured experimentally. Some parameters such as, ρ and N can be easily estimated. However, for some internal parameters such as, yarn-to-yarn friction coefficient and normal load in the crossing points, there are no empirical approaches. These parameters are related to the internal mechanical properties of the plain woven fabrics, have not been measured directly yet. As an alternative, the force balance model, which has been introduced by Valizadeh et al, is employed to estimate these parameters [10].

Force balance model is capable of predicting the internal mechanical parameters of the plain woven fabrics in a similar yarn pullout test based on the force distribution concepts. These parameters are yarn-to-yarn friction coefficient, normal load at crossovers, lateral forces, lateral strain, weave angle variations, and pullout force. Crimp angle of yarns within the fabric (θ in (Figure 3)), its elastic modulus and linear density and fabric deformation data (α in (Figure 1(a))) are the required empirical factors of force balance model. The main equations of this model are presented in (Table 1) [10]:

TABLE 1 Discriminating equations of force balance model.

Variation of crimp angle for the opposed yarn due to fabric deformation by angle α during pullout test (Figure 1a) θ is opposed yarn crimp angle in relaxed condition	$\theta' = \arctan(\tan\theta \ \cos^2\alpha)$	(7)
Opposed yarns strain during pullout test	$\varepsilon_y = \dfrac{\cos\theta - \cos\alpha\cos\theta'}{\cos\alpha\cos\theta'}$	(8)
Yarn-to-yarn friction coefficient	$\mu = \sin\alpha \ \cot g\theta'$	(9)
Lateral force in opposed yarn direction	$T_{yf} = E_{yf}.\rho\dfrac{\cos\theta - \cos\theta'\cos\alpha}{\cos\theta'\cos\alpha}$	(10)
Normal load in each crossover	$F_N = 2T_{yf}\sin\theta'$	(11)
Pullout force for a sample with N crossovers	$F = \mu \ N \ F_N$	(12)

To specify material properties, anisotropy, and non linear behavior of textile structures have to be considered. These assumptions make the assessments quite complex. However, previous studies showed that the simplifying linear elastic material could lead to reliable results in the modeling of yarn pullout [1, 5, 10].

14.3 EXPERIMENTAL PROCEDURE

Five plain woven fabrics, two polyester fabrics, and three cotton fabrics were examined in a yarn pullout test, as schematically shown in (Figure 1(a)). Characteristics of these fabrics are presented in (Table 2). Sample PET1 is consisted of twisted multi filaments of polyester as weft and warp yarns. Sample PET2 has intermingled polyester multi filaments as warp and textured polyester multi filaments as weft yarns. The COT1 is a scoured cotton fabric with ringspun yarns as wefts and warps. The COT2 is a bleached cotton fabric with ringspun weft and warp yarns, and finally COT3 is a bleached cotton fabric with open-end spun wefts and warps.

The sample is clamped in a relaxed condition along the warp direction. Therefore, the warp yarns are the opposed yarns. The middle weft yarn with a 10 mm free end was pulled in the pullout test. The pulling velocity was 10 mm/min and it kept fixed until the pulled yarn left the weave completely. The pullout force was measured by a Zwick tensiometer "model 1440-60" (made in Germany). The test was repeated five times for each fabric.

TABLE 2 Fabrics characteristics.

	COT1	COT2	COT3	PET1	PET2
Dimensions of rectangular sample (mm^3)	**0.3×10í 50**	**0.29×10í50**	**0.33×10í50**	**0.15×10í50**	**0.19í8.2í50**
Ends/cm	25	25	28	45	34
Picks/cm	24	22	27	25	30
Linear density of warp yarns (tex)	25	22.5	30	8.9	17
Linear density of weft yarns (tex)	24	22	30	17	8.9
Number of crossovers (N)	24	24	28	44	28

Fabric deformation angle α and warp crimp angle θ within the fabric (in relaxed condition) were measured by image processing method. Cross sectional microscopic image of fabric in warp direction was used to measure θ (example of which is represented in (Figure 3)). To estimate α, Yarn pullout test was video recorded using a digital camera (Sony DCR-PC120E). The deformation angles were calculated using an image processing program in Matlab software.

The maximum deformation angle at static point is referred as α_S. Static parameters such as, μ_S and F_{NS}, are related to α_S. Similarly, the dynamic parameters are related to

α_D. Whereas α_D varies in dynamic zones, the mean value of these angles (α_Ds), in the second zone has been considered as the dynamic deformation angle.

E_{yf} is the elastic modulus of the opposed yarns but due to the internal junctions, the yarns could not reveal their intrinsic properties. Therefore, Force balance model suggests Equation (13), which calculates yarn elastic modulus considering fabric elastic modulus E_f. E_f in warp yarn direction was measured by the same Zwick tensiometer.

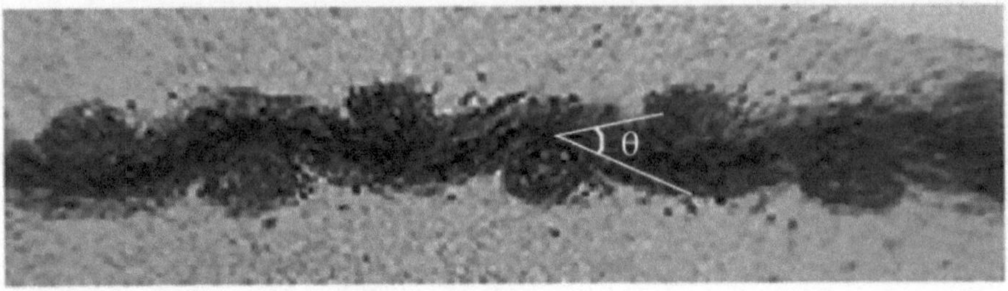

FIGURE 3 Cross section of a plain woven fabric.

$$E_{yf} = \frac{fabric \;\; modulus \;\; in \;\; opposed \;\; direction \times sample \;\; thickness \times sample \; width}{\rho \times number \;\; of \;\; yarns \;\; in \;\; sample \; width} \tag{13}$$

By measuring θ, E_f in warp direction, yarn linear density, and fabric deformation angle in pullout test, μ_S, μ_D, F_{NS}, and F_{ND} of oscillation model can be computed using Equations (711). The last required parameter of the oscillation model is the spring stiffness constant of the pulled yarn (K). This factor was considered as the slop of force extension curve of the pulled yarn in a simple tensile test [1]. However, because of the junctions between warp and weft yarns the pulled yarn, which is settled in the weave construction, would not exhibit such spring stiffness constant. Hence, the pulled yarn has been assumed as a series of springs jointed together along the length of this yarn. Number of these connections is equal to the number of the crossovers. Therefore, the equivalent spring constant of such a system (K_{eq}) has been calculated by the spring in series law:

$$\frac{1}{K_{eq}} = \sum_{i=1}^{N} \frac{n_i}{K_i} \tag{14}$$

In this case the equivalent spring constant will be:

$$K_{eq} = \frac{K}{N} \tag{15}$$

It should be noted that when the number of crossovers decreases during the third step of the pullout test, K_{eq} should be estimated simultaneously according to the number of the remained crossovers ($K_{eq} = \frac{K}{N-i}$).

14.4 DISCUSSION AND RESULTS

Theory of the oscillation model considers two terms in the principals:

1. Contact and friction in crossing points of the weave
2. The stick-slip oscillatory behavior of the yarn pullout test upon the vibration hypothesis.

As the following report will show, the oscillation model responses seem to be consistent with the behavior of fabric in such a loading condition, but it obviously needs some additional aspects.

Constrained yarns in the fabric or other textile structures reveal mechanical responses, which are different in compare with the free yarns. Therefore, their intrinsic behavior would not be displayed perfectly. This is actually the reason of representation of Equations (13) and (15). Table 3 reveals fabric modulus along the warp direction E_f, the modified warp yarns modulus E_{yf}, the spring stiffness constant, and the equivalent spring stiffness constant for the weft yarn of each fabric.

Oscillation model predicts the force displacement profiles of the yarn pullout tests using the parameters suggested by the force balance model, the modified warp yarns modulus and the equivalent spring constant of the weft yarns, as are shown in (Figure 4(a)4(e)). These diagrams reveal the comparison between the experimental results and the model predictions for the cotton and the polyester fabrics. Presented profiles show that the oscillation model operates more efficient on polyester fabrics comparing to the cotton fabrics.

TABLE 3 Structural and mechanical properties of fabrics.

	COT1	COT2	COT3	PET1	PET2
$(E_f)_{warp}$ (N/mm^2)	83.67	93.12	202.17	456.6	184.75
$(E_{yf})_{warp}$ (N/tex)	0.416	0.48	0.79	1.67	0.6
K (N/m)	82.5	57.5	190.12	90	132.6
K_{eq} *(N/m)*	3.3	2.3	6.79	2	3.9

Regarding to the fact that the cotton fabrics consist of uneven staple yarns, this is quite reasonable that their behavior under pullout test could not perfectly comply with discriminating equations of the oscillation model. On the other hand, the polyester fabrics consist of more uniform multifilament yarns; therefore, the prediction of their behavior through the oscillation model eventuates in results, more consistent with actual pullout behavior. However, PET1, which consists of twisted polyester multi filaments, reveals a better conformity with the experimental results toward the textured, intermingled multi filaments of PET2.

It should be noted that, if in any circumstances the pullout test get involved in a complicacy, for example if the end of the pulled yarn get trapped in the weave, the oscillation model does not have the ability to predict it.

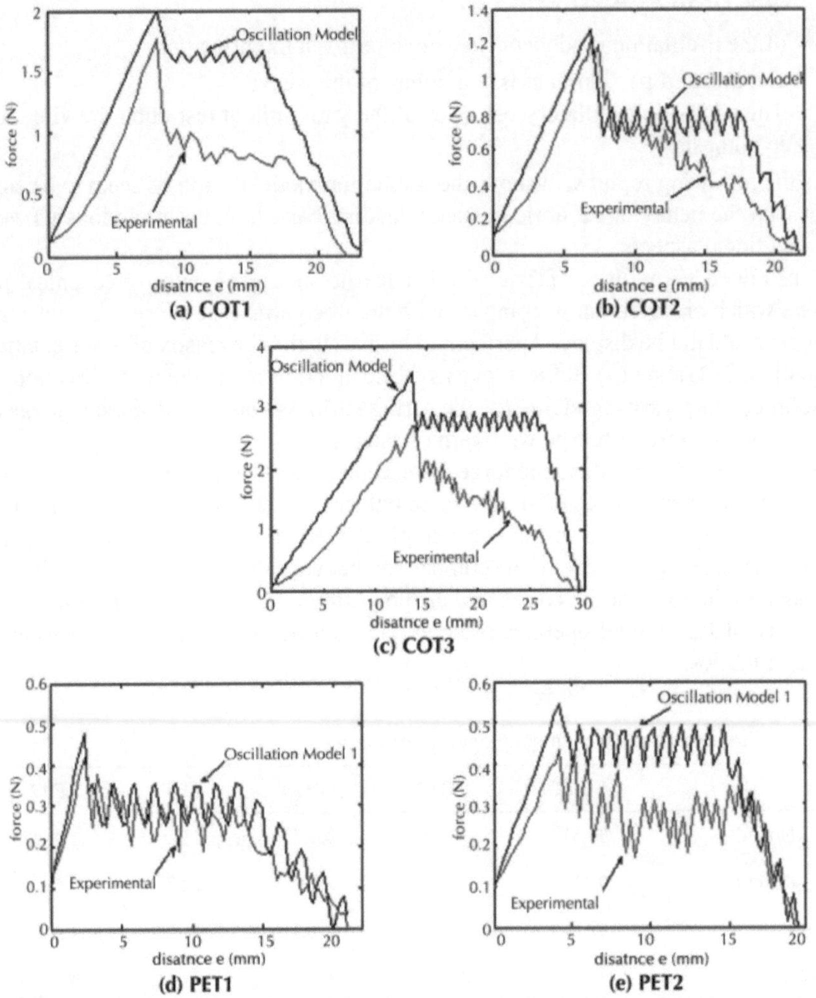

FIGURE 4 Experimental yarn pullout force displacement profiles vs. oscillation method.

Furthermore, it seems that the applied mean dynamic friction coefficient (μ_D) for cotton fabrics is an overestimation of the real μ. Here the two last parts of the pullout profile has less conformity with the experimental results. Hence, the oscillation model is a more appropriate method to predict the pullout behavior of the fabrics composed of filament or even yarns.

14.5 CONCLUSION

Oscillation model, based on the force balance model, is an analytical method, capable of predicting the pullout behavior of the yarns within the plain woven fabrics, fixed from two opposite sides. This method emphasizes on the vibratory behavior of the yarns during the pullout test. Force balance model was used to determine the internal frictional parameters of oscillation model. Simple equations were reported to estimate the equivalent yarn spring stiffness constant and its tensile modulus within the fabric. Results of oscillation model and experimental measurements were compared together. Although, the anticipated force displacement profiles are clearly consisted in the shape, the experimental investigations showed that oscillation model is suitable for fabrics, made from filament or more even yarns.

KEYWORDS

- **Coulombs law**
- **Internal mechanical parameters**
- **Oscillation model**
- **Pullout behavior**
- **Yarn-to-yarn friction coefficient**

REFERENCES

1. Badrossamay, M. R., Hosseini Ravandi, S. A., and Morshed, M. Fundamental parameters affecting yarn pullout behavior. *Journal of Textile Institute*, **92**, 280287 (2001).
2. Hong, J. and Jayaraman, S. Friction in textiles. *Textile progress*, **34**(12) (2003).
3. Kirkwood, J. E., Kirkwood, K. M., Lee, Y. S., Egres Jr., R. G., Wagner, N. J., and Wetzel, E. D. Yarn pullout as a mechanism for dissipating ballistic impact energy in kevlar® km-2 fabric, part ii: Predicting ballistic performance. *Textile Research Journal*, **74**, 939948 (2004).
4. Kirkwood, K. M., Kirkwood, J. E., Lee, Y. S., Egres Jr., R. G., Wagner, N. J., and Wetzel, E. D. Yarn pullout as a mechanism for dissipating ballistic impact energy in kevlar® km-2 fabric, part i: Quasi static characterization of yarn pullout, *Textile Research Journal.*, **74**, 920928 (2004).
5. Motamedi, F., Baily, A. I., Briscoe, B. J., and tabor, D. Theory and practice of localized fabric deformation. *Textile Research Journal*, **59**, 160172 (1989).
6. Pan, N. and Young Youn, M. Behavior of yarn pullout from woven fabrics: theoretical and experimental. *Textile Research Journal*, **63**, 629637 (1993).
7. Pan, N. Theoretical modeling and analysis of fiber pullout behavior from a bonded fibrous matrix: The elastic-bond case. *Journal of Textile Institute*, **84**, 472485 (1993).
8. Sebastian, S. A. R. D., Baily, A. I., Briscoe, B. J., and tabor, D. Extension, displacements and forces associated with pulling a single yarn from a fabric. *Journal of Physics D. Applied Physics*, **20**, 130139 (1987).
9. Seto, W. W. *Theory and problems of mechanical vibration*, Tehran, Jazil (1992).
10. Valizadeh, M., Hosseini Ravandi, S. A., Salimi, M., and Sheikhzadeh, M. Determination of internal mechanical characteristics of woven fabrics based on the force balance analysis of yarn pullout test. *Journal of Textile Institute*, **99**, 3755 (2008).

15 Some New Aspects of Polymeric Nanofibers: Part I

A. K. Haghi and G. E. Zaikov

CONTENTS

15.1 Introduction 141
15.2 Experimental 142
15.3 Discussion and Results 145
15.4 Conclusion 147
Keywords 147
References 148

15.1 INTRODUCTION

Fibrous materials used for filter media provide advantages of higher filtration efficiency and lower air resistance, which are closely associated with fiber fineness [1, 2]. Filtration efficiency is one of the most important concerns for filter performance [1-5]. There are various methods to produce ultra fine fibers [1, 2]. Recently, much attention is being directed towards electrospinning as a unique technique for the fabrication of nanofibers [1-9]. In electrospinning, a high voltage is applied to a capillary containing polymer solution. At a sufficient voltage to overcome surface tension forces, a charged fluid jet is ejected from the needle tip. The jet is stretched and elongated before it reaches the target, then dried and collected as randomly oriented structures in form of nonwoven mat. Electrospinning provides an ultra thin mat of extremely fine fibers with very small pore size and high porosity, which makes them unique candidates for use in filtration, and possibly protective clothing applications [1-5].

These nanowebs have good aerosol particle obstruction and comparatively low air resistance. Recently, the filtration properties of electrospun mats have been studied [3-8]. To provide appropriate mechanical properties, nanofiber webs have been applied onto various substrates. Substrates are often chosen to resemble conventional filter materials [8].

Gibson et al [6, 7] have reported some properties of electrospun mats. They compared performances of electrospun fiber mats with properties of textiles and membranes currently used in protective clothing systems and showed that electrospun layers are extremely efficient for trapping airborne particles. Also, they reported that the air flow resistance and aerosol filtration properties are affected by the coating weight. It was shown that an extremely thin layer of electrospun nanofibers eliminated particle penetration through the layer [6-7]. Transport properties of electrospun nylon6 mats were investigated by Ryu et al [9]. They found that concentration of polymer solution affected the fiber diameter, pore size, Brunauer-Emmett-Teller (BET) surface area, and gas transport properties of mats. It was shown that the filtration efficiency of Nylon6 nanofilters is superior to the commercialized high efficiency particulate air (HEPA) filter for 0.3 μm test particles [4]. Li et al [10] found that the pore size and pore size distribution of electrospun polylacticacid (PLA) membranes are strongly associated with fiber mass, fiber diameter, and fiber length.

Researchers found that the electrospun mats provide good aerosol particle protection, without a significant change in moisture vapor transport. It was shown that materials used in protective clothing must provide a combination of high barrier performance and thermal comfort [6, 11]. It has been recognized that the heat and moisture transport behavior of textile materials is one of the most important factors influencing the dynamic comfort and performance of clothing in normal use [12]. Significant theoretical and experimental investigation has been done in this field by various researchers [12, 13]

Despite the many excellent studies which have been done on the properties of nanofiber mats, due to insufficient information on heat and moisture transfer of nanofiber mats, more studies in these fields are required.

In this study, bulky nanofiber mats were produced by using a modified electrospinning method. Bulk porosity of nanofiber mats was measured and the effect of bulk porosity on the air permeability and moisture transfer of nanofiber mats was evaluated.

15.2 EXPERIMENTAL

Polymer solution with concentration 15 wt. % was prepared by dissolving PAN in DMF. The PAN solution in DMF was loaded into a 1 ml syringe. A modified electrospinning system consists of two syringe pumps was used to produce bulky nanofiber mats. The experimental setup of electrospinning has been shown in Figure 1.

Polymer solution fed at speed of 2.8 μl/min through needles with an outer diameter of 0.7 mm. A 9 kv voltage applied between the two needles while nozzle distance was 10 cm. The tip to collector distance was 15 cm.

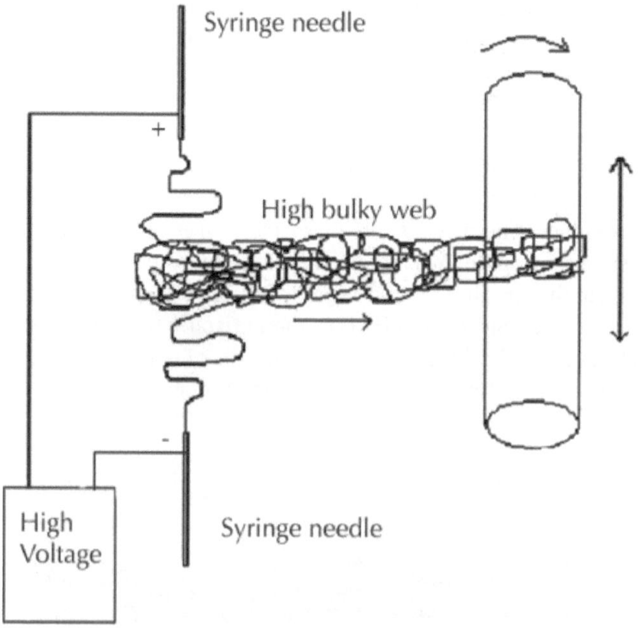

FIGURE 1 Schematic illustration of the modified electospinning system setup.

The rotation speed of the drum was changed from 31.5 to 103.5 RPM. The bulk porosity of nanofiber mats was measured by the following equations:

$$\varepsilon = 1 - \frac{\rho_{mat}}{\rho_f} \quad (1)$$

$$\rho_{mat} = \frac{W_{mat}}{L \times A} \quad (2)$$

Where ε, ρmat, ρf, W_{mat} ,L, and A are bulk porosity, nanofiber mat density (g/cm^3), nanofiber density (1.16g/cm^3), weight of nanofiber mat (g), nanofiber mat thickness (0.2 cm), and nanofiber mat area (cm^2), respectively.

The morphology of electrospun mats was observed by a scanning electron microscope (SEM, Philips XL-30). Thickness of nanofiber mats was measured by measuring of collector diameter before and after electrospinning using digital images as shown in Figure 2. Microstructure measurement software was used to calculate the diameter of PAN nanofibers from SEM images at magnification of 5,000. A minimum of 100 fibers were used to calculate the mean values of the fibers diameter.

Air permeability of PAN nanofiber mats was measured by using a Shirley permeation analyzer at 65% relative humidity (RH), 25 $^\circ C$, and 100 Pa.

In order to investigate the dynamic moisture transfer of nanofiber mats, an experimental apparatus was designed and constructed which was able to simulate the sweating of human body (Figure 3). It consists of a controlled environmental chamber, sweating guarded hot plate, and data acquisition system. The guarded hot plate, maintained at 37°C and used as a heat source, was housed in a chamber with ambient conditions of 25°C, and 65% RH. The diffusion cell consists of a water container, a piece of animal skin for simulating human skin, humidity sensors. One side of nanofiber mats faces the sweating skin but does not contact it, whereas another side is exposed to the controlled environment. The driving forces for the movement of moisture vapor are the temperature and vapor gradients maintained between the points where the moisture vapor emerges from the simulated skin (37°C and 90% RH) and the ambient environment controlled at 25°C and 65% RH. For comparison between different samples, each sample is tested for 10 min. The one way analysis of variance at the 0.05 level of significance was used to comparisons among the treatments means.

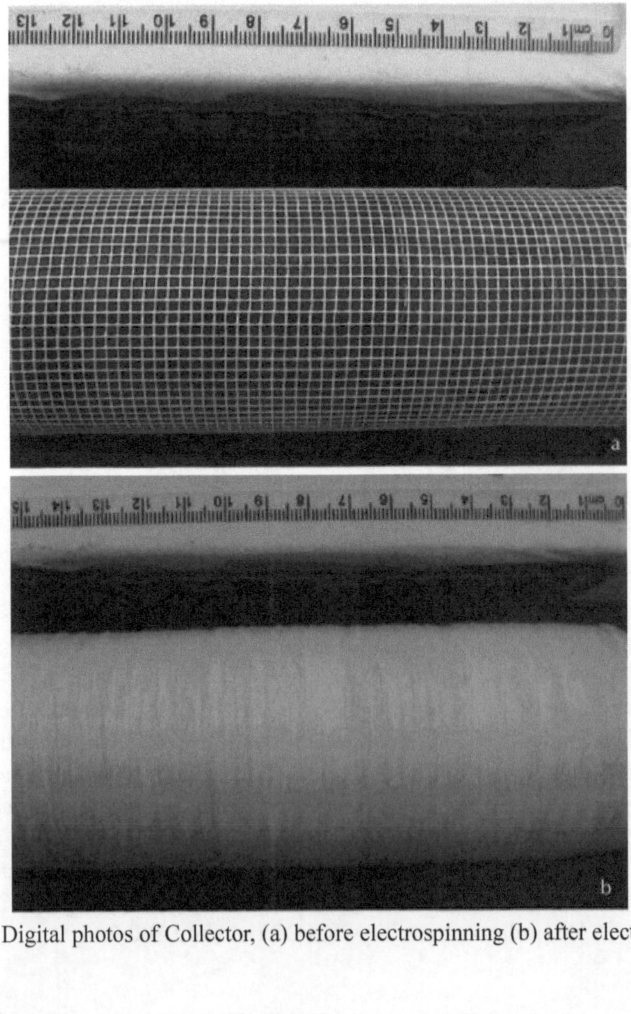

FIGURE 2 Digital photos of Collector, (a) before electrospinning (b) after electrospinning.

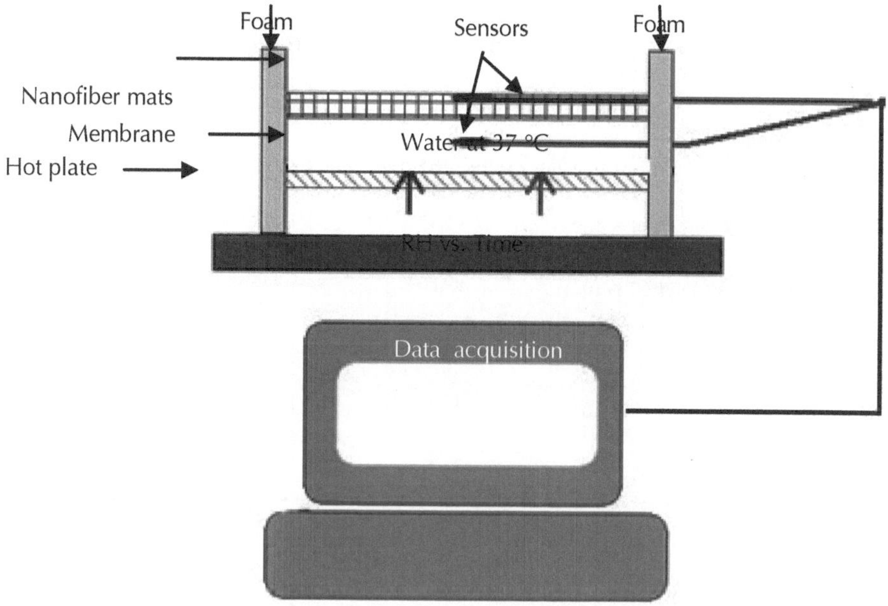

FIGURE 3 Schematic design of the instrument for measurement of moisture transfer.

15.3 DISCUSSION AND RESULTS

Present experimental investigation was carried out for a wide range of rotation speed of collector. Table 1 represents the characteristics and transport properties of bulky nanofiber mats. Results indicate that bulk nanofiber mats have high porosity. Increase in speed rotation of collector increases the density of nanofiber mats and decreases bulk porosity of nanofiber mats. Statistical analysis stresses on the significant reduction of the bulk porosity of nanofiber mats.

The based on the results in Table 1 and Figure 4, the air permeability of nanofiber mats increases by increasing in porosity and decreasing density of nanofiber mats. Reduction in mat density results in increase of the bulk porosity and pore size of the mat. Also results show that the moisture transfer of nanofiber mats is high which reduces by increasing the density and decreasing the bulk porosity of nanofiber mats as shown in Figure 5.

Statistical analysis shows the transport properties significantly reduce as bulk porosity of nanofiber mats decreases.

Comparing the nanofiber mats with different porosities, shows that the nanofiber mat with higher porosity and air permeability has the higher moisture transfer. This result indicates that the high resistance to the air permeability of nanofiber mats dose not impeded the moisture transfer through the bulk nanofiber mats. This difference is related to the size of water vapor and air molecules.

TABLE 1 Characteristics and transport properties of bulk nanofiber mats.

Rotation speed of collector *(RPM))*	Density of nanofiber mats *(g/cm³)*	Bulk porosity (%)	Air permeability *(l/m²sec)*	Moisture transfer (%)
substrate	0	~100	4000 >	100
31.5	0.84×10^{-3}	99.92	477.5	98.4
49.5	1.65×10^{-3}	99.86	245.5	82.5
67.5	3.1×10^{-3}	99.73	126.7	74.7
85.5	4.62×10^{-3}	99.6	58.7	64.8
103.5	6.64×10^{-3}	99.43	46.9	60.6

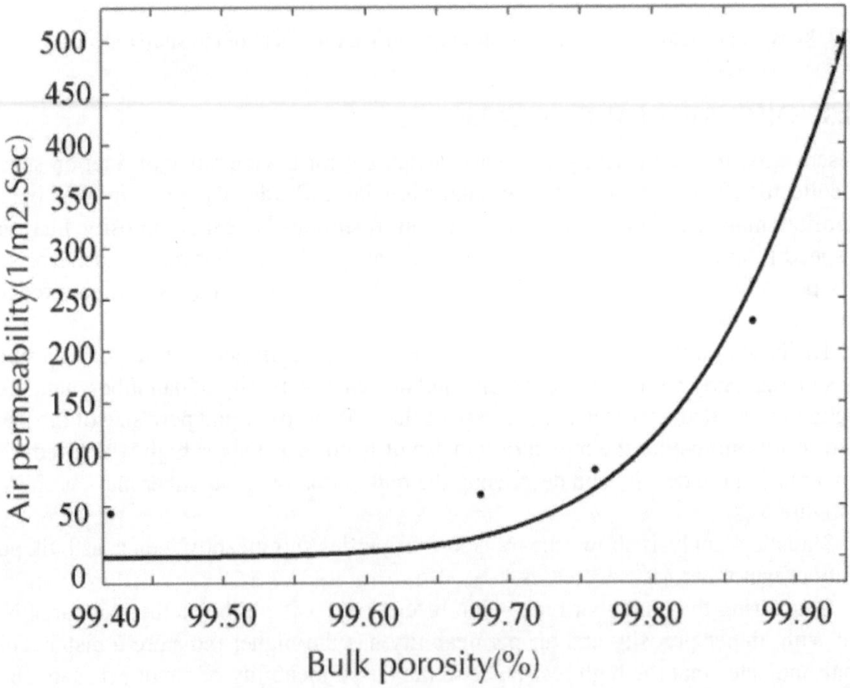

FIGURE 4 Air permeability versus bulk porosity of nanofiber mats.

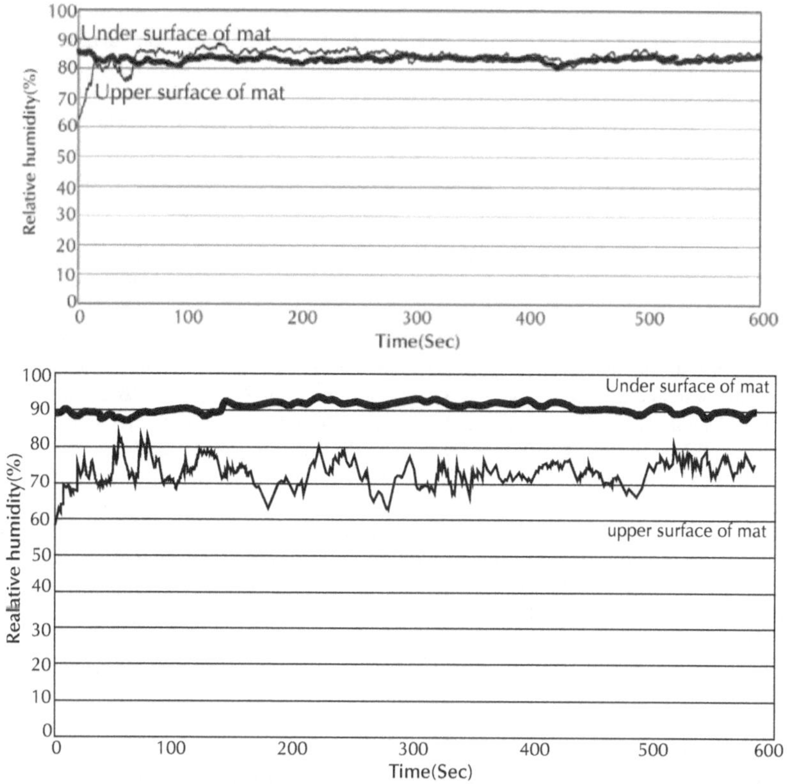

FIGURE 5 Experimental results of the relative humidity of the nanofiber mats with different porosity, (a) 99.92% and (b) 99.42%.

15.4 CONCLUSION

Electrospun nanofiber mats by modified electospinning system have high porosity. Results show that the porosity of nanofiber mats increases by decreasing speed rotation of drum. Air permeability and moisture transfer of nanofiber mats were reduced with reduction the porosity of nanofiber mats. Based on experimental results, moisture transfer behavior is influenced by nanofiber mats porosity. Bulk nanofiber mats with high rates of water vapor diffusion and low air permeability are promising candidates for protective clothing applications.

KEYWORDS

- **Brunauer-Emmett-Teller**
- **Electrospinning**
- **Electrospun polylacticacid**
- **Nanofiber mats**
- **Nylon 6**

REFERENCES

1. Huang, Z. M., Zhang, Y. Z., Kotaki, M., and Ramakrishna, S. A Review on Polymer Nanofibers by Electrospinning and their Applications in Nanocomposites. *Composites Science and Technology*, **63**, 2223-2253 (2003).
2. Ramakrishna, S., Fujihara, K., Teo, W. E., Lim, T. C., and Ma, Z. Introduction to Electrospinning and Nanofibers. *World Scientific Publishing Co. Pte.Ltd. ISBN* 981-256-415-2 (2005).
3. Gibson, P. and Gibson, H. S. *Use of Electrospun Nanofibers for Aerosol Filtration in Textile Structures,* Proceeding of 24th Army Science Conference (ASC2004) Orland, FL, (20 November-December, 2004).
4. Ahn, Y. C., Park, S. K., Kim, G. T., Hwang, Y. J., Lee, C. G., Shin, H. S., and Lee, J. K. Development of High Efficiency Nanofilters Made of Nanofibers. *Current Applied Physics,* **6**, 1030-1035 (2006).
5. Graham, K., Ouyang, M., Raether, T., Grafe, T., McDonald, B., and Knauf, P. Polymeric Nanofibers in Air Filtration Application, *Fifteenth Annual Technical conference and Expo of the American Filtration and Separations Society,* Galveston, Texas, pp. 9-12 (April, 2002).
6. Gibson, P., Gibson, H. S., and Rivin, D. Transport Properties of Porous Membranes Based on Electrospun Nanofiber, *Colloids and Surfaces A,* **187-188**, 469-481 (2001).
7. Gibson, P., Gibson, H. S., and Rivin, D. Electrospun Fiber Mats: Transport Properties. *AIChE Journal*, **45**(1), 190-195 (1999).
8. Grafe, T., Gogins, M., Barris, M., Schaefer, J., and Canepa, R. *Nanofibers in Filtration Applications in Transportation, International Conference and Exposition of the INDA,* Chicago, Illinois, pp. 1-15 (December, 3-5, 2001).
9. Ryu, Y. J., Kim, H. Y., Lee, H. K., Park, H. C., and Lee, D. R. Transport Properties of Electrospun Nylon6 Nonwoven Mats. *European Polymer Journal*, **39**, 1883-1889 (2003).
10. Li, D., Frey, M. W., and Joo, Y. L. Characterization of Nanofibrous Membranes with Capillary Flow Porometry, *J. of Membrane Science*, , **286**, 104 (2006).
11. Graham, K., Gogins, M., and Gibson, H. S. Incorporation of Electrospun Nanofibers into Functional Structures, *INJ, 21-27, summer,* (2004).
12. Li, Y. and Wong, A. S. W. Clothing Biosensory Engineering. *The Textile Institute*, England (2006).
13. Wang, Z., Li,Y., Zhu Q. Y., and Lue, Z. X. Radiation and Conduction Heat Transfer Coupled with Liquid Water Transfer, Moisture Sorption, and Condensation in Porous Polymer Materials. *Journal of Applied Polymer Science*, **89**, 2780-2790 (2003).

16 Some new Aspects of Polymeric Nanofibers: Part II

A. K. Haghi and G. E. Zaikov

CONTENTS

16.1 Introduction 150
16.2 Experimental 151
16.2.1 Materials 151
16.2.2 Electrospinning 151
16.2.3 Characterization 151
16.2.4 Experimental Design 152
16.3 Discussion and Results 153
16.3.1 Morphological Analysis of Nanofibers 158
16.3.2 Analysis of Variance (Anova) 162
16.3.3 Response Surfaces for Average Fiber Diameter (AFD) 162
16.3.3.1 Solution Concentration 162
16.3.3.2 Applied Voltage 163
16.3.3.3 Tip to Collector Distance 163
16.3.3.4 Volume Flow Rate 163
16.3.4 Response Surfaces for Contact Angle (Ca) of Electrospun Fiber Mat 163
16.3.4.1 Solution Concentration 164
16.3.4.2 Applied Voltage 165
16.3.4.3 Tip to Collector Distance 165
16.3.4.4 Volume Flow Rate 160
16.3.5 Determination of Optimal Conditions 160
16.3.6 Relationship between AFD and CA of Electrospun Fiber Mat 167
16.4 Conclusion 167
References 167

16.1 INTRODUCTION

In recent years, electrospinning as a simple and effective method for preparation of nanofibrous materials have attracted increasing attention [1]. Electrospinning process, unlike the conventional fiber spinning systems (melt spinning, wet spinning, etc.), uses electric field force instead of mechanical force to draw, and stretch a polymer jet [2]. This method is providing nonwoven mat with individual fiber diameter ranged from micrometer to nanometer. Due to their high specific surface area, high porosity, and small pore size, the unique nanofibers have been suggested as excellent candidate for many applications including filtration [3], multifunctional membranes [4], biomedical agents [5], tissue engineering scaffolds [6,7], wound dressings [8], full cell [9] and protective clothing [10].

The typical electrospinning apparatus consists of three components to fulfill the process including syringe filled with a polymer solution, a high voltage supplier to provide the required electric force for stretching the liquid jet and a grounded collection plate to hold the nanofiber mat. In electrospinning the electrical forces applied between the nozzle and collector draws the polymer solution toward the collection plate as a jet. Usually, voltages range from 5 to 30 kV, sufficient to overcome the surface tension forces of the polymer solution. During the jet movement to the collector, the solvent evaporates and dry fibers are randomly deposited on the surface of a collector [11-16]. A schematic representation of electrospinning is shown in Figure 1.

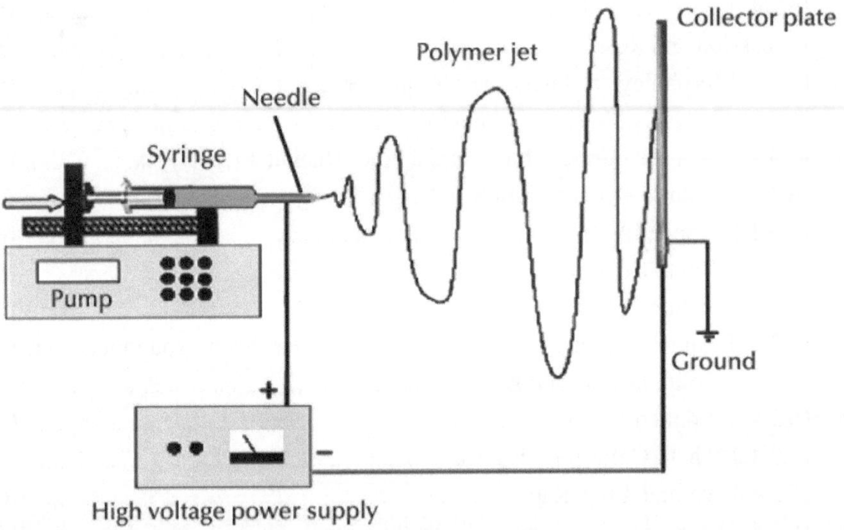

FIGURE 1 Schematic of electrospinning set up.

The morphology and the diameter of the electrospun nanofibers can be affected by many electrospinning parameters including solution properties (the concentration, liquid viscosity, surface tension, and dielectric properties of the polymer solution), processing conditions (applied voltage, volume flow rate, tip to collector distance, and the strength of the applied electric field), and ambient conditions (temperature, atmospheric pressure and humidity) [17-20].

The wettability of solid surfaces is a very important property of surface chemistry, which is controlled by both the chemical composition and the geometrical microstructure of surface [21-23]. When a liquid droplet contacts a solid surface, it will spread or remain as droplet with the formation of angle between the liquid and solid phases. Contact angle (CA) measurements are widely used to characterize the wettability of solid surface. Surface with a water CA greater than 150° is usually called superhydrophobic surface. On the other hand, when the CA is lower than 5°, it is called superhydrophilic surface. Fabrication of these surfaces has attracted considerable interest for both fundamental research and practical studies [23-25].

In this work, we investigate the effect of four electrospinning parameters (solution concentration, applied voltage, tip to collector distance, and volume flow rate) on the average fiber diameter (AFD) and CA of electrospun PAN nanofiber mat. The aim of the present study is to establish quantitative relationship between electrospinning parameters and AFD and CA of electrospun fiber mat by response surface methodology.

16.2 EXPERIMENTAL

16.2.1 Materials

Polyacrylonitrile (PAN) powder was purchased from Polyacryle Co. (Iran). The weight average molecular weight (M_w) of PAN was approximately 100,000 g/mol. The solvent *N-N*, dimethylformamide (DMF) was obtained from Merck Co. (Germany). These chemicals were used as received.

16.2.2 Electrospinning

In experiment, the PAN powder was dissolved in DMF and gently stirred for 24 hr at 50°C. Therefore, homogenous PAN/DMF solution was prepared in different concentration ranged from 10 to 14 wt.%.

Electrospinning was set up in a horizontal configuration. The electrospinning apparatus consist of 5 ml plastic syringe connected to a syringe pump and a rectangular grounded collector (aluminum sheet). A high voltage power supply (capable to produce 0–40 kV) was used to apply a proper potential to the metal needle. It should be noted that all electrospinnings were carried out at room temperature.

16.2.3 Characterization

The morphology of the gold-sputtered electrospun fibers were observed by scanning electron microscope (SEM, Philips XL-30). The AFD and distribution was determined from selected SEM image by measuring at least 50 random fibers. The wettability of electrospun fiber mat was determined by CA measurement. The CA measurements were carried out using specially arranged microscope equipped with camera and PCTV vision software as shown in Figure 2. The droplet used was distilled water and was 1 μl in volume. The CA experiments were carried out at room temperature and were repeated five times. All CAs measured within 20 s of placement of the water droplet on the electrospun fiber mat.

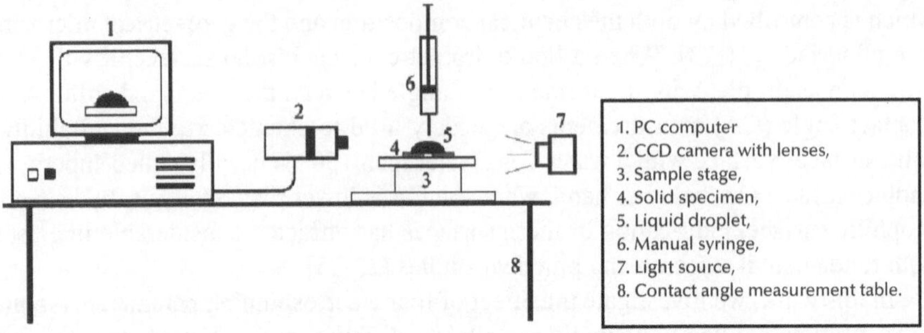

FIGURE 2 Schematic of CA measurement set up.

16.2.4 Experimental Design

The response surface methodology (RSM) is a combination of mathematical and statistical techniques used to evaluate the relationship between a set of controllable experimental factors and observed results. This optimization process is used in situations where several input variables influence some output variables (responses) of the system. The main goal of RSM is to optimize the response, which is influenced by several independent variables, with minimum number of experiments. The central composite design (CCD) is the most common type of second-order designs that used in RSM and is appropriate for fitting a quadratic surface [26, 27].

In the present study, CCD was employed to establish relationships between four electrospinning parameters and two responses including the AFD and the CA of electrospun fiber mat. The experiment was performed for at least three levels of each factor to fit a quadratic model. Based on preliminary experiments, polymer solution concentration (X_1), applied voltage (X_2), tip to collector distance (X_3), and volume flow rate (X_4) were determined as critical factors with significance effect on AFD and CA of electrospun fiber mat. These factors were four independent variables and chosen equally spaced, while the AFD and the CA of electrospun fiber mat were dependent variables (responses). The values of –1, 0, and 1 are coded variables corresponding to low, intermediate, and high levels of each factor respectively. The experimental parameters and their levels for four independent variables are shown in Table 1.

TABLE 1 Design of experiment (factors and levels).

Factor	Variable	Unit	Factor level		
			-1	**0**	**1**
X_1	Solution concentration	(wt.%)	10	12	14
X_2	Applied voltage	(kV)	14	18	22
X_3	Tip to collector distance	(cm)	10	15	20
X_4	Volume flow rate	(ml/hr)	2	2.5	3

The following quadratic model, which also includes the linear model, was fitted to the data.

$$Y = \beta_0 + \sum_{i=1}^{k} \beta_i . x_i + \sum_{i=1}^{k} \beta_{ii} . x_i^2 + \sum \sum_{i<j=2}^{k} \beta_{ij} . x_i x_j + \varepsilon \quad (1)$$

where Y is the predicted response, x_i and x_j are coded variables, β_0 is constant coefficient, β_i is the linear coefficients, β_{ii} is the quadratic coefficients, β_{ij} is the second order interaction coefficients, k is the number of factors, and ε is the approximation error [26, 27].

The experimental data were analyzed using design-expert software including analysis of variance (ANOVA). The values of coefficients for parameters (β_0, β_i, β_{ii}, β_{ij}) in Equation (1), p-values, the determination coefficient (R^2) and adjusted determination coefficient (R^2_{adj}) were calculated by regression analysis.

16.3 DISCUSSION AND RESULTS

16.3.1 Morphological Analysis of Nanofibers

The PAN solution in DMF were electrospun under different conditions, including various PAN solution concentrations, applied voltages, volume flow rates and tip to collector distances, to study the effect of electrospinning parameters on the morphology and diameter of electrospun nanofibers.

Figure 3 shows the SEM images and fiber diameter distributions of electrospun fibers in different solution concentration as one of the most effective parameters to control the fiber morphology. As observed in Figure 3, the AFD increased with increasing concentration. It was suggested that the higher solution concentration would have more polymer chain entanglements and less chain mobility. This causes the hard jet extension and disruption during electrospinning process and producing thicker fibers.

The SEM image and corresponding fiber diameter distribution of electrospun nanofiber in different applied voltage are shown in Figure 4. It is obvious that increasing the applied voltage cause an increase followed by a decrease in electrospun fiber diameter. As demonstrated by previous researchers [17, 30], increasing the applied voltage may decrease, increase or may not change the fiber diameter. In one hand, increasing the applied voltage will increase the electric field strength and higher electrostatic repulsive force on the jet, favoring the thinner fiber formation. On the other hand, more surface charge will introduce on the jet and the solution will be removed more quickly from the tip of needle. As a result, the AFD will be increased [29-30].

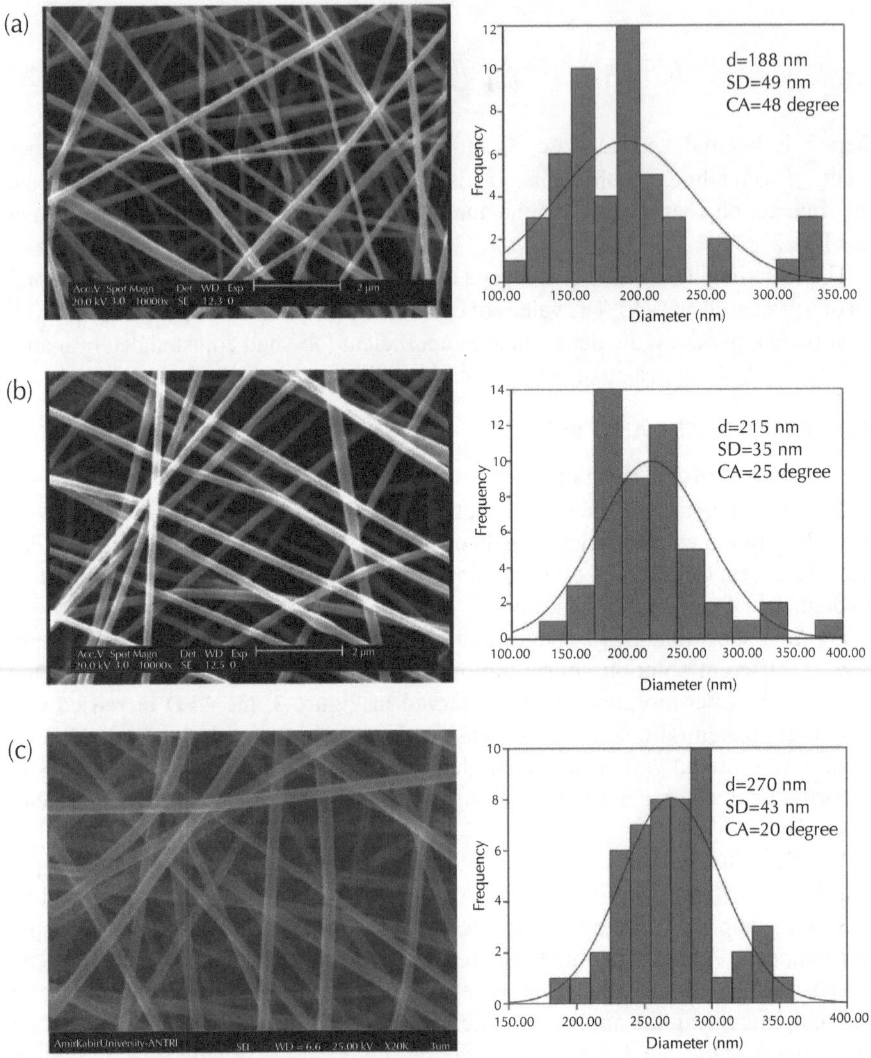

FIGURE 3 The SEM images and fiber diameter distributions of electrospun fibers in solution concentration of (a) 10 wt.%, (b) 12 wt.% and (c) 14 wt.%.

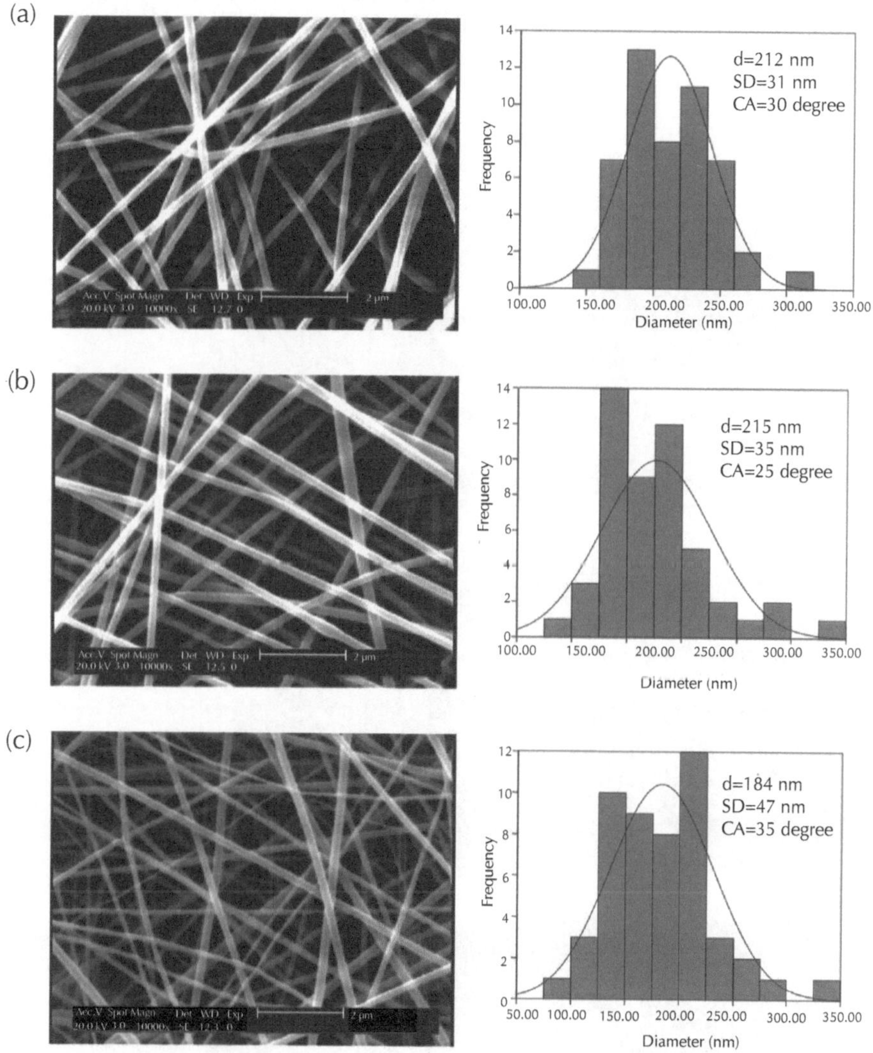

FIGURE 4 The SEM images and fiber diameter distributions of electrospun fibers in applied voltage of (a) 14 kV, (b) 18 kV, and (c) 22 kV.

Figure 5 represents the SEM image and fiber diameter distribution of electrospun nanofiber in different spinning distance. It can be seen that the AFD decreased with increasing tip to collector distance. Because of the longer spinning distance could give more time for the solvent to evaporate, increasing the spinning distance will decrease fiber diameter [30-31].

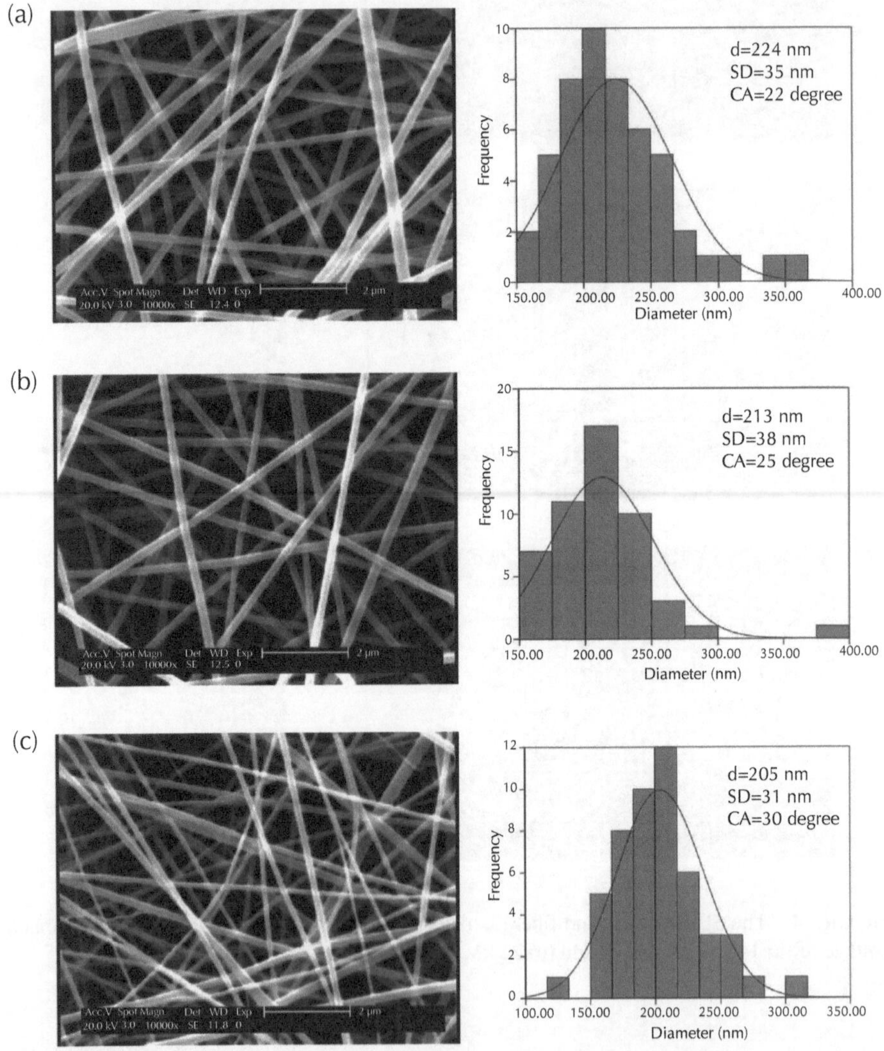

FIGURE 5 The SEM images and fiber diameter distributions of electrospun fibers in tip to collector distance of (a) 10 cm, (b) 15 cm, and (c) 20 cm.

The SEM image and fiber diameter distribution of electrospun nanofiber in different volume flow rate are illustrated in Figure 6. It is clear that increasing the volume flow rate cause an increase in AFD. Ideally, the volume flow rate must be compatible with the amount of solution removed from the tip of the needle. At low volume flow rates, solvent would have sufficient time to evaporate and thinner fibers were produced but at high volume flow rate, excess amount of solution fed to the tip of needle and thicker fibers result [28-31].

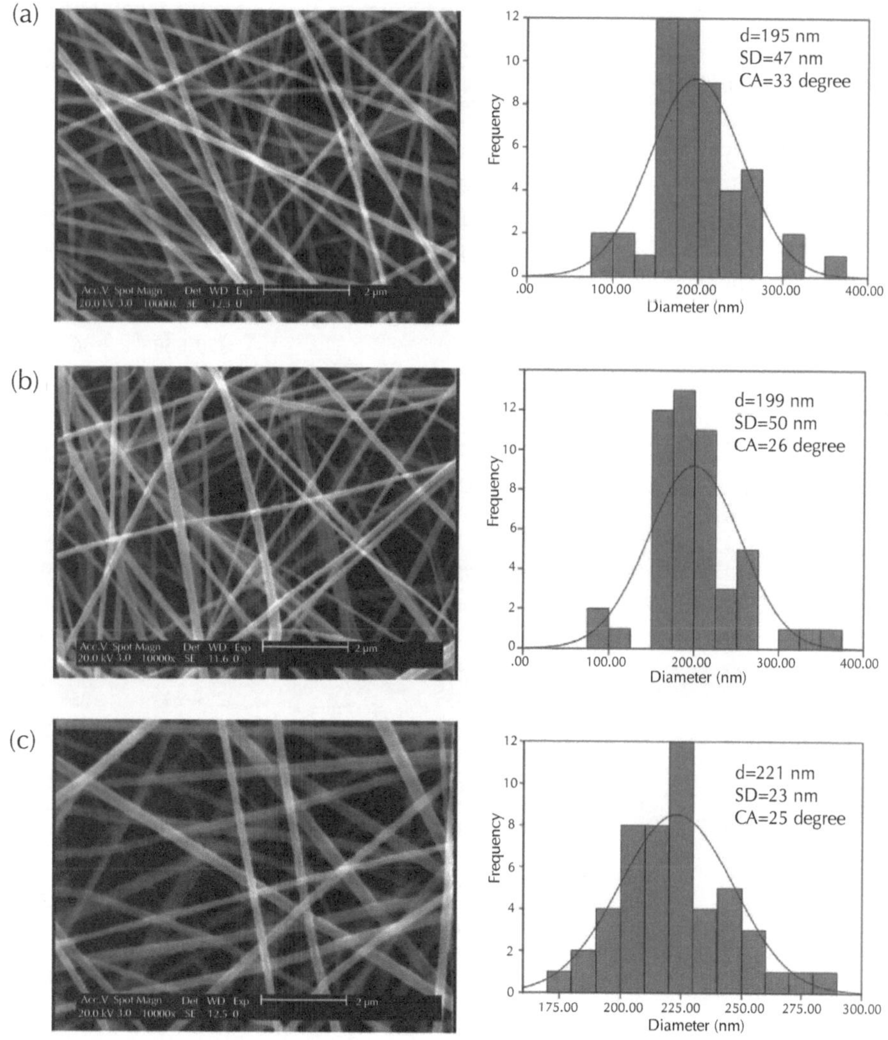

FIGURE 6 The SEM images and fiber diameter distributions of electrospun fibers in volume flow rate of (a) 2 ml/hr, (b) 2.5 ml/hr, and (c) 3 ml/hr.

16.3.2 Analysis of Variance (ANOVA)

All 30 experimental runs of CCD were performed as described in Table 2. A significance level of 5% was selected; that is statistical conclusions may be assessed with 95% confidence. In this significance level, the factor has significant impact on response if the p-value is less than 0.05, when p-value is greater than 0.05, it is concluded the factor has no significant effect on response.

TABLE 2 The actual design of experiments and responses for AFD and CA.

No.	Electrospinning parameters				Responses	
	X_1 Concentration	X_2 Voltage	X_3 Distance	X_4 Flow rate	AFD (nm)	CA (°)
1	10	14	10	2	206 ± 33	44 ± 6
2	10	22	10	2	187 ± 50	54 ± 7
3	10	14	20	2	162 ± 25	61 ± 6
4	10	22	20	2	164 ± 51	65 ± 4
5	10	14	10	3	225 ± 41	38 ± 5
6	10	22	10	3	196 ± 53	49 ± 4
7	10	14	20	3	181 ± 43	51 ± 5
8	10	22	20	3	170 ± 50	56 ± 5
9	10	18	15	2.5	188 ± 49	48 ± 3
10	12	14	15	2.5	210 ± 31	30 ± 3
11	12	22	15	2.5	184 ± 47	35 ± 5
12	12	18	10	2.5	214 ± 38	22 ± 3
13	12	18	20	2.5	205 ± 31	30 ± 4
14	12	18	15	2	195 ± 47	33 ± 4
15	12	18	15	3	221 ± 23	25 ± 3
16	12	18	15	2.5	199 ± 50	26 ± 4
17	12	18	15	2.5	205 ± 31	29 ± 3
18	12	18	15	2.5	225 ± 38	28 ± 5
19	12	18	15	2.5	221 ± 23	25 ± 4
20	12	18	15	2.5	215 ± 35	24 ± 3
21	12	18	15	2.5	218 ± 30	21 ± 3
22	14	14	10	2	255 ± 38	31 ± 4
23	14	22	10	2	213 ± 37	35 ± 5
24	14	14	20	2	240 ± 33	33 ± 6
25	14	22	20	2	200 ± 30	37 ± 4
26	14	14	10	3	303 ± 36	19 ± 3
27	14	22	10	3	256 ± 40	28 ± 3
28	14	14	20	3	283 ± 48	39 ± 5
29	14	22	20	3	220 ± 41	36 ± 4
30	14	18	15	2.5	270 ± 43	20 ± 3

The results of ANOVA for AFD and CA of electrospun fiber mat are shown in Table 3 and Table 4 respectively. Equations (2) and (3) are the calculated regression equation.

TABLE 3 ANOVA for average fiber diameter (AFD).

Source	Sum of squares	F-value	p-value
Model	31004.72	28.67	< 0.0001
X_1	17484.50	226.34	< 0.0001
X_2	4201.39	54.39	< 0.0001
X_3	2938.89	38.04	< 0.0001
X_4	3016.06	39.04	< 0.0001
X_1X_2	1139.06	14.75	0.0016
X_1X_3	175.56	2.27	0.1524
X_1X_4	637.56	8.25	0.0116
X_2X_3	39.06	0.51	0.4879
X_2X_4	162.56	2.10	0.1675
X_3X_4	60.06	0.78	0.3918
X_1^2	945.71	12.24	0.0032
X_2^2	430.80	5.58	0.0322
X_3^2	0.40	0.005	0.9433
X_4^2	9.30	0.12	0.7334
Lack of Fit	711.41	0.8	Not significant

$R^2 = 0.9640$; $R^2_{adj} = 0.9303$.

TABLE 4 ANOVA for CA of electrospun fiber mat.

Source	Sum of squares	F-value	p-value
Model	4175.07	32.70	< 0.0001
X_1	1760.22	193.01	< 0.0001
X_2	84.50	9.27	0.0082
X_3	338.00	37.06	< 0.0001
X_4	98.00	10.75	0.0051
X_1X_2	42.25	4.63	0.0116
X_1X_3	42.25	4.63	0.0116
X_1X_4	42.25	4.63	0.0116
X_2X_3	12.25	1.34	0.2646
X_2X_4	6.25	0.69	0.4207
X_3X_4	2.25	0.25	0.6266
X_1^2	161.84	17.75	0.0008
X_2^2	106.24	11.65	0.0039
X_3^2	0.024	0.003	0.9597
X_4^2	21.84	2.40	0.1426
Lack of Fit	95.30	1.15	Not significant

$R^2 = 0.9683$; $R^2_{adj} = 0.9387$.

$$\begin{aligned} AFD = {} & 212.11 + 31.17X_1 - 15.28X_2 - 12.78X_3 + 12.94X_4 \\ & - 8.44X_1X_2 + 3.31X_1X_3 + 6.31X_1X_4 + 1.56X_2X_3 - 3.19X_2X_4 - 1.94X_3X_4 \\ & + 19.11X_1^2 - 12.89X_2^2 - 0.39X_3^2 - 1.89X_4^2 \end{aligned} \quad (2)$$

$$\begin{aligned} CA = {} & 25.80 - 9.89X_1 + 2.17X_2 + 4.33X_3 - 2.33X_4 \\ & - 1.63X_1X_2 - 1.63X_1X_3 + 1.63X_1X_4 - 0.88X_2X_3 - 0.63X_2X_4 + 0.37X_3X_4 \\ & + 7.90X_1^2 + 6.40X_2^2 - 0.096X_3^2 + 2.90X_4^2 \end{aligned} \quad (3)$$

From the p-values presented in Table 3 and Table 4, it can be concluded that the p-values of terms X_3^2, X_4^2, X_2X_3, X_1X_3, X_2X_4, and X_3X_4 in the model of AFD and X_3^2, X_4^2, X_2X_3, X_2X_4, and X_3X_4 in the model of CA is greater than the significance level of 0.05, therefore, they have no significant effect on corresponding response. Since, the terms had no significant effect on AFD and CA of electrospun fiber mat, these terms were removed and fitted the equations by regression analysis again. The fitted equations in coded unit are given in Equation (4) and (5).

$$\begin{aligned} AFD = {} & 211.89 + 31.17X_1 - 15.28X_2 - 12.78X_3 + 12.94X_4 \\ & - 8.44X_1X_2 + 6.31X_1X_4 \\ & + 18.15X_1^2 - 13.85X_2^2 \end{aligned} \tag{4}$$

$$\begin{aligned} CA = {} & 26.07 - 9.89X_1 + 2.17X_2 + 4.33X_3 - 2.33X_4 \\ & - 1.63X_1X_2 - 1.63X_1X_3 + 1.63X_1X_4 \\ & + 9.08X_1^2 + 7.58X_2^2 \end{aligned} \tag{5}$$

Now, all the p-values are less than the significance level of 0.05.

The predicted *versus* actual plots for AFD and CA of electrospun fiber mat are shown in Figures 7 and 8 respectively. Actual values are the measured response data for a particular run, and the predicted values evaluated from the model. These plots have determination coefficient (R^2) of 0.9640 and 0.9683 for AFD and CA respectively. It can be observed that experimental values are in good agreement with the predicted values.

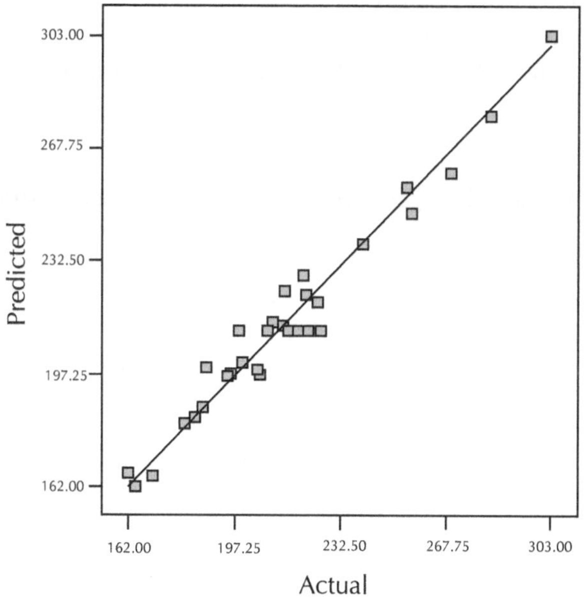

FIGURE 7 The predicted versus actual plot for AFD of electrospun fiber mat.

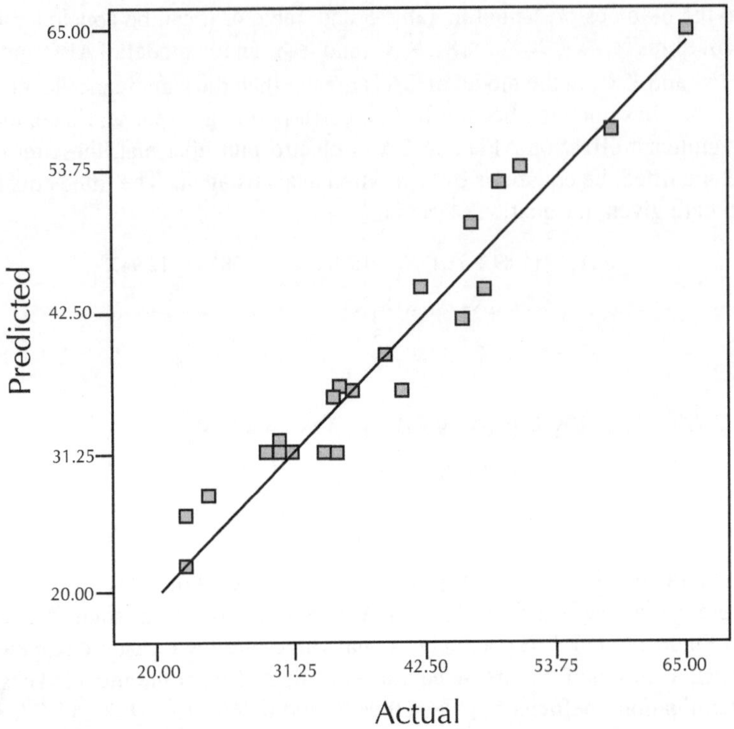

FIGURE 8 The predicted versus actual plot for CA of electrospun fiber mat.

16.3.3 Response Surfaces for Average Fiber Diameter (AFD)

16.3.3.1 Solution Concentration

Generally, a minimum solution concentration is required to obtain uniform fibers from electrospinning. This concentration, polymer chain entanglements are insufficient, and a mixture of beads and fibers is obtained. As the solution concentration increases, the shape of the beads changes from spherical to spindle-like [17].

In the present study, the AFD increased with solution concentration as shown in Figure 9(a), (b), and (c) that is agreement with previous observations [28-30]. Figure 9(a) shows the effect of changing solution concentration and applied voltage at fixed spinning distance and volume flow rate. It can be seen the increase in AFD with increase in solution concentration at any given voltage. As shown in Figure 9(b), no interaction was observed between solution concentration and spinning distance. This means that the function of solution concentration was independent from spinning distance for AFD. The effect of solution concentration on AFD was influenced by volume flow rate (Figure 9(c)) and this agrees the presence of the term X_1X_4 in the model of AFD.

16.3.3.2 Applied Voltage

Figure 9(a), (d), and (e) show the effect of applied voltage on AFD. In this work, the AFD suffer an increase followed by a decrease with increasing the applied voltage.

The surface plot in Figure 9(a) indicated that there was a considerable interaction between applied voltage and solution concentration and this is in agreement with the presence of the term X_1X_2 in the model of AFD. In the present study, applied voltage influenced AFD regardless of spinning distance and volume flow rate as shown in Figure 9(d) and (e). The absence of X_2X_3 and X_2X_4 in the model of AFD proves this observation.

16.3.3.3 Tip to Collector Distance

The tip to collector distance was found to be another important processing parameter as it influences the solvent evaporating rate and deposition time as well as electrostatic field strength. Figure 9(b), (d), and (f) represents the decrease in AFD with spinning distance. It was founded that spinning distance influences AFD independent from solution concentration, applied voltage and volume flow rate. This agrees the absence of X_1X_3, X_2X_3, and X_3X_4 in the model of AFD.

16.3.3.4 Volume Flow Rate

The effect of volume flow rate on AFD is shown in Figure 9(c), (e), and (f) and indicated the increase in AFD with volume flow rate. Figure 9(c) shows the surface plot of interaction between volume flow rate and solution concentration. It can be seen that at fixed applied voltage and spinning distance, the increase in volume flow rate and solution concentration result the higher AFD. As depicted in Figure 9(e) and (f), the effect of volume flow rate on AFD was independent from applied voltage and spinning distance. This observation confirms the absence of X_2X_4 and X_3X_4 in the model of AFD.

16.3.4 Response Surfaces for Contact Angle (CA) of Electrospun Fiber Mat

16.3.4.1 Solution Concentration

Figure 10(a), (b), and (c) show the effect of solution concentration on CA of electrospun fiber mat. In this work, the CA of electrospun fiber mat decrease with increasing the solution concentration.

Figure 10(a) shows the surface plot of interaction between solution concentration and applied voltage. It is obvious that at fixed spinning distance and volume flow rate, the increase in applied voltage and decrease in solution concentration result the higher CA. As shown in Figure 10(b), there was a considerable interaction between solution concentration and spinning distance and this is in agreement with the presence of the term X_1X_3 in the model of CA. The surface plot in Figure 10(c) shows the interaction between solution concentration and volume flow rate at fixed applied voltage and spinning distance. It can be seen that at any given flow rate, CA of electrospun fiber mat will increase as solution concentration decreases.

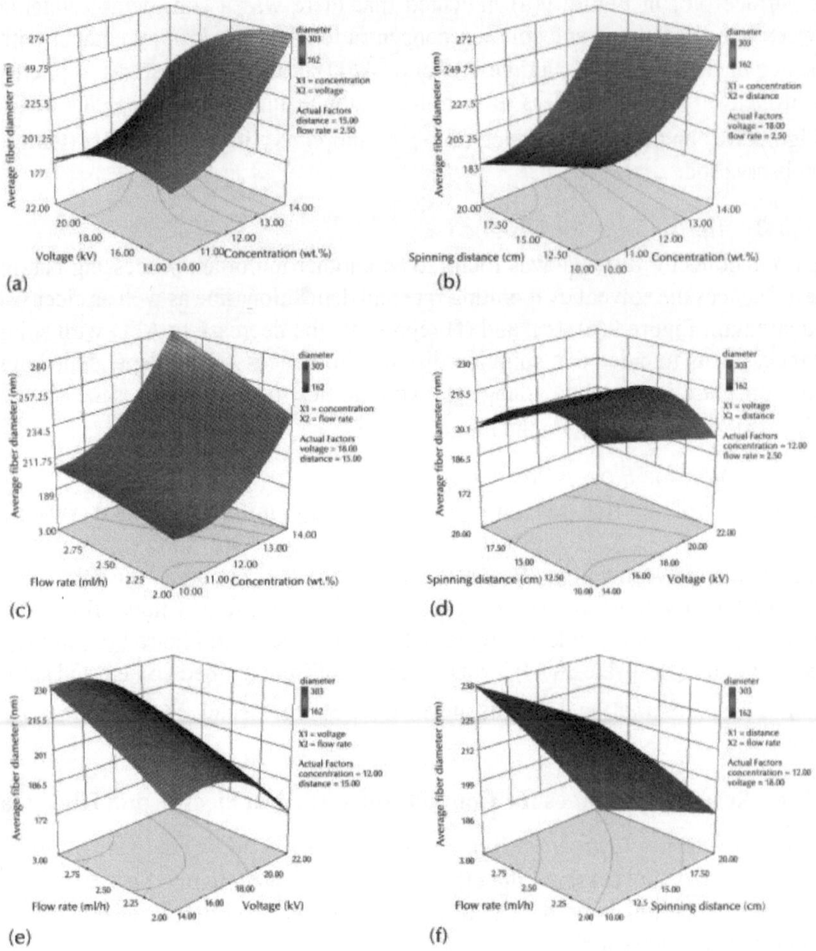

FIGURE 9 Response surfaces plot for AFD showing the effect of: (a) solution concentration and applied voltage, (b) solution concentration and spinning distance, (c) solution concentration and volume flow rate, (d) applied voltage and spinning distance, (e) applied voltage and flow rate, and (f) spinning distance and volume flow rate.

16.3.4.2 Applied Voltage

The effect of applied voltage on CA of electrospun fiber mat is shown in Figure 10(a), (d), and (e). It can be seen that the CA suffer a decrease followed by an increase with increasing the applied voltage. As depicted in Figure 10(a), the impact of applied voltage on CA of electrospun fiber mat will change at different solution concentration. Figure 10 (d) shows that there was no combined effect between applied voltage and spinning distance. Also no interaction was observed between applied voltage and volume flow rate (Figure 10 (e)). Therefore, applied voltage had interaction with solution concentration which had been confirmed by the existence of the term X_1X_2 in the model of CA.

16.3.4.3 Tip to Collector Distance

The impact of spinning distance on CA of electrospun fiber mat is illustrated in Figure 10(b), (d), and (f). Increasing the spinning distance causes the CA of electrospun fiber mat to increase. As demonstrated in Figure 10(b), low solution concentration cause the increase in CA of electrospun fiber mat at large spinning distance. Spinning distance affected CA of electrospun fiber mat regardless of applied voltage and volume flow rate (Figure 10(d) and (f)) as could be concluded from the model of CA. This means that no interaction exists between these variables.

16.3.4.4 Volume Flow Rate

The surface plot in Figure 10(c), (e), and (f) represented the effect of volume flow rate on CA of electrospun fiber mat. Figure 10(c) shows the interaction between volume flow rate and solution concentration. As illustrated in Figure 10(e) and (f), the effect of volume flow rate on CA of electrospun fiber mat was independent from applied voltage and spinning distance.

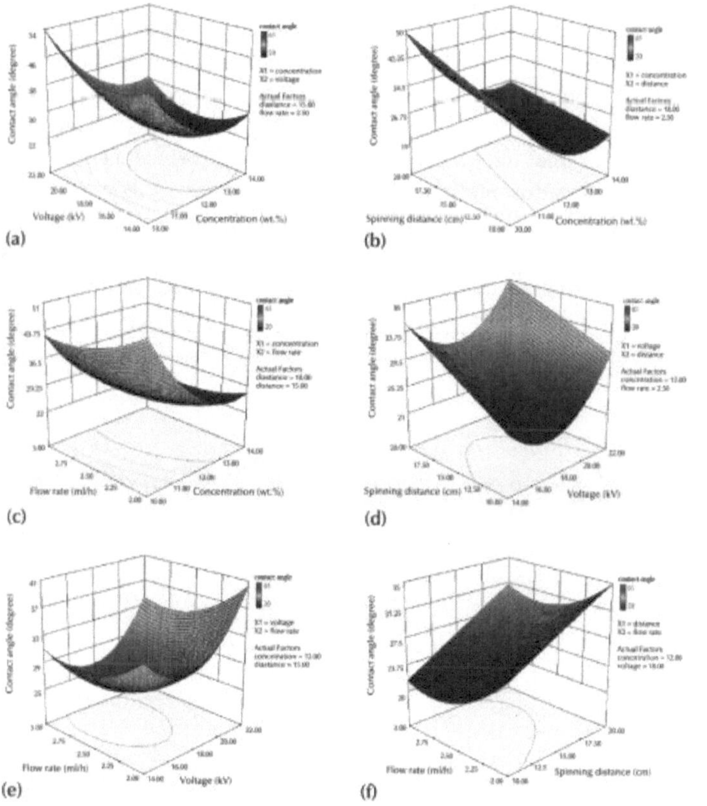

FIGURE 10 Response surfaces plot for CA of electrospun fiber mat showing the effect of: (a) solution concentration and applied voltage, (b) solution concentration and spinning distance, (c) solution concentration and volume flow rate, (d) applied voltage and spinning distance, (e) applied voltage and flow rate, and (f) spinning distances and volume flow rate.

16.3.5 Determination of Optimal Conditions

The optimal conditions were established by desirability. Independent variables namely solution concentration, applied voltage, spinning distance, and volume flow rate were set in range and dependent variable (contact angle), was fixed at minimum. The optimal conditions in the tested range for minimum CA of electrospun fiber mat are shown in Table 5.

TABLE 5 Optimum values of the process parameters for minimum CA of electrospun fiber mat.

Parameter	Optimum value
Solution concentration (wt. %)	13.2
Applied voltage (kV)	16.5
Spinning distance (cm)	10.6
Volume flow rate (ml/hr)	2.5

16.3.6 Relationship Between AFD and CA of Electrospun Fiber Mat

The wettability of surface controlled both by the surface chemistry and surface roughness. The morphology and structure of electrospun fiber mat, such as the nanoscale fibers and interfibrillar distance, increases the surface roughness as well as the fraction of contact area of droplet with the air trapped between fibers. It is proved that the CA can provide valuable information about surface roughness. Figure 11 shows that the variation of CA with AFD. The CA is observed to decrease with the increase in AFD, which is in good agreement with other report [32]. It can be concluded that the thinner fibers, due to their high surface roughness, have higher CA than the thicker fibers.

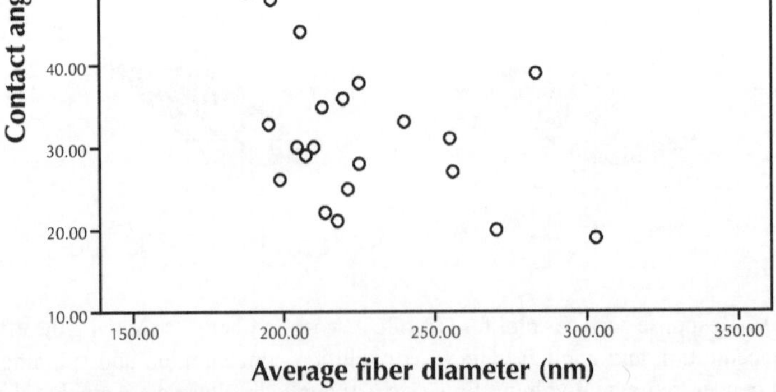

FIGURE 11 Variation of CA with AFD.

16.4 CONCLUSION

In this work, the effects of four electrospinning parameters (solution concentration, applied voltage, tip to collector distance, and volume flow rate) on AFD and CA of PAN nanofiber mat were investigated using RSM. The CCD was used to establish a quantitative relationship between factors and corresponding responses. The results showed that polymer solution concentration was the most significant factor impacting the AFD and CA of electrospun fiber mat. The RSM was successfully applied to find out the optimum level of the factors. The optimum solution concentration, applied voltage, spinning distance, and flow rate were found to be 13.2 wt.%, 16.5 kV, 10.6 cm, and 2.5 ml/hr, respectively, for minimum CA of electrospun fiber mat. The values obtained from the model were in good agreement with the experimental values and a good determination coefficient (R^2) of 0.9640 and 0.9683 was obtained for AFD and CA respectively. This study also suggests that the thin fibers exhibit high surface roughness as well as high CA.

KEYWORDS

- **Average fiber diameter**
- **Central composite design**
- **Contact angle**
- **Dimethylformamide**
- **Nanofibers**
- **Polyacrylonitrile**

REFERENCES

1. Shams Nateri, A. and Hasanzadeh, M. *J. Comput. Theor. Nanosci.*, **6**, 1542–1545 (2009).
2. Kilic, A., Oruc, F., and Demir, A. *Text. Res. J.*, **78**, 532–539 (2008).
3. Dotti, F., Varesano, A., Montarsolo, A., Aluigi, A., Tonin C., and Mazzuchetti, G. *J. Ind. Text.*, **37**, 151–162 (2007).
4. Lu, Y., Jiang, H., Tu, K., and Wang, L. *Acta Biomater.*, **5**, 1562–1574 (2009).
5. Lu, H., Chen, W., Xing, Y., Ying D., and Jiang, B. *J. Bioact. Compat. Pol.*, **24**, 158–168 (2009).
6. Nisbet, D. R., Forsythe, J. S., Shen, W., Finkelstein, D. I., and Horne, M. K. *J. Biomater. Appl.*, **24**, 7–29 (2009).
7. Ma, Z., Kotaki, M., Inai, R., and Ramakrishna, S. *Tissue Eng.*, **11**, 101–109 (2005).
8. Hong, K. H. *Polym. Eng. Sci.*, **47**, 43–49 (2007).
9. Zhang, W. and Pintauro, P. N. *Chem. Sus. Chem.*, **4**, 1753–1757 (2011).
10. Lee, S. and Obendorf, S. K. *Text. Res. J.*, **77**, 696–702 (2007).
11. Reneker, D. H. and Chun, I. *Nanotechnology*, **7**, 216–223 (1996).
12. Shin, Y. M., Hohman, M. M., Brenner, M. P., and Rutledge, G. C. *Polymer*, **42**, 9955–9967 (2001).
13. Reneker, D. H., Yarin, A. L., Fong H., and Koombhongse, S. *J. Appl. Phys.*, **87**, 4531–4547 (2000).
14. [14]. Zhang, S., Shim, W. S., and Kim, J. *Mater. Design*, **30**, 3659–3666 (2009).
15. Yördem, O. S., Papila, M., and Menceloğlu, Y. Z. *Mater. Design*, **29**, 34–44 (2008).
16. Chronakis, I. S. *J. Mater. Process. Tech.*, **167**, 283–293 (2005).
17. Haghi, A. K. and Akbari, M. *Phys. Status. Solidi. A.*, **204**, 1830–1834 (2007).

18. Zhu, M., Zuo, W., Yu, H., Yang, W., and Chen, Y. *J. Mater. Sci.*, **41**, 3793–3797 (2006).
19. Ding, B., Kim, H., Lee, S., Lee, D., and Choi, K. *Fiber Polym.*, **3**, 73–79 (2002).
20. Kanafchian, M., Valizadeh, M., and Haghi, A. K. *Korean J. Chem. Eng.*, **28**, 445–448 (2011).
21. Miwa, M., Nakajima, A., Fujishima, A., Hashimoto, K., and Watanabe, T. *Langmuir*, **16**, 5754–5760 (2000).
22. Öner, D. and McCarthy, T. J. *Langmuir*, **16**, 7777–7782 (2000).
23. Abdelsalam, M. E., Bartlett, P. N., Kelf, T., and Baumberg, J. *Langmuir*, **21**, 1753–1757 (2005).
24. Nakajima, A., Hashimoto, K., Watanabe, T., Takai, K., Yamauchi, G., and Fujishima, A. *Langmuir*, **16**, 7044–7047 (2000).
25. Zhong, W., Liu, S., Chen, X., Wang, Y., and Yang, W. *Macromolecules*, **39**, 3224–3230 (2006).
26. Myers, R. H., Montgomery, D. C., and Anderson-cook, C. M. *Response surface methodology: Process and product optimization using designed experiments*, 3rd ed., John Wiley and Sons, USA (2009).
27. Gu, S. Y., Ren, J., and Vancso, G. J. *Eur. Polym. J.*, **41**, 2559–2568 (2005).
28. Zhang, S., Shim, W. S., and Kim, J. *Mater. Design*, **30**, 3659–3666 (2009).
29. Zhang, C., Yuan, X., Wu, L., Han, Y., and Sheng, J. *Eur. Polym. J.*, **41**, 423–432 (2005).
30. Ziabari, M., Mottaghitalab, V., and Haghi, A. K. In *Nanofibers: Fabrication, Performance, and Applications*. By W. N. Chang (Ed.), Nova Science Publishers, USA, Chapter 4, pp. 153–182 (2009).
31. Ramakrishna, S., Fujihara, K., Teo, W. E., Lim, T. C., and Ma, Z. *An Introduction to Electrospinning and Nanofibers*, National University of Singapore, World Scientific Publishing, Singapore (2005).
32. Ma, M., Mao, Y., Gupta, M., Gleason, K. K., and Rutledge, G. C. *Macromolecules*, **38**, 9742–9748 (2005).

17 Mathematical Model of Nanofragment Cross-linked Polymers

N. V. Ulitin, R. Ya. Deberdeev, E. V. Samarin, T. R. Deberdeev, I. I. Nasyrov, and G. E. Zaikov

CONTENTS

17.1 Introduction 169
17.2 Topological Structure of Epoxy Amine Polymers 170
17.3 Theoretical Estimation of Mathematical Model of Cross-Linked Polymers Compliance's Constants 171
17.3.1 Glass Transition Temperature and Coefficients of Thermal Expansion 171
17.3.2 Hyperelastic State's Constant 173
17.3.3 Weighting Coefficient 176
17.3.4 Spectrum of A-Relaxation Times and Parameters 179
17.4 Numerical Experiment and Realized Thermomechanical Behavior of Experimental Objects 179
17.5 Conclusion 180
Keywords 180
References 180

17.1 INTRODUCTION

Information about the quantity of cross-linked polymers strain properties in their one or another physical state with known assessment of mechanical tensions is nesessary, for choice of the optimal cross-linked polymer matrix and high strength fiberglass products. In the network of search solutions of this problem was developed and confirmed by experiments with cross-linked epoxy amine polymers ,a new mathematical account of nanofragment cross-linked polymers pliability [1]. A purpose of this research is the theoretical estimation of constants, for this mathematical model of the same experimental objects and the display of a computer is simulation of their deformative behavior for the dersived values reason.

17.2 TOPOLOGICAL STRUCTURE OF EPOXY AMINE POLYMERS

For modeling of experimental objects topological structure write their repeated fragments of meshes in the one for all forms (Figure 1). In the topology mark out following elementary units: nitrogen atom surrounded by three methylene groups that are elastic effective mesh points (designated their general quantity in meshes' structure = N_{3f}, mole); tetramethylene units (N_{σ}, mole); fragments DGEBA[1] (N_{π}, mole); fragments AH (1-aminohexane) with two methylene groups close to nitrogen atom (N_{2f}, mole).

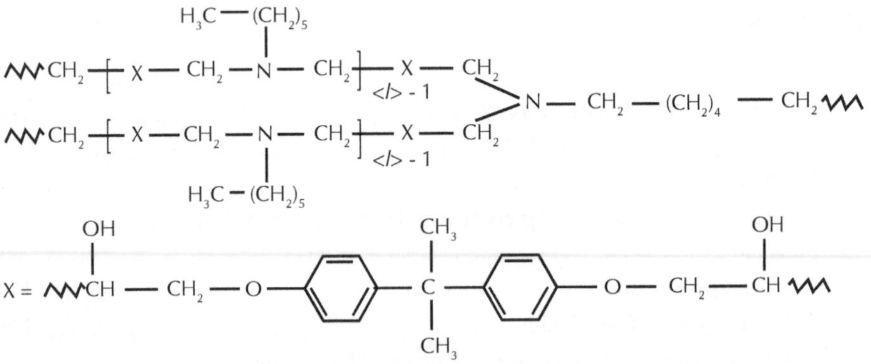

FIGURE 1 General repeated fragment of topological structure for all experimental objects.

For reasons of stehiometry it possible to write down following expressions.

$$N_{2f} = xn(HMDA),\ N_{3f} = 2n(HMDA),\ N_{\sigma} = n(HMDA),\ N_{\pi} = (2+x)n(HMDA)$$

$$N_{tot} = N_{2f} + N_{3f} + N_{\sigma} + N_{\pi} = (5+2x)n(HMDA)$$

For the purpose of further use in accounts and interpretation of derived data express amounts of elastic effective mesh points and fragments of DGEBA as static parameters.

$$n_{3f} = N_{3f} / N_{tot} = 2/(5+2x),\quad n_{\pi} = N_{\pi} / N_{tot} = (2+x)/(5+2x)$$

[1]Splits, and their signs are given in [1].

Key parameter of meshes' structure, which describes their composition and structure of repeated fragments accurate within defects is the number average degree of polymerization inter nodal chain $<l> = n_{\pi} / n_{3f} = 1 + 0.5x$.

17.3 THEORETICAL ESTIMATION OF MATHEMATICAL MODEL OF CROSS-LINKED POLYMERS COMPLIANCE'S CONSTANTS

17.3.1 Glass Transition Temperature and Coefficients of Thermal Expansion

Describe glass transition temperature (T_g, K) of experimental objects according to the formula [2]:

$$T_g = \left(\sum_i \Delta V_i\right)_{r.f.} \Big/ \left(\left(\sum_i a_i \Delta V_i + \sum_j b_j\right)_l + \left(\sum_i K_i \Delta V_i\right)_y\right) \quad (1)$$

where $\left(\sum_i \Delta V_i\right)_{r.f.}$ = Van der Waals's volume of repeated fragment, Å^3; a_i, b_j = increments, which characterize energy of poor dispersion and strong specific intermolecular interactions, respectively, K^{-1} и Å^3/K; ΔV_i = Van der Waals's volume of *i*th atom, Å^3; $\left(\sum_i a_i \Delta V_i + \sum_j b_j\right)_l$ = increments set for linear chains, which are part of the repeated fragment, Å^3/K; K_i = parameter, which is input for atoms of mesh point and is depended on energy of chemical bonds, , K^{-1}; $\left(\sum_i K_i \Delta V_i\right)_y$ = increments set for mesh point, Å^3/K.

Describe Van der Waals's volume of experimental objects' repeated fragment as [2]:

$$\left(\sum_i \Delta V_i\right)_{r.f.} = (<l> - 1)\left(\sum_i \Delta V_i\right)_{l,1} + \left(\sum_i \Delta V_i\right)^*_{l,1} + \frac{1}{2}\left(\sum_i \Delta V_i\right)_{l,2} + \left(\sum_i \Delta V_i\right)_y$$

where $\left(\sum_i \Delta V_i\right)_{l,1}$ = Van der Waals's volume of repeated fragment of sewn together chains' linear fragments, Å^3; $\left(\sum_i \Delta V_i\right)^*_{l,1}$ = the Van der Waals's volume of repeated link of sewn together chains' linear fragments, which is joined to mesh point, Å^3; $\left(\sum_i \Delta V_i\right)_{l,2}$= Van der Waals's volume of linear fragment's repeated link sewn together bridges, Å^3; $\left(\sum_i \Delta V_i\right)_y$ = the Van der Waals's volume of mesh point, Å^3.

$$\left(\sum_i \Delta V_i\right)_{r.f.} = 458.9 <l> - 57.4 \text{ Å}^3$$

So long as, the oxygen atom in fragment -O-C_6H_4- causes shift of e-density in aromatic ring and plays the role of negative part of dipole in intermolecular dipole-dipole interaction, it is participation in the last, was allowed with multiplier 0.5. Then:

$$\left(\sum_i a_i \Delta V_i + \sum_j b_j\right)_l = (1437.54 < l > - 478.76)10^{-3}\ \mathring{A}^3/K$$

Increments set of mesh point $\left(\sum_i K_i \Delta V_i\right) = 82.179 \cdot 10^{-3}\ \mathring{A}^3/K$.

The final expression after substitution of all considered components in Equation (1) looks as following form (calculated values are given in Table 1):

$$T_g = ((458.9 < l > - 57.4)/(1437.54 < l > - 396.581)) \cdot 10^3 - 273.15,\ ^\circ C$$

Define coefficient of thermal expansion in glassy state (α_g, K^{-1}, or degree^{-1}) [2]:

$$\alpha_g = \left(\left(\sum_i \alpha_i \Delta V_i + \sum_j \beta_j\right)_l + \left(\sum_i K_i \Delta V_i\right)_y\right) \Big/ \left(\sum_i \Delta V_i\right)_{r.f.} \qquad (2)$$

where α_i = the partial coefficient of volumetrically thermal expansion due to poor dispersion interaction *i*th atom with next atoms, K^{-1}; β_j = increment reflecting contribution of different types of intermolecular interactions to coefficient of thermal expansion, $\mathring{A}^3/K$.

On the same principles of that in account $\left(\sum_i a_i \Delta V_i + \sum_j b_j\right)_l$ have

$$\left(\sum_i \alpha_i \Delta V_i + \sum_j \beta_j\right)_l = (137.615 < l > - 45.855)10^{-3}\ \mathring{A}^3/K$$

The final calculating formula derived from Equation (2):

$$\alpha_g = ((137.615 < l > + 36.324)/(458.9 < l > - 57.4)) \cdot 10^{-3} \qquad (3)$$

It is possible to calculate the coefficient of thermal expansion in hyperelastic state according to the Simkhi-Boyer's equation for cross-linked polymers:

$$\alpha_\infty = (0.106/T_g) + \alpha_g \qquad (4)$$

Values of coefficients of thermal expansion in glassy and hyperelastic states are defined respectively to the Equations (3) and (4) were average out for all experimental objects 3.9 × 10^{-4} degree^{-1} и 6.8 × 10^{-4} degree $^{-1}$ and were close by experimental data — 4.3 × 10^{-4} degree^{-1} and 7.0 × 10^{-4} degree^{-1}, — relative discrepancy of theoretical value as compared with experimental value were 9 and 3%, respectively.

TABLE 1 Structural glass transition temperature (Theoretical and experimental values).

$x = n(AH)/n(HMDA)$	$<l>$	T_g,°C theor.	T_g,°C exper. [1]	ε, %*
0.0	1.00	113	109	4
0.5	1.25	95	99	4
1.0	1.50	85	88	3
1.5	1.75	79	77	3
2.0	2.00	74	71	4

* Hereinafter relative discrepancy

$\varepsilon = \left|(\text{experimental value-calculated value})/\text{experimental value}\right| \cdot 100, \%$.

17.3.2 Hyperelastic State's Constant

A theoretical value of the hyperelastic state's constant for experimental objects estimate according to the Equation [3]:

$$A_\infty = 1\Big/\left(\frac{f}{2}C_{tot}n_{3f}RF\right), \text{K/Mpa} \tag{5}$$

where f = functionality of elastic effective mesh points (f = 3 for use objects); C_{tot} = concentration of elementary links at the structural glass transition temperature, mole/cm^3; R = gas constant, J/(mole×K); F = the front factor.

According to formulation

$$C_{tot} = N_{tot}\Big/V_{tot}\left(T_g\right) \tag{6}$$

where N_{tot} = general amount of elementary links, mole; $V_{tot}\left(T_g\right)$ = volume, which elementary links occupy at the structural glass transition temperature, cm^3/mole.

A volume, which elementary links occupy at the structural glass transition temperature, is equated to

$$V_{tot}(T_g) = N_\sigma V_\sigma(T_g) + N_{3f}V_{3f}(T_g) + N_{2f}V_{2f}(T_g) + N_\pi V_\pi(T_g)$$

where $V_\sigma(T_g)$, $V_{3f}(T_g)$, $V_{2f}(T_g)$, and $V_\pi(T_g)$ = molar volumes of respective elementary links at the structural glass transition temperature, cm³/mole.

Express molar volumes of respective elementary links Through molar masses and substitute them to Equation (6):

$$C_{tot} = \left((5+2x)d(T_g)\right)/\left(M_\sigma + 2M_{3f} + xM_{2f} + (2+x)M_\pi\right) \tag{7}$$

where $d(T_g)$ = density of experimental objects at their structural glass transition temperatures, g/cm³; M_σ, M_{3f}, M_{2f}, and M_π = molar masses of respective elementary links, g/mole.

Find components of Equation (7). Temperature dependence of density for cross-linked polymers looks like this [2]:

$$d(T) = k_g M_{r.f.} \Big/ \left(10^{-24}\left[1+\alpha(T-T_g)\right]N_A\left(\sum_i \Delta V_i\right)_{r.f.}\right), \quad \alpha = \begin{cases}\alpha_g, & T<T_g\\ \alpha_\infty, & T>T_g\end{cases}$$

where $d(T)$ = density of polymer at the temperature T, g/cm³; $k_g \approx 0.681$ = universal value of the molecular packing's coefficient for cross-linked polymers at T_g; $M_{r.f.}$ = molar mass of repeated mesh's fragment, g/mole; 10^{-24} = conversion factor Å³ to cm³; $N_A = 6.023 \cdot 10^{23}$ mole^{-1} = Avogadro constant.

Taking into account that $M_{r.f.} = 441 < l > - 43$ $g/mole$, transform Equation (7) to the form:

$$d(T) = 1.13(441<l>-43)\Big/\left(\left[1+\alpha(T-T_g)\right](458.9<l>-57.4)\right), \alpha = \begin{cases}\alpha_g, & T<T_g\\ \alpha_\infty, & T>T_g\end{cases} \tag{8}$$

Equation (8) makes it possible to determine values of density at every temperature (for check: $d_{theor.}(25) = 1.14$ g/sm^3, and $d_{exp.}(25) = 1.11$ g/sm^3, difference is 3%).

Molar masses of relevant elementary links:

$$M_\sigma = \underbrace{4M_C + 8M_H}_{-(CH_2)_4-} = 56 \text{ g/mole}; \qquad M_{3f} = \underbrace{M_N + 3M_C + 6M_H}_{N(CH_2)_3-} = 56 \text{ g/mole};$$

$$M_{2f} = \underbrace{M_N + 8M_C + 17M_H}_{-(CH_2)_2N-CH_2-(CH_2)_4-CH_3} = 127 \text{ g/mole}$$

$$M_\pi = 2\left(\underbrace{M_C + M_O + 2M_H}_{-CH-OH} + \underbrace{M_C + 2M_H}_{-CH_2-} + \underbrace{M_O}_{-O-} + \underbrace{6M_C + 4M_H}_{-C_6H_4-}\right) + \underbrace{3M_C + 6M_H}_{>C(CH_3)_2} =$$

$= 314 \ g/\text{mole}$

Substitute in Equation (7) molar masses of relevant elementary links and temperature function of density for derived values and expression (8) respectively:

$$C_{tot} = 1.13(5+2x)(441<l>-43)/((796+441x)(458.9<l>-57.4)) \quad (9)$$

In view of the fact that front factor's values for cross-linked polymers grow linearly with increase of the mesh points' concentration within limits 0.650.85 [3], will find the following calculating equation:

$$F = 1.125n_{3f} + 0.4 \quad (10)$$

Substitution of expressions (9) and (10) to the Equation (5) will give:

$$A_\infty = (796+441x)(458.9<l>-57.4)/(1.695(1.125n_{3f}+0.4)R(5+2x)(441<l>-43)n_{3f}) \quad (11)$$

Values F and A_∞, defined according to Equations (10) and (11) are given in Table 2.

TABLE 2 The Hyperelastic state's constant and front factor (Theoretical and experimental values)

x	$<l>$	n_{3f}	A_∞, *K/MPa*		ε,%	*F*		ε,%
			calc.	*exper. [1]*		*calc.*	*exper.*	
0.0	1.00	0.4000	33.5	35.0	4	0.8500	0.8540	5
0.5	1.25	0.3333	47.3	53.0	11	0.7750	0.7164	8
1.0	1.50	0.2857	62.1	69.3	10	0.7214	0.6636	9
1.5	1.75	0.2500	77.7	81.6	5	0.6813	0.6608	3
2.0	2.00	0.2222	93.9	93.5	4	0.6500	0.6613	2

It is shown that experimental data are satisfactorily described by theoretical values A_∞ (Figure 2(a)). In accordance with Table 2 and Figure 2(b), analytical expression (10) reproduces experimental values of front factor in general.

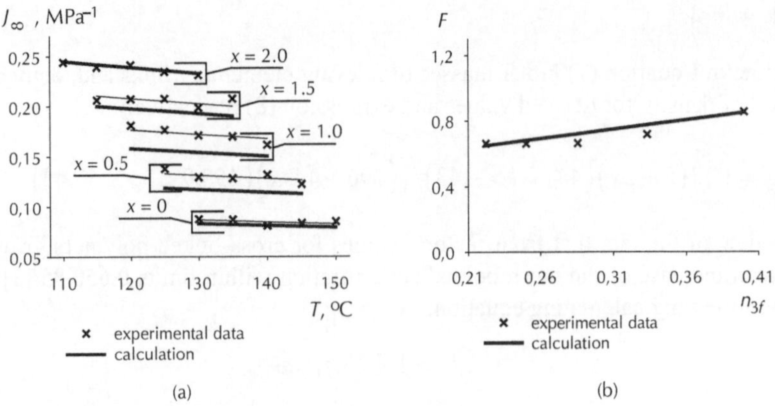

FIGURE 2 Temperature dependence of the equilibrium shear pliability (a); Front factor's dependence of n_{3f} (b).

17.3.3 Weighting Coefficient

Weighting coefficient $w_{J,\beta}$ will appreciate by the formula [1]:

$$w_{J,\beta} = \left(J_{\beta,\infty}T\right)/A_\infty \tag{12}$$

where $J_{\beta,\infty}$ = glass transition state's shear pliability ($T < T_g - 15\ ^\circ\text{C}$), MPa-1. Temperature of experiment was T = 25°C, therefore, will find theoretical value $J_{\beta,\infty}$ at this temperature. With this purpose will consider difference between main values of dielectric permeability's tensor [3, 4].

$$\varepsilon_1 - \varepsilon_2 = (1/2)Ja_1\left(\tau_1 - \tau_2\right) \tag{13}$$

where J = relaxational operator of shear pliability, MPa-1; a_1 = any coefficient; τ_1, τ_2 = main values of transverse strain, MPa.

Considering equality of inductive capacity and square refractive index $\varepsilon \approx n^2$ [4], receive

$$\varepsilon_1 - \varepsilon_2 = n_1^2 - n_2^2 = \left(n_1 - n_2\right)\left(n_1 + n_2\right) = \left\{\begin{matrix} n_1 - n_2 = \Delta n \\ \left(n_1 + n_2\right)/2 \approx n_0 \quad [5] \end{matrix}\right\} = 2n_0\Delta n \tag{14}$$

where Δn = strain electromagnetic anisotropy; n_0 = refractive index of unstrained polymer object in glassy state.

After substitution of expression (14) in Equation (13)

$$\Delta n = \left(Ja_1/\left(4n_0\right)\right)\Delta\tau \tag{15}$$

In fact Equation (15) is equation of Bruster-Vertgame's Law [2], then

$$J_{\beta,\infty} = 4C_{\beta,\infty} n_0 / a_1 \tag{16}$$

where $C_{\beta,\infty}$ = strain electromagnetic susceptibility of glassy state (MPa^{-1}), which at 25°C it is possible to determine by expression [2]:

$$C_{\beta,\infty} = \left(\left(\sum_i C_i \right)_{r.f.} \Big/ \left(10^{-24} N_A \left(\sum_i V_i \right)_{r.f.} \right) \right) + \mathrm{P} \tag{17}$$

In Equation (17): $\left(\sum_i C_i \right)_{r.f.}$ = increments set of repeated mesh's fragment, which reproduces summary contribution of each atom and type of intermolecular interaction in strain electromagnetic susceptibility of glassy state, $cm^3/(MPa \cdot mole)$; $P = 3.61387 \cdot 10^{-4}$ MPa^{-1}— universal parameter.

Arising under the influence of load in glassy state polymers electromagnetic anisotropy is caused only by deformation of valence angles and change of interatomic distances, further in calculation $\left(\sum_i C_i \right)_{r.f.}$ increment C_h for fragment >CH-OH was inputted with multiplier 0.5. Then

$$\left(\sum_i C_i \right)_{r.f.} = (11.4065 - 86.8863 < l >) 10^{-3} \ \mathrm{sm}^3 / (\mathrm{MPa} \cdot \mathrm{mole})$$

From Equation (17) will get final calculating equation:

$$C_{\beta,\infty} = \left(\left((11.4065 - 86.8863 < l >) \cdot 10^{-3} \right) \Big/ \left(0.6023 (458.9 < l > - 57.4) \right) \right) + 3.61387 \cdot 10^{-4}$$

Relative difference between derived theoretical value $C_{\beta,\infty} = 4.9 \cdot 10^{-5}$ MPa^{-1} and experimental $5.3 \cdot 10^{-5}$ MPa^{-1} was 7%.

So as $\varepsilon \approx n^2$, refractive index at 25°C will find from following formula [2]:

$$\left((\varepsilon_0 - 1)/(\varepsilon_0 + 2) \right) \left(M_{r.f.} / d(T) \right) = \left(\sum_i (R_D)_i + \sum_j (\Delta R_D)_j \right)_{r.f.} \tag{18}$$

where $(R_D)_i$ = atomic refractions of atom by Eisenlor, $cm^3/mole$; $(\Delta R_D)_j$ = amendment to orientation of dipoles.

Substitute into Equation (18) the dependence of density from temperature and will express inductive capacity of unstressed polymer in glassy state:

$$\varepsilon_0 = \frac{10^{-24} \left[1 + \alpha_g (T - T_g) \right] N_A \left(\sum_i \Delta V_i \right)_{r.f.} + 2k_g \left(\sum_i (R_D)_i + \sum_j (\Delta R_D)_j \right)_{r.f.}}{10^{-24} \left[1 + \alpha_g (T - T_g) \right] N_A \left(\sum_i \Delta V_i \right)_{r.f.} - k_g \left(\sum_i (R_D)_i + \sum_j (\Delta R_D)_j \right)_{r.f.}} \tag{19}$$

Incoming in Equations (18) and (19) increments set $\left(\sum_i (R_D)_i + \sum_j (\Delta R_D)_j\right)_{r.f.}$ for usable experimental objects are equal to $131.476 < l > -14.954 \text{ sm}^3 / \text{mole}$.

After conversion final expression of refractive index at 25°C looks as

$$n_0 = \sqrt{\frac{0.6023\left[1+3.9\cdot10^{-4}\left(298-T_g\right)\right]\left(458.9<l>-57.4\right)+1.362\left(131.476<l>-14.954\right)}{0.6023\left[1+3.9\cdot10^{-4}\left(298-T_g\right)\right]\left(458.9<l>-57.4\right)-0.681\left(131.476<l>-14.954\right)}}$$

Equation for coefficient a_1 (glassy state), looks as:

$$a_1 = -6\left(\partial\delta\varepsilon/\partial T\right)\left(K/\alpha_g\right)$$

where $\partial\delta\varepsilon/\partial T$ = one of summands, which is decomposed the derivative of inductive capacity by temperature and charged with appearance of electromagnetic anisotropy in polymer when the impact of load, K^{-1}, or degree^{-1}; K = multiplier, value of which depends on the polymer's topology (for example, for linear polymers is universal value, not determined by the nature of polymer and is equal to 1).

$\partial\delta\varepsilon/\partial T$ makes, it possible to determine method of increments [2]:

$$\partial\delta\varepsilon/\partial T = \left(\sum_i \delta C_i \Delta V_i + \sum_i \delta C_{i,imi}\right)_{r.f.} \Big/ \left(\sum_i \Delta V_i\right)_{r.f.} \tag{20}$$

As with the putting into C_h for fragment >CH-OH and increment δC_h for it is multiplied by 0.5. Then

$$\left(\sum_i \delta C_i \Delta V_i + \sum_i \delta C_{i,imi}\right)_{r.f.} = \left(-146.6962<l>-5.309\right)10^{-6}\ \mathring{A}^3/\text{K}$$

For calculation K [5] was received next dependence:

$$K_{cr.}/K_{lin.} = u_{0,lin.}^{1.0}\Big/u_{0,cr.}^{1.0}$$

where $K_{cr.}$ = quantity of multiplier for cross-linked polymer; $K_{lin.}$ = quantity of multiplier for hypothetical linear polymer, which consists of repeated fragments of cross-linked polymer (in case of epoxy-amine polymers this multiplier is equal to similar multiplier for epoxide resin); $u_0^{1.0}$ = worth of stripe of relevant material by deformations. Average values in glassy state for epoxide resins is $u_{0,lin.}^{1.0} \approx 15.4\cdot10^{-3}$ and for cross-linked polymers on their basis $u_{0,cr.}^{1.0} \approx 7.0\cdot10^{-4}$ [5]. On the basis of these data by Equation (21) $K_{cr.} = 22$.

Final expression for a_1 is obtained after substitution of derived values $K_{cr.}$ and α_g and expression for $\partial\delta\varepsilon/\partial T$ in initial equation (formula is not shown because it is inconvenience). Derived by Equation (16) in the issue of substitution of parameters a_1 and n_0 in it for their expressions and averaged by all experimental objects a value of shear pliability at 25°C was equal to 2.5 × 10^{-3} MPa^{-1}, and it is respective experimental value — 2.6 × 10^{-3} MPa^{-1} (relative difference 4%). Define by Equation (12) and experimental values of weighting coefficient look in Table 3.

TABLE 3 Weighting coefficient (Theoretical and experimental values).

x	$< l >$	$w_{J,\beta}$		ε, %
		calc.	exper. [1]	
0.0 1.00		0.0220	0.0180	22
0.5	1.25	0.0157	0.0150	5
1.0	1.50	0.0120	0.0140	14
1.5	1.75	0.0096	0.0120	20
2.0	2.00	0.0080	0.0080	0

17.3.4 SPECTRUM OF α-RELAXATION TIMES AND PARAMETERS

For theoretical estimation of average **α**-relaxation times at different temperatures will employ derivative in research [1] and temperature dependence of **α**-relaxation times. Given the fact that cross-site chains of experimental objects have narrow molecular mass distribution, take average value $\Xi_{J,\alpha} = 0.5$ as a spectrum width.

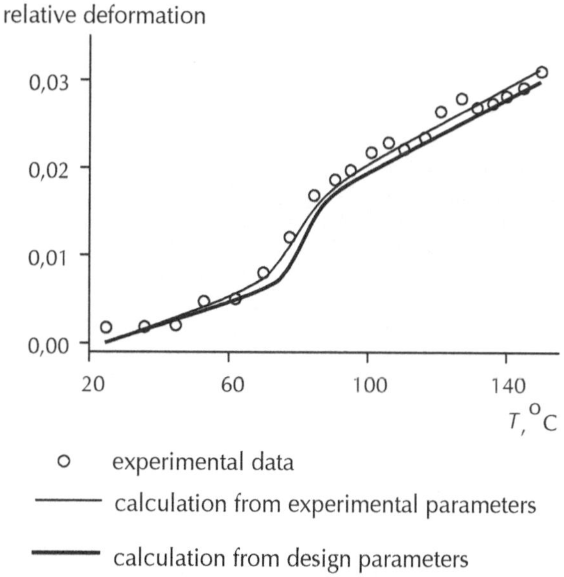

FIGURE 3 Thermomechanical curve for the experimental object with composition x = 2 (heating velocity 0.9°C/min, load 720 g).

17.4 NUMERICAL EXPERIMENT AND REALIZED THERMOMECHANICAL BEHAVIOR OF EXPERIMENTAL OBJECTS

As an example, showing that on the base of the developed model and put in it calculated theoretically values of constants, it is possible to carry out good conformed to experimental results numerical experiments of the evaluation of cross-linked polymers' deformative behavior, which progresses without destruction of their chemical structure, were predicted results of thermomechanical experiment of experimental objects (Figure 3).

17.5 CONCLUSION

Uniqueness of introduced mathematical model of cross-linked polymers' pliability lies in the fact that is constants may be determined not only by the results of short-term tests [1], but how it was shown in this chapter, may be estimated theoretically by the way of convergence to the known incremental calculating schemes. Exactly this peculiarity of mathematical formalism will be useful in the selection of it is base cross-linked polymer matrixes with necessary strain properties in specified ranges of temperatures and tensions for high strength fiberglass products.

KEYWORDS

- **Cross-linked polymer**
- **Glass transition temperature**
- **Hyperelastic state's constant**
- **Mesh point**
- **Weighting coefficient**

REFERENCES

1. Ulitin, N. V., Deberdeev, T. R., and Deberdeev, R. Ya. Viscoelastic compliance of cross-linked polymers with nanoscale cross-site chains, (2012).
2. Askadskii, A. A. and Kondrashenko, V. I. *Computer science of materials of polymers*. Vol.1. Atomic-molecular level. M. Scientific world, p. 544 (1999).
3. Bajenov, S. L., Berlin, A. A., Kulkov, A. A., and Oshmyan, V. G. *Polymer composite materials: Solidity and technology*. Dolgoprudnii, Intellect, p. 352 (2010).
4. Blythe, T. and Bloor, D. *Electrical properties of polymers*. Cambridge University Press, Cambridge, p. 492 (2005).
5. In *The photoelasticity's method vol. 3.* G. L Khesin(Ed.). M. Gostechizdat, **3** (1975).

18 Internal Mechanical Properties of the Plain Woven Polymeric Fabrics: An Analytical Approach

A. K. Haghi and G. E. Zaikov

CONTENTS

18.1 Introduction 181
18.2 Analytical Approach 182
18.3 Conclusion 184
Keywords 184
References 184

18.1 INTRODUCTION

Pullout test is a conventional and suitable method to investigate the effects of yarn properties and the structural characteristics of weave on fabric mechanical behavior. Frictional specifications of the fabric yarns influence the fabric strength and efficiency and its ability to absorb energy. This chapter is concerned with formulating an analytical model on yarn pullout force in plain-woven fabrics. The model can predict variations in the internal mechanical parameters of woven fabrics based on a force-balance analysis. These parameters are yarn-to-yarn friction coefficient, normal load at cross-overs, lateral forces, lateral strain, weave angle variations, and pullout force. These parameters were predicted using fabric deformation data, which were measured by image processing method during a yarn pullout test and the information of weave angle of fabric, its modulus and density, which were obtained experimentally. Yarn pullout force was calculated through Amonton's friction law to evaluate the efficiency of the presented model. This study demonstrates that the force balance model is correlated quantitatively with the experimental yarn pullout results.

In this chapter an attempt is made to introduce a mechanistic model using the force-balance analysis, and to calculate some fabric mechanical parameters that is yarn-to-yarn friction coefficient, normal load, lateral forces, weave angle variations, and the yarn pullout force. These parameters are useful in determination of the internal mechanical characteristics of the woven fabrics, and could be employed to simulate yarn pullout behavior of such fabrics.

To evaluate the competency of the theoretical controversies, suggested equations of the force-balance model have been used to estimate the amount of the force required to pullout a single yarn from the weave in a yarn pullout test, the efficiency of the force-balance model has been examined by calculation of the correlation coefficient between experimental data and that of the model results.

18.2 ANALYTICAL APPROACH

In yarn pullout test a rectangular fabric sample (for simplicity a plain pattern is considered here) is clamped to the two opposite side of a U frame. The pullout force is applied to a single yarn (for instance at the middle) with a specific free end length. The yarn is pulled out in 3-steps, the detail of which will be given in the discussion.

As is readily inferred, there are two kinds of stresses in such an experiment:

I. Shear stresses happening in the fabric plane parallel to the pulled yarn direction.
II. Tensile stresses occurring (a) in the pulled yarn and (b) in the longitudinal direction of the fabric, parallel to the opposed yarn direction.

First of all the structure of the weave and its variations during the test has been considered. During the test, the weave angle θ decreases to θ' due to the extension. If the height and the volume variations are considered to be negligible (this assumption could be reinforced by selecting a densely weaved fabric), θ' would be calculated as follows:

$$V = V' \Rightarrow t.h.L = t'.h'.L'$$
$$Or\ t.h.p = t'.h'.p' = h(t - \Delta t)(p + \Delta p)$$
$$t' = t\frac{p}{p'} \Rightarrow \tan\theta' = \frac{t'}{2p'} = \frac{t.p}{2p'^2} \quad (1)$$

$\tan\theta = \dfrac{t}{p}$, and on the other hand during pullout $p' = \dfrac{p}{\cos\alpha}$, consequently:

$$\tan\theta' = \frac{t\cos^2\alpha}{2p}$$

$$Or\ \theta' = \mathrm{Arc}\tan(\tan\theta.\cos^2\alpha) \quad (2)$$

Strain of the opposed yarns $\varepsilon_y = \frac{\Delta x}{x}$ are estimated as follows:

$$x = \frac{p}{\cos\theta} \quad , \quad x' = \frac{p'}{\cos\theta'} = \frac{p}{\cos\alpha\cos\theta'}$$

and

$$\varepsilon_y = \frac{x' - x}{x} = \frac{\cos\theta - \cos\alpha\cos\theta'}{\cos\alpha\cos\theta'} \tag{3}$$

Note that when α is small θ' is close to θ whereupon ε is small too.

By determination of the strain and the weave angle variations during the yarn pull-out test, an account is given to the force-balance. Effects of shear stresses have been implied in the friction force which is equal to the pullout force F according to Equation (4) (the Amounton's law):

$$F = N.\mu.F_N \tag{4}$$

The pullout force is normalized by the number of crossovers N in the direction of the pulled yarn, to represent the pullout force per each crossover. that is

$$f = \frac{F}{N} = \mu.F_N \tag{5}$$

It is clear that during the process four main forces are formed: the normal load F_N, the normalized pullout force f and two lateral forces T_f in the fabric plane. It should be noted that T_f is the projection of the lateral forces $T_{y,}$ propagating in the opposed yarn direction.

We have:

$$F_N = 2T_y \sin\theta' \tag{6}$$

$$f = 2T_f \sin\alpha \tag{7}$$

$$T_f = T_y \cos\theta' \tag{8}$$

With regard to Equations (5–8) a simple function for friction coefficient is obtained:

$$\mu = \frac{f}{F_N} = \frac{\sin\alpha}{\tan\theta'} \tag{9}$$

With regards to the fact that α and θ (or θ') are ever salient angles, so $\sin\alpha$ and $\tan\theta'$ and consequently yarn-to-yarn friction coefficient μ are ever positive too.

$$T_y = E_y.\varepsilon_y.\rho$$

Hence:

$$T_y = E_y.\rho\frac{\cos\theta - \cos\theta'\cos\alpha}{\cos\theta'\cos\alpha} \tag{10}$$

And

$$F_N = 2E_y.\rho\frac{\cos\theta - \cos\theta'\cos\alpha}{\cos\theta'\cos\alpha}.\sin\theta' \quad (11)$$

So that:

$$T_f = E_y.\rho(\frac{\cos\theta}{\cos\alpha} - \cos\theta') \quad (12)$$

In order to evaluate the efficiency of the force-balance model, the yarn pullout force was simulated by Amonton's Law (Equation (6)). Therefore, the yarn-to-yarn friction coefficient and the normal load in crossovers were required to be known, these values were calculated through the force-balance model. Therefore, the fabric modulus and its density along the warp direction, the number of crossovers and the weave angle of the fabric should be given. In this study, these parameters were determined experimentally. The quantities of the fabric deformation angle α, and the variations of the weave angle θ' during a yarn pullout test at any instance should be known too. These factors were provided using the fabric deformation information, which were acquired by the image processing method.

18.3 CONCLUSION

In this study the yarn pullout test is applied to investigate internal mechanical properties of the plain woven fabrics. In the first step an analytical model was developed, inputs of which employs simple mechanical properties such as the fabric modulus, the weave angle and the fabric deformation angles during the pullout test. This model predicts important mechanical parameters such as the weave angle variations, the yarn-to-yarn friction coefficient, the normal load in crossovers, the lateral forces, and the opposed yarn strain within the fabric. This approach may be extended to other types of the woven fabrics.

KEYWORDS

- **Analytical approach**
- **Analytical model**
- **Force-balance analysis**
- **Internal mechanical properties**
- **Plain-woven fabrics**

REFERENCES

1. Badrossamay, M. R., Hosseini Ravandi, S. A., and Morshed, M. Fundamental Parameters Affecting Yarn Pullout Behavior. *J. Tex. Inst.*, **92** 280–287 (2001).
2. De Jong, S. and Postle, R. An Energy Analysis of Woven-Fabric Mechanics by Mean of Optimal-Control Theory Part I: Tensile Properties. *J. Text. Inst.*, **68** 350–361 (1977).

3. Hearle, J. W. S. and Shanahan, W. J. An Energy Method for Calculations in fabric Mechanics Part I: Principle of The Method. *J. Text. Inst.*, **69** 81–91 (1978).
4. Kirkwood, K. M. et al. Yarn Pullout as a Mechanism for Dissipating Ballistic Impact energy in Kevlar® KM-2 Fabric, Part I: Quasi Static Characterization of Yarn Pullout. *Tex. Res. J*, **74** 920–928 (2004).
5. Kirkwood, J. E. et al. Yarn Pullout as a Mechanism for Dissipating Ballistic Impact energy in Kevlar® KM-2 Fabric, Part II: Predicting Ballistic Performance. *Tex. Res. J*, **74** 939–948 (2004).
6. Love, L. Graphical Relationships in Cloth Geometry for Plain, twill and Sateen Weaves. *Tex. Res. J*, **24** 1073–1083 (1954).
7. Motamedi, F., Baily, A. I., Briscoe, B. J. and Tabor, D. Theory and Practice of Localized Fabric Deformation. *Textile Res. J.*, **59** 160–172 (1989).
8. Pan, N. and Young Youn, M. Behavior of Yarn Pullout from Woven Fabrics: Theoretical and Experimental. *Textile Res. J.*, **63** 629–637 (1993).
9. Painter, E. V. Mechanics of Elastic Performance of Textile Materials, Part VIII: Graphical Analysis of Fabric Geometry. *Tex. Res. J*, **22** 153–169 (1952).
10. Peirce, F. T. The Geometry of Cloth Structure, *J. Text. Inst.*, **28** T45–69 (1937).
11. Scelzo, W. A., Backer, S., and Boyce, M. C. Mechanistic Role of Yarn and Fabric Structure in Determining Tear Resistance of Woven Cloth. Part II: Modeling Tongue Tear. *Tex. Res. J*, **64** 321–329 (1994).
12. Sebastian, S. A. R. D., Baily, A. I., Briscoe, B. J., and Tabor, D. Extension, Displacements and Forces Associated with Pulling a Single Yarn from a Fabric. *J. Phys. D. Appl. Phys.*, **20** 130–139 (1987).
13. Seo, M. H. et al. Mechanical Properties of Fabric Woven from Yarns Produced by Different Spinning Technologies: Yarn Failure in Woven Fabric. *Textile Res. J.*, **63** 123–134 (1993).
14. Realff, M. L. et al. Mechanical Properties of Fabric Woven from Yarns Produced by Different Spinning Technologies: Yarn Failure as a Function of Gauge Length. *Textile Res. J.*, **61** 517–530 (1991).
15. Taylor, H. M. Tensile and Tearing Strength of Cotton Cloths. *J. Text. Inst.*, **50** T161–188 (1959).

19 Effect of Insufficient Watering and Melamine Salt of Bis(Oxymethyl)phosphonic Acid (Melaphen) on the Fatty Acid Composition

G. E. Zaikov, I. V. Zhigacheva, T. B. Durlakova, T. A. Misharina, M. B. Terenina, N. I. Krikunova, I. P. Generozova, A. P. Shugaev, and S. G. Fattakhov

CONTENTS

19.1 Introduction 187
19.2 Materials and Methods 189
19.2.1 Pea Seeds Germination 189
19.2.2 Statistics 190
19.2.3 Reagents 190
19.3 Discussions and Results 190
19.4 Conclusion 195
Keywords 196
References 196

19.1 INTRODUCTION

In this work we studied the effect of insufficient watering (IW) and melamine salt of bis(oxymethyl)-phosphonic acid (melaphen) on the fatty acid composition and the energy of 5 day etiolated pea seedling mitochondria (*Pisum sativum).* It has been shown that IW results in alteration of fatty acid composition in mitochondrial membranes of seedlings. The ratio of the content of C_{18}—unsaturated fatty acids to the stearic acid content decreases by 1.5 times. Significant changes are observed in content of fatty acids with 20 carbon atoms: the ratio of unsaturated fatty acids to saturated fatty acids decreases by 3.3 times. The changes in fatty acid composition of mitochondrial membranes are in correlation with changes in maximum rates of NAD-dependent substrates oxidation (the Pearson's coefficient of correlation for C_{18} fatty acids is 0.676; for C_{20} fatty acids – 0.963). We discuss the regulatory role of the C_{18} and C_{20} unsaturated fatty acids in mitochondrial energy metabolism of pea seedlings.

Cell membranes are among cell components damaged under the conditions of water deficit. Water deficit modifies the cell and organelle membranes and thus alters their functions and cell metabolism [1]. The alterations occur at the fatty acid composition of biological membranes. The content of saturated fatty acid increases while the content unsaturated fatty acid decreases [2]. These changes reflect on the activity of enzymes associated with chloroplast and mitochondrial membranes and result in decreasing the functional activity of these organelles [3]. As known from the literature, regulators of plant growth and development improve their tolerance to biotic and abiotic stresses, to water deficit in particular [4]. One of such growth regulators is melaphen (MF)—a melamine salt of bis(oxymetyl)phosphonic acid [5]:

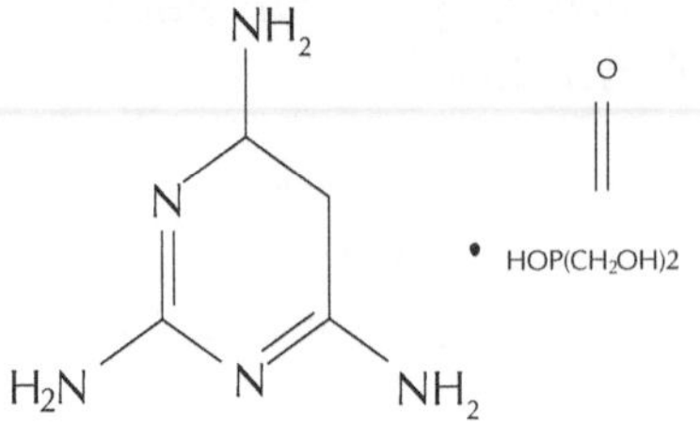

The aim of this work is study the effect of IW and the plant growth regulator—MF on the fatty acids composition lipid fraction of mitochondrial membranes and bioenergetical function of 5 day pea seedling mitochondria. The development and, probability, survival of plant in any case are more dependent on availability of water than on any other environmental factors. At present, numerous data accumulated demonstrating that even weak water deficit affected plant metabolism and thus their growth and development [6]. Metabolism of plants survived even short-term strong drought could not be recovered [7]. Water deficit modified cell membranes, which affected their functions and disturbed cell metabolism [1]. The alterations occur at the level of glycolipids, monogalactosyl-diacyl-glycerol and digalactosyl-diacyl- glycerol [2, 8]. The content of unsaturated fatty acids decreases in these lipids which results in the decreasing the membrane "fluidity", alteration in the lipid–protein ratio, and eventually in the activity changes of the enzymes associated with membrane, first of all enzymes which enter into complex of electron-transport chain of mitochondria and chloroplasts [3]. The energy metabolism plays a significant role in adaptive response of the organism. Mitochondria play a key role in the energy, redox, and metabolic processes in cell

[9]. As the plant grows regulators enhance the resistance of plant to different kinds of stress and MF is such a preparation, it was interesting to find out if a preliminary treatment of seeds with MF would exert a protective effect under the conditions of water deficit.

19.2 MATERIALS AND METHODS

The study was carried out on mitochondria isolated from pea seedlings (*Pisum sativum*) obtained in standard conditions and in the conditions of IW.

19.2.1 Pea Seeds Germination

The seeds from the control group were washed with soap solution and 0.01% $KMnO_4$ solution and left in water for 60 min. The seeds from the experimental group were placed in the 2 x 10^{-12} M MF solution for 60 min. After 1 day exposure, half of the seeds from the control group and half of the seeds treated with MF were placed onto a dry filter paper in open cuvettes. Two days later the seeds were placed into closed cuvettes with periodically watered filter paper and left for 2 day. On the 5th day the amount of germinated seeds was calculated and mitochondria isolated.

Isolation of mitochondria from 5 day sprouts epicotyls was performed by a method of [8] in our modification. The epicotyls having a length of 1.5–5 cm (20–25 g) were placed into a homogenizer cup, poured with an isolation medium in a ratio of 1:2, and then were rapidly disintegrated with scissors and homogenized with the aid of a press. The isolation medium comprised: 0.4 M sucrose, 5 mM EDTA, 20 mM KH_2PO_4 (pH 8.0), 10 mM KCl, 2 mM dithioerythritol, and 0.1% BSA (free of fatty acids).The homogenate was centrifugated at 25,000 g for 5 min. The precipitate was re-suspended in 8 ml of a rinsing medium and centrifugated at 3,000 g for 3 min. The suspension medium comprised: 0.4 M sucrose, 20 mM KH_2PO_4, 0.1% BSA (free of fatty acids) (pH 7.4). The supernatant was centrifuged for 10 min at 11,000 g for mitochondria sedimentation. The sediment was re-suspended in 2–3 ml of solution contained: 0.4 M sucrose, 20 mM KH_2PO_4 (pH 7.4), 0.1% BSA (without fatty acids) and mitochondria were precipitated by centrifugation at 11,000 g for 10 min. The suspension of mitochondria (about 6 mg of protein/ml) was stored in ice.

The rate of mitochondria respiration was measured with the aid of Clarke oxygen electrodes and LP-7 polarograph (Czechia). Mitochondria were incubated in a medium containing 0.4 M sucrose, 20 mM HEPES-Tris buffer (pH 7.2), 5 mM KH_2PO_4, 4 mM $MgCl_2$, 5 mM EDTA and 0.1% BSA, 10 mM mlate + glutamate, pH 7.4. The rate of respiration was expressed in ng-atom O/(mg protein min).

Fatty acid methyl esters (FAMEs) were produced by acidic methanolysis of mitochondrial membrane lipids [11] or using one-step methylation of fatty acids, excluding the extraction of lipids [12]. The MEFA were purified by the method of thin layer chromatography on the silica plates and hexanol elution. For a quantitative control of the methanolysis process. An internal standard—pentadecane was used.

The FAME identification was performed by chromato-mass-spectrometry (GC-MS) using a Hewlett-Packard-6890 spectrophotometer with a HP-5972 mass-selective detector and after the retention times [13]. The FAME were separated in the HP-5MS capillary column (30 m x 0.25 mm, phase layer = 0.25 μm) at programmed temperature increase from 60 to 285°C at the rate of 5°C /min. Evaporator temperature is 250°C, detector temperature is 289°C. Mass spectra were obtained in the regime of electron impact ionization at 70 eV and the scan rate of 1 sec. per mass decade in the scan mass range of 40–450 a.u.m.

The FAME quantification was performed using a Kristall 2,000 M chromatograph (Russia) with flame-ionization detector and quarts capillary column SPB-1 (50 m x 0.32 mm, a nonpolar phase layer = 0.25 μm). FAME analysis was performed at programmed temperature increase from 120 to 270°C at the rate of 4°C/min. Temperature of injector and detector—270°C; the helium carrier gas rate was 1.5 ml/min. Each sample contained 2 μl of the hexane extract. The FAME content in samples was calculated as the ratio of peak area of a corresponding acid to the sum of peak areas of all found FAMEs.

Unsaturation index was calculated as a total percentage of unsaturated fatty asids with a certain number atoms multiplied by the number of double bonds, and divided by 100%. For example, for fatty acids with 18 carbon atoms unsaturation index is equal to (18:2ω6) x 2 + (18:3ω3) x 3 + 18:1ω9 + 18:1ω7/100.

Lipid peroxidation (LPO) activity was assessed by fluorescent method [14]. Lipids were extracted by the mixture of chloroform and methanol (2:1). Lipids of mitochondrial membranes (3–5 mg of protein) were extracted in the glass homogenizer for 1 min at 10°C. Thereafter, equal volume of distilled water was added to the homogenate, and after rapid mixing the homogenate was transferred into 12 ml centrifuge tubes. Samples were centrifuged at 600 g for 5 min. The aliquot (3 ml) of the chloroform (lower) layer was taken, 0.3 ml of methanol was added, and fluorescence was recorded in 10 mm quartz cuvettes with a spectrofluorometer (FluoroMaxHoribaYvon, Germany). The excitation wavelength was 360 nm, the emission wavelength was 420–470 nm. The results were expressed in arbitrary units per mg protein. The using of this method permits recording both fluorescence of 4-hydroxynonenals and the fluorescence of MDA. The emission wavelength depends on the nature of the Schiff's bases: the Schiff's bases formed by 4-hydroxynonenals have fluorescence wavelength 430–435 nm; those formed by MDA, 460–470.

19.2.2 Statistics

Tables and figures present means and their standard deviations (M + m). In Figures 2 and 3, correlations between unsaturation coefficients of C 18 and C 20 fatty acids and the rates of NAD-dependent substrate oxidation are presented; they were calculated using Statistica v. 6 software for Windows.

19.2.3 Reagents

The following reagents were used: potassium carbonate, methanol, chloroform (Merck, Germany), hexane (Panreac, Spain), acetyl chloride (Acros, Belgium), sucrose, Tris, EDTA, FCCP, malate, glutamate, FA-free BSA (Sigma, United States), and Hepes (MB Biomedicals, Germany).

19.3 DISCUSSIONS AND RESULTS

The IW resulted in 3-fold increase in content of LPO products in pea seedling mitochondrial membranes (Figure 1).

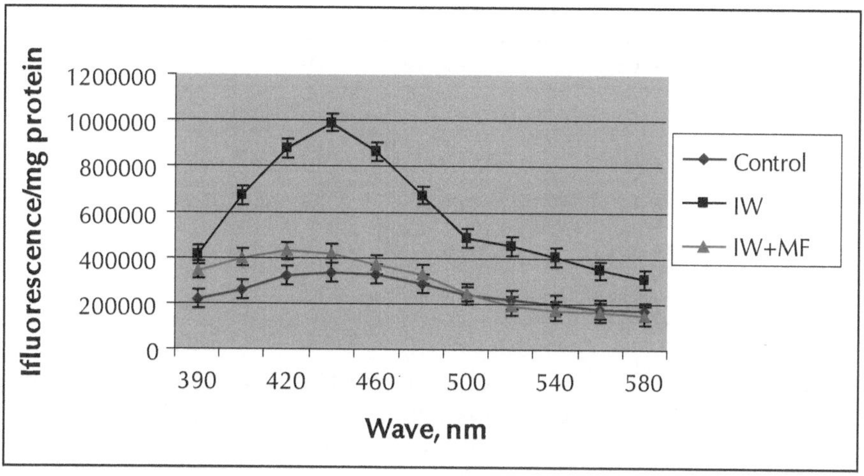

FIGURE 1 Fluorescence spectra of LPO products in mitochondrial membranes of 5 day pea seedlings under condition of IW and the treatment of seeds with melaphen (IW + MF).

The treatment of seeds with a 2 x 10^{-12} M MF solutions decreased the content of LPO products to the control values. The IW promoted LPO accompanied by modification of the fatty acid composition of pea seedling mitochondrial membranes. Water deficit led to the increase in the relative content of saturated and a decrease in the content of unsaturated fatty acids in mitochondrial membranes of pea seedlings (Table 1).

TABLE 1 Effects of IW and MF on the relative fatty acids content in mitochondrial membrane lipids of pea seedlings,%.

Fatty acid	Control	Control+MF	IW	IW+MF
12:0	0.34 ± 0.03	0.34 ± 0.01	0.94 ± 0.30	0.34 ± 0.20
14:0	0.68 ± 0.03	0.64 ± 0.02	0.67 ± 0.20	0.69 ± 0.20
16:1ω7	0.36 ±0,03	0.36 ± 0.004	0.47 ± 0.13	.42 ± 0.005
16:0	18.64 ± 0.75	18.63 ± 0.05	20.74 ± 0.11	18.96 ± 0.50
17:0	0.45 ± 0.05	0.78 ± 0.12	0.66 ± 0.10	0.45 ± 0.16
18:2ω6	50.72 ± 0.80	50.74 ± 0.40	45.22 ± 0.10	50.65 ± 0.01
18:3ω3	11.3 ± 0.02	10.67 ± 0.01	9.18 ± 0.30	10.81 ± 0.09
18:1ω9	5.27 ± 0.40	5.25 ± 0.37	6.77 ± 0.20	5.22 ± 0.01
18:1ω7	0.81 ± 0.10	0.79 ± 0.24	0.61 ± 0.03	0.73 ± 0.05
18:0	4.10 ± 0.18	4.14 ± 0.32	5.83 ± 0.38	4.10 ± 0.15
20:2ω6	0.82 ± 0.01	0.80 ± 0.02	0.30 ±0.05	0.82 ± 0.01

TABLE 1 *(Continued)*

Fatty acid	Control	Control+MF	IW	IW+MF
20:1ω 9	2.22 ± 0.01	2.79 ± 0.01	1.57 ± 0.01	2.6 ± 0.03
20:1ω7	1.45 ± 0.01	1.14 ± 0.01	1.00 ± 0.01	1.52 ± 0.01
20:0	1.23 ± 0.03	1.20 ± 0.03	2.52 ± 0.20	1.30 ± 0.05
22:0	1.23 ± 0.11	1.20 ± 0.03	2.52 ± 0.20	1.04 ± 0.05
24:0	0.37 ± 0.02	0.55 ± 0.005	0.98 ± 0	0.35 ± 0.10

The relative content of linoleic acid was reduced by 11%, that of linolenic acid by 29%. The content of stearic acid increased by 41%, which resulted in the decrease in the total content of C_{18} unsaturated fatty acids relative to the content of stearic acid from 16.61 ± 0.30 to 10.59 ± 0.20. Similar effect of the water deficit on the fatty acid composition of the mitochondrial membranes from maize, potato, and leaves of *Arabidopsis thaliana* and apricot was observed earlier [2, 15-18]. The authors detected a considerable decrease of the levels of linoleic and linolenic acids and an increase of the level of stearic acid in the membranes.

Substantial changes occurred also in the relative content of fatty acids with 20 carbon atoms. The pool of 20:2ω6 reduced by 2.7 times, 20:1ω–91.3 times. At the same time, the content of eicosanoic acid (20:0) increased more than 2-fold. As a result, the ratio of pool unsaturated fatty acids containing 20 carbon atoms (20:1ω7 + 20:1ω 9) + (20:2ω6) x 2 to eicosanic acid in mitochondrial membrane lipids decreased from 3.65 ± 0.03 to 1.20± 0.16.

The observed alterations possibly influence lipid–protein relation and thus alter the activity of the enzymes associated with the membrane. Indeed, IW results in a decrease of the maximal rates of NAD-dependent substrates oxidation. The rate of the pair glutamate + malate oxidation in the presence of uncoupling agent (FCCP) drops from 70.0 ± 4.6 down to 48.9 ± 3.2 ng oxygen atom/mg of protein min and the respiratory control rate (RCR) decreases from 2.27 ± 0.1 to 1.7 ± 0.2 (Table 2).

The treatment of seeds with a 2 x10^{-12} M MF solution before germination prevents the alteration of the oxidative phosphorylation efficiency caused by IW. Besides, the preliminary treatment with MF reduces the rates of NAD-dependent substrates oxidatio in the presence of ATP or FCCP to the control values (Table 2). Apparently, the described alterations are related with the physicochemical state of mitochondrial membranes.

Indeed the treatment with MF protects the unsaturated fatty acid from LPO and prevents thereby from changes in the fatty acid composition of seedling membranes in condition of IW (Table 1). In the group of seedlings subjected to IW combined with MF treatment (group IW + MF) concentrations of such saturated fatty acids as lauric, palmitic, and stearic acids are lower by 65, 7.5, and 30%, respectively, than in seedlings subjected to IW only (group IW). The level of C_{18}—unsturated fatty acids playing an important role in plant resistance to the effect of adverse factors of environment [19] increased. Thus, the level of linoleic acid increases by 12% and linolenic acid, by 15% (Table 1). The ratio of the sum of unsaturated C_{18} acids to saturated $C_{18:0}$ acids increases 1.5-fold in comparison with the group of IW. The level of C_{20} fatty acids also changed. The level of 20:2 ω 6 acid increases 2.73 times and that of 20:1ω9 acid, 2.28 times. The level of eicosanic acid decreases 1.92 times. As a result, the ratio of the sum of unsaturated C_{20} acids to saturated $C_{20:0}$ acids returns to the control values. The changes in the fatty acid composition of mitochondrial membranes were accompanied by changes in maximum rates of NAD-dependent substrates oxidation. A decrease in unsaturation coefficient of fatty acids in mitochondrial membranes led to decreasing the rates of NAD-dependent substrates oxidation and efficiency of oxidative phosphorylation.

TABLE 2 Effects of IW and MF on the rate of NAD-dependent substrate oxidation by mitochondria isolated from pea seedlings, ng-atom/(mg protein min).

Treatment	V_0	V_3	V_4	V_3/V_4	FCCP
Control	20.0 ± 1.5	68.0 ± 4.1	30.0 ± 2.0	2.27 ± 0.1	60.0 ± 4.6
IW	14.0 ± 2.0	48.6 ± 3.0	40.2 ± 1.0	1.7 ± 0.2	41.9 ± 3.2
IW+MF	19.8 ± 3.0	66.0 ± 2.4	27.5 ± 1.3	2.4 ± 0.2	60.3 ± 5.2

Notes: Incubation medium contained 0.4 M sucrose, 20 mM Hepes--Tris, 5 mM KH2PO4, 2 mM MgCl2, 5 mM EDTA, 10 mM malate + glutamate, pH 7.4. ADP (200 μM) and FCCP (0.5 μM) were added. V0 = the rate of oxygen decay in the presence of 10 mM malate + glutamate as substrate; V3 = the rate of substrate oxidation in the presence of ADP (state 3, 200 μM ADP); V4 = the rate of substrate oxidation after added ADP consumption (state 4); FCCP = the rate of substrate oxidation in the presence of FCCP (carbonylcyanide –p-trifluoromethoxyphenylhydrazone). The results of 10 experiments are presented, M ± m.

On the basis of presented data, it may be supposed that a prevention of unsaturated fatty acids peroxidation, in particular C_{18} and C_{20} acids in membranes of plant tissues leads to enhancement of plant resistance to IW. In fact, a close correlation was observed between the unsaturation coefficient of C_{18} fatty acids in mitochondrial membranes (∑unsaturated C_{18} fatty acids/ $C_{18:0}$) and maximum rates of NAD-dependent substrate oxidation (the correlation coefficient r = 0.765) (Figure 2).

An even greater correlation is observed between the unsaturation coefficient of C_{20} fatty acids (20:2 ω6) x 2 + 20:1 ω 9+ 20:1 ω 7/20:0) and maximum rates of NAD-dependent substrate oxidation (r = 0.964) (Figure 3).

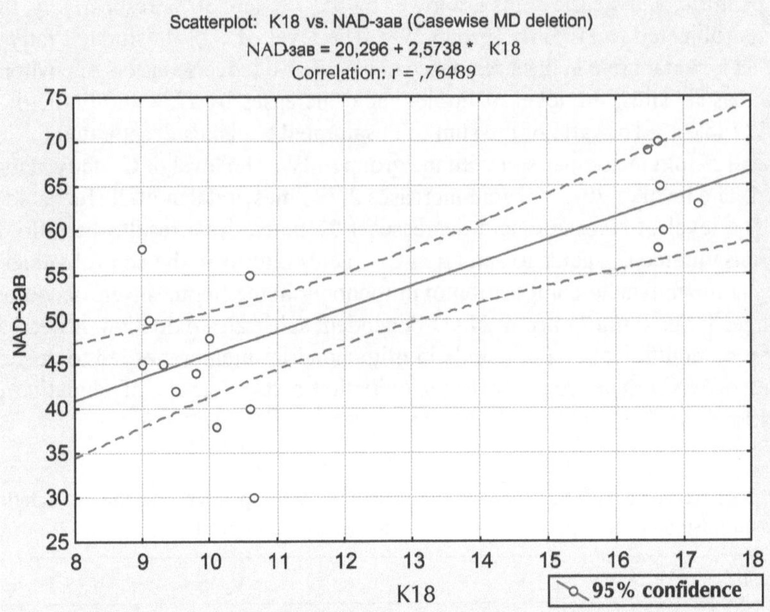

FIGURE 2 Correlation between the unsaturation coefficient of C20 fatty acids and maximum rates of NAD-dependent substrate oxidation. Y-axis shows the maximum rates of NAD-dependent substrate oxidation; X-axis- unsaturation coefficient of C18 fatty acids.

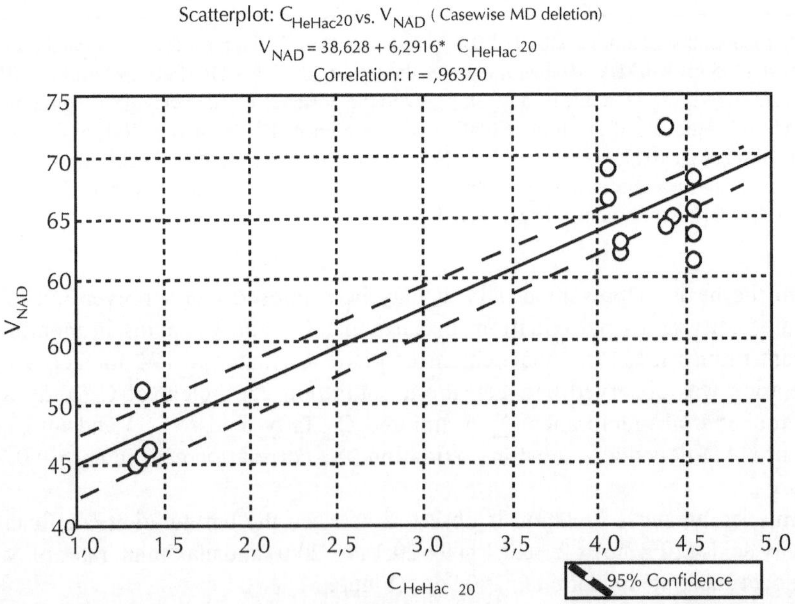

FIGURE 3 Correlation between the unsaturation coefficient of C20 fatty acids and maximum rates of NAD-dependent substrate oxidation. Y-axis shows the maximum rates of NAD-dependent substrate oxidation; X-axis- the ratio (20:2 ω6) x 2+ 20:1 ω 9+ 20:1 ω 7/20:0).

Changes in physical and chemical properties of mitochondrial membranes resulting in changes in the energy metabolism affected also physiological indices, for example, seedling growth. As evident from Figure 4, pea seed treatment with MF stimulated root growth by 5 times and sprouts growth by 3.5 times under conditions of water deficit. Observed stimulation of seedling root growth under IW has a great adaptive significance.

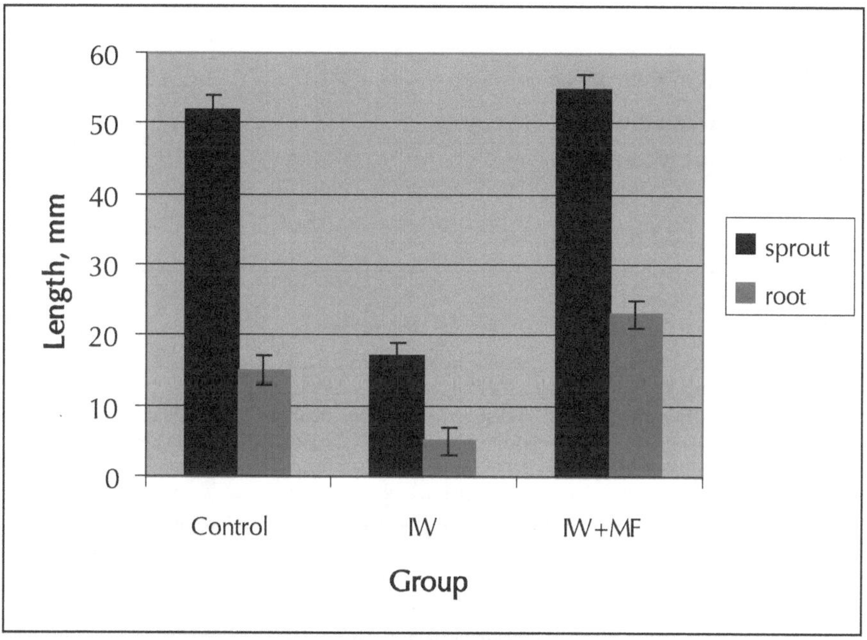

FIGURE 4 The lengths of sprouts and roots of 5 days pea seedling under condition of IW and treatment of pea seeds in this condition with melaphen (IW + MF).

Thus, under condition of IW, MF decreased the intensity of LPO in mitochondrial membranes. As a result, the pool of unsaturated fatty acids containing 18 and 20 carbon atoms in the lipid phase of mitochondrial membranes remained unchanged. The prevention of changes in fatty acid composition of mitochondrial membranes affected the bioenergetic indices: there was maintained a high activity of the NADH-dehydrogenase complex of the respiratory chain of mitochondria.

19.4 CONCLUSION

On the basis of data obtained, it is possible to suggest that tolerance to water stress is determined by the cell antioxidant system protecting unsaturated C_{18} fatty acids and unsaturated very-long-chain fatty acids against modifications induced by the oxidative stress, for example activation of free radical processes [20]. Changes in the C_{20} fatty acids in the mitochondrial membrane lipids are noted for the first time. Just the unsaturation coefficient of C_{20} fatty acids was correlated with the highest rates of

NAD-dependent substrate oxidation. At the same time, mitochondria of storage organs and seeds are characterized by relatively low rates of oxidation of NAD-dependent substrates. The result of maintenance high activity of NAD-dependent dehydrogenases is the support the energy processes in cell that promotes the resistance of plant to varying environmental conditions. Under conditions of IW, protective effect of MF is apparently determined by maintenance in the content of unsaturated fatty acids with 18 and 20 carbon atoms in lipid phase of mitochondrial membranes.

KEYWORDS

- **Biological membranes**
- **Fatty acid composition**
- **Insufficient watering**
- **Lipid peroxidation**
- **Mitochondrial membranes**

REFERENCES

1. Kizis, D., Lumbreras, V., and Pages, M. Role of AP2/EREBP Transcription Factors in Gene Regulation during Abiotic Stress. *FEBS Lett.*, **498**, 187–189 (2001).
2. Junior, R. R. M., Oliveira, M. S. C., Baccache, M. A., and de Paula, F. M. Effect of Water Deficit and Rehydratation on the Polar Lipid and Membrane Resistance Leaves of Phaseolus vulgaris L. cv. Perola. *Brazilian Arch. Biol. and Technol.*, **5** 361–367 (2008).
3. Shugaeva, N. A., Vyskrebentseva, E. I., orekhova, S. O., and Shugaev, A. G. Effect of water deficiency on respiration of conducting bundles of shuger-beet chard. *Plant Physiol.* (Rus.), **54**(3) 373–380 (2007).
4. Zhirmunskaya, N. M. and Shapovalov, A. A. Physiological aspect of use of growth regulators for enhancement of drought-resistance of plant. *Agrokhimiya*, **5** 102–119 (1987).
5. Fattakhov, S. G. Reznik, V. S., and Konovalov, A. I. Melamine salt of bis(ox methyl)phosphoric acid (MELAPHEN) as a new generation of regulator of plant grows. *Proceedings of the 13th International Conference on Chemistry of Phosphorus Compounds*. Saint-Petersburg., p. 80 (2002).
6. Ribas-Carbo, M., Tailor, N. L., Giles, L, Busquets, S., Finnegan, P. M., and Day, D. A. et al. Effects of Water Stress on Respiration in Soybean Leaves. *Plant Physiol.*, **139**, p. 466–473 (2005).
7. Boyer, J. S. Plant Productivity and the Environment. *Science*, **218** 443–448 (1982).
8. Sahsah, Y, Campos, P., Gareil, M. et al. Enzymatic degradation of polar lipid in Vigna uniquiculata leaves and influence of drought stress. *Plant. Physiol.*, **104** 577–586 (1988).
9. Atkin, O. K. and Macherel, D. The Crucial Role of Plant Mitochondria in Orchestrating Drought Tolerance. *Ann. Bot.*, **103** 581–59 (2009).
10. Popov, V. N., Ruuge, E. K., and Starkov, A. A. Influence ingibitors of electron transport on formation of active forms of oxygen at oxidation of succinate by mitochondria. *Biohimiya.*, **68**(7) 910–916 (2003).
11. Carreau, J. P. and Dubacq, J. P. Adaptation of Macroscale Method to the Microscale for Fatty Acid Methyl Transesterification of Biological Lipid Extracts. *J. Chromatogr.*, **1516** 384–390 (1979).
12. Wang, J., Sunwoo, H, Cherian, G., and Sim, I. S. Fatty Acid Determination in Chicken Egg Yolk. A Comparison of Different Methods Poultry. *Science*, **79** 1168–1171 (2000).

13. Golovina, R. V. and Kuzmenko, T. E. Thermodynamic Evaluation Interaction of Fatty Acid Metyl Esters with Polar and Nonpolar Stationary Phases, Based on Their Retention Indices Chromatographia. *Chromatogr.*, **10** 545–546 (1977).
14. Fletcher, B. I., Dillard, C. D., and Tappel, A. L. Measurement of Fluorescent Lipid Peroxidation Products in Biological Systems and Tissues. *Ann. Biochem.*, **52** 1–9 (1973).
15. Makarenko, S. P., Konstanyinov, Yu. M., Khotimchenko, S. V., Konenkina, T. A., and Arzev, A. S. Fatty acid composition of mitochondria membrane lipids of *Zea mays* and *elymus sibiricus. Plant Physiol.* (Rus), **50** 487–492 (2003).
16. Gigon, A., Matos, A. R., Laffray, D., Fodil, Y. Z., Thi Anh-Thu, Pham. Effect of Drought Stress on Lipid Metabolism in the Leaves of *Arabidosis thaliana* (Ecotype Columbia). *Ann. Bot.*, **94** 345–351 (2004).
17. Leone, A., Costa, A., Grillo, S., Tucci, M., Horvarth, I., and Vigh, L. Acclimation to Low Water Potential Determines Changes in Membrane Fatty Acid Composition and Fluidity in Potato Cells. *Plant Cell Environ.*, **19** 1103–1109 (1996).
18. Guo Yun-ping and Li Jia-rui. Changes of Fatty Acids Composition of Membrane Lipids, Ethylene Release and Lipoxygenase Activity in Leaves of Apricot under Drought. *J. Zhejiang Univ (Agricalt. and Life Sci).*, **28** 513–517 (2002).
19. Demin, A. N., Deryabin, A. N., Sinkevich, M. S., and Trunova, T. I. Introduction of gene *desA* *Δ12-acyl-lipid desaturase* of cyanobacterium enhange the resistance of potapo plants to oxidative stress caused by hypothermia. *Plant Physiol.* (Rus), **55** 710–720 (2000).
20. Torres-Franklin, M. L., Repellin, A., Van-Biet Huynh, and d'Arcy-Lameta A. Omega-3 Fatty Acid Desaturase (*FAD3*, *FAD7*, *FAD8*) Gene Expression and Linolenic Acid Content in Cowpea Leaves Submitted to Drought and After Rehydration. *Environm. Experiment. Bot.*, **65** 162–169 (2009).

20 Meniscus Height Changing in the Capillary and Ultraviolet Spectra of Ascorbic Acid and Paracetamol High-diluted Solutions

F. F. Niyazy, N. V. Kuvardin, E. A. Fatianova, and G. E. Zaikov

CONTENTS

20.1 Introduction 199
Keywords 207
References 207

20.1 INTRODUCTION

During last decades there is a tendency of the growing interest to the study of high diluted solutions of bioactive substances. Besides, concentration ranges under study are related to the category of supersmall or, in other words, "illusory" concentrations. Such solutions, unlike more saturated ones, but with pretherapeutic content of active substance, may possess high biological activity.

Use of bio-objects to reveal supersmall doses (SSD) effect in the substances is the most exact method today allowing not only to define the effect existence and to find out how it shows itself, but to determine concentration ranges of its action. However, the use of this method will entail great difficulties. In this connection it is necessary to search for other methods, including physico-chemical ones, allowing to define presence of SSD effect in the compounds if only at the stage of preliminary tests. Study of physico-chemical bases of SSD effect display is one of the most interesting questions in the given sphere of research and attracts attention of many scientists [1-4].

Antineoplastic and antitumorous agents, radioprotectors, neutropic preparations, neupeptides, hormones, adaptogenes, immunomodulators, antioxidants, detoxicants, stimulants, and inhibitors of plants growth and so on are included into the group of bioactive substances possessing SSD effect. Study of high diluted solutions of bioactive compounds was carried out on one-component solution that is ones containing only one solute. But now, mainly multicomponent medical preparations, possessing several therapeutic actions, are used in medicine. So, preparations of analgesic-antipyretic action are possibly used at sharp respiratory illnesses, accompanied by muscular pain and rise of temperature. It is possible that effects of multicomponent medical preparation in supersmall concentrations and its separate components will differ.

We studied some physico-chemical properties of high diluted aqueous solutions of paracetamol and ascorbic acid with the purpose of finding out peculiarities of their change at solutions dilution and also of definition of possible concentration ranges of SSD effect action. Paracetamol and ascorbic acid are the components of combined analgesic-antipyretic and anti-inflammatory preparations [5]. Ascorbic acid is used as fortifying and stimulating remedy for immune system. Paracetamol (acetominophene) has anaesthetic and febrifugal effect.

There have been prepared one-component solutions of ascorbic acid and paracetamol, two-component solutions of paracetamol with ascorbic acid (relation of dissoluted compounds in solutions is 1:1), in the following concentrations of dissoluted substances (mole/l): 10^{-1}, 10^{-3}, 10^{-5}, 10^{-7},10^{-9}, 10^{-11}, 10^{-13}, 10^{-15}, 10^{-17}, 10^{-19}, 10^{-21}, and 10^{-23}. Water cleansed by reverse osmosic was used as solvent. Solutions were prepared by successive dilution by 100 times using classical methods. Initial solution was 0.1M one. Before choosing of solution portion for the following dilution the sample was subjected to taking antilogs.

Prepared solutions were studied by cathetometric method of substances screening, the ones acting in supersmall concentrations, and also by method of electronic spectroscopy.

Cathetometric method of substances screening is based on the study of the change of solution meniscus height in capillary [6]. Results of measuring meniscus height of ascorbic acid, paracetamol, paracetamol with ascorbic acid solutions are given in Figure 1, Figure 2, and Figure 3, accordingly.

Meniscus height of dilution with ascorbic acid concentration of 10^{-3} mole/l was 0.8 mm (Figure 1). During further dilution value of meniscus height in the capillary reduces, but changes are not uniquely defined. The most lowering of meniscus height is observed in samples in which content of ascorbic acid is 10^{-9}, 10^{-13}, 10^{-15}, and 10^{-17} mole/l.

Meniscus height reduces on an average by 13.75%. Lowering of meniscus height by 23.7%, in comparison with more concentrated solution, is also observed for the sample with ascorbic acid content of 10^{-23} mole/l.

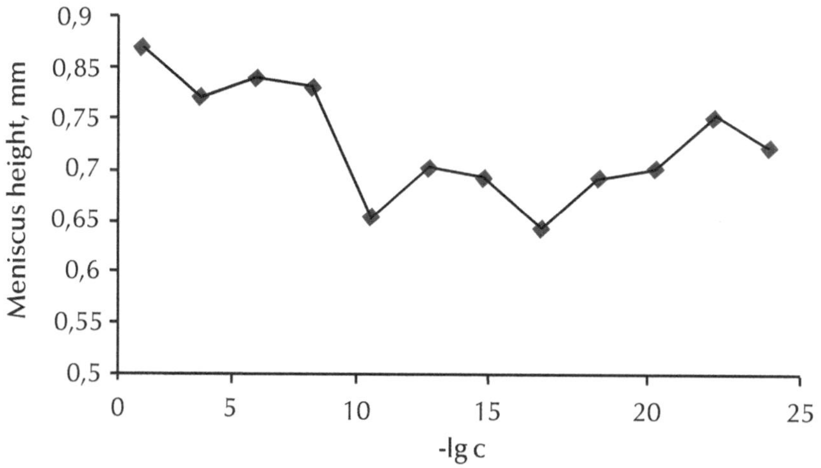

FIGURE 1 Values of meniscus height in the capillary of ascorbic acid solutions (concentration, mole/l).

Equivalent lowering of meniscus in the capillary has been stated for solutions with ascorbic acid concentration of 10^{-9} and 10^{-13}–10^{-17} mole/l between, these concentration ranges there is concentration range in which there are no essential changes. Being based on literature data and also on cathetometric method for screening of substances activity in supersmall doses, we can assume that dilutions of ascorbic acid in concentrations of 10^{-9} mole/l and 10^{-13}–10^{-17} mole/l show biological activity regarding bio-objects.

While studying solutions of paracetamol by cathetometric method it can be observed that in first dilutions by 100 times (paracetamol concentrations being 10^{-3} – 10^{-7} mole/l) height of meniscus reduces slightly, maximum 5.7%, regarding meniscus height of the first dilution. This slight lowering of meniscus height in the capillary is caused by rather large dose of active substance in these dilutions. However, sudden lowering of meniscus height in the capillary up to 0.71 mm, which is by 19.3% lower than meniscus height of the first dilution, is observed at diluting paracetamol solution up to the concentration of 10^{-9} mole/l (Figure 2).

The same dependence is observed for dilution of paracetamol solution with concentration of 10^{-15} mole/l. So, at this concentration meniscus height is 0.7 mm.

Growth of meniscus height in the capillary is observed further for solutions with the following dilution. This process is motivated by very high dilution that is the solution, according to its composition and properties, tries to attain the state of pure solvent.

These changes are polymodal dependence effect–concentration, described for different substances and different properties in domestic and world scientific literature. From the Figure and its description it is clearly seen that there are concentrations of paracetamol solution of 10^{-9} mole/l and 10^{-15} mole/l, for which there has been stated essential lowering of meniscus in the capillary regarding other concentrations. Between these concentration ranges there is a concentration range in which there are no essential changes, this range being the so called "dead zone".

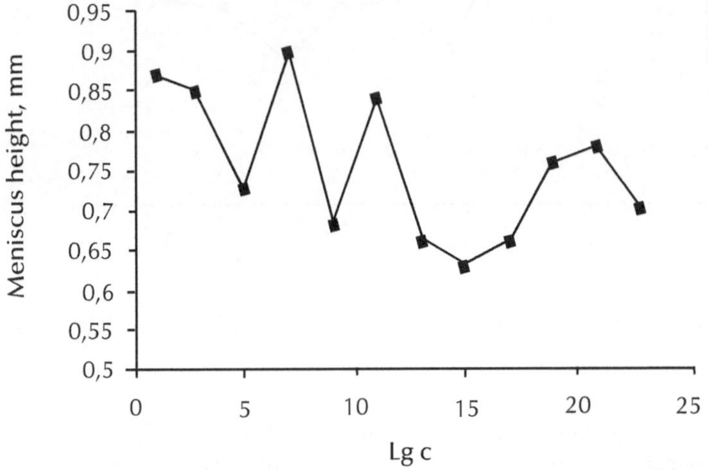

FIGURE 2 Values of meniscus height in the capillary of paracetamol solution (concentration, mole/l).

We have studied solutions of ascorbic acid and paracetamol mixture. It is necessary to note that not uniquely defined change of meniscus height is observed in solutions containing simultaneously two active substances. Meniscus of one-component solutions is narrower than that of water, but in two-component solutions both reduction and increase of meniscus height values are possible in comparison with water (Figure 3).

Lowering of meniscus height is observed in solutions with concentration of paracetamol with ascorbic acid 10^{-5}, 10^{-9}, 10^{-13}, 10^{-15}, and 10^{-17} mole/l on 11.5, 17.5, 20, and 23.6% correspondingly. It is possible to distinguish two concentration ranges in the field of super-small concentrations where reduction of meniscus height in the capillary takes place: this is–10^{-9} mole/l and 10^{-13}–10^{-17} mole/l. Received concentration ranges coincide with data of cathetometric studies of ascorbic acid and paracetamol solutions (Table 1).

Increase of meniscus height in comparison with the solvent is observed in solutions with concentrations of diluted compounds of 10^{-1}, 10^{-3}, 10^{-7}, and 10^{-11} mole/l. Growth of meniscus height above such value of the solvent can be explained by the change in SSD effect display.

We have analyzed ultra-violet spectra of solutions of ascorbic acid, paracetamol, paracetamol with ascorbic acid. All the spectra were read from spectrophotometer Cary 100, UV-visible Spectrophotometer in the range of 200–350 nm.

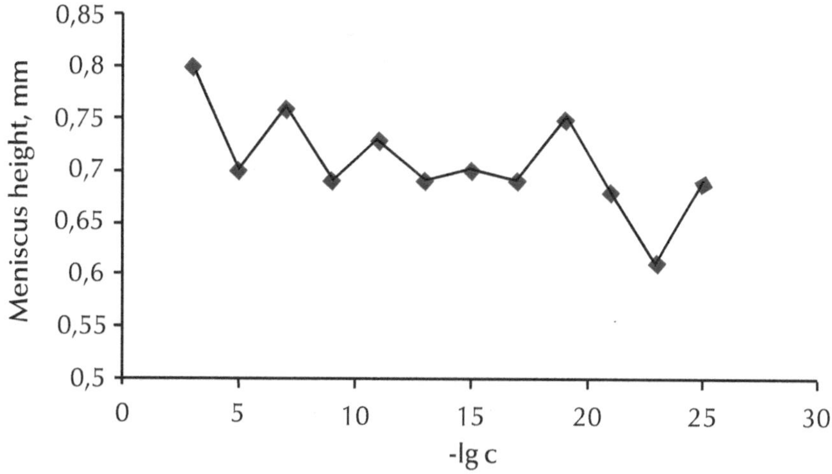

FIGURE 3 Values of meniscus height in the capillary of paracetamol solutions with ascorbic acid.

The most absorption in all solutions takes place in short wave part of the effective wave band. More concentrated solutions (10^{-1}, 10^{-3} mole/l) of ascorbic acid have maximum absorption in the length interval of 220–280 nm, of paracetamol–220–310 nm, paracetamol with ascorbic acid–200–320 nm. Tendency to narrowing of absorption field, reduction of optical density value and quantity of tops up to one or two is registered in all solutions, under study, as content of diluted substance reduces.

TABLE 1 Results of study of ascorbic acid, paracetamol, paracetamol with ascorbic acid solutions of wide concentration range by method of electronic spectroscopy and cathetometric method.

Method of study	Concentration ranges, mole/l		
	Ascorbic acid solutions	Paracetamol solutions	Solutions of paracetamol with ascorbic acid
Cathetometric method of screening	10^{-9}, 10^{-13} – 10^{-17}, 10^{-23}	10^{-9}, 10^{-15}	10^{-9}, 10^{-13} – 10^{-17}
Electronic spectroscopy	10^{-9}, 10^{-15}, 10^{-19}	10^{-9}, 10^{-13}, 10^{-15}, 10^{-21}	10^{-9}, 10^{-13}, 10^{-15}, 10^{-21}

Solutions spectra with ascorbic acid concentration of 10^{-15} and 10^{-19} mole/l are equal in form and differ from spectra of other dilutions. Lowering of ascorbic acid concentration is not accompanied by uninterrupted reduction of optical density value. Irregular growth of absorption in comparison with more concentrated solutions is observed in solutions with ascorbic acid concentrations of 10^{-9}, 10^{-15}, and 10^{-19} mole/l (Table 2).

TABLE 2 Maximum values of absorption of ascorbic acid solutions in ultra-violet field.

Concentration, mole/l	Absorption maximum	
	Wave length, nm	A
10^{-3}	230–270	0,79
10^{-5}	265	0,04
10^{-7}	220–230	0,012
10^{-9}	225	0,03
10^{-11}	235	0,01
10^{-13}	220	0,015
10^{-15}	220	0,04
10^{-17}	220–240	0,005
10^{-19}	220	0,03
10^{-21}	220	0,005
10^{-23}	220	0

Community in given structures may be assumed taking into account coincidence of waves length and values of optical density for solutions with concentrations 10^{-9}, 10^{-15}, and 10^{-19} mole/l.

Gradual reduction of paracetamol concentration in solutions is not accompanied by the same reduction of optical density value. Growth of absorption in comparison with more concentrated solutions is observed for solutions with concentrations 10^{-9}, 10^{-13}, 10^{-15}, and 10^{-21} mole/l (Table 3). All this allows to assume the rise of structural changes in these solutions.

TABLE 3 Maximum values of absorption of paracetamol solutions in ultra-violet field.

Concentration, mole/l	Absorption maximum	
	Wave length, nm	A
10^{-1}	230	0,58
10^{-3}	270	0,53

TABLE 3 *(Continued)*

Concentration, mole/l	Absorption maximum	
	Wave length, nm	A
10^{-5}	245	0,15
10^{-7}	240	0,02
10^{-9}	235	0,015
10^{-11}	240	0,007
10^{-13}	245	0,007
10^{-15}	220–245	0,01
10^{-17}	235	0,003
10^{-19}	290	0,001
10^{-21}	225	0,02
10^{-23}	230	0,005

Spectrum of ascorbic acid and paracetamol solution in concentrations of 10^{-1} mole/l is characterized by wide absorption band in the field of 200–320 nm. Dilution of 0.1 M solution by 100 times is accompanied by reduction of absorption field width up to 200–280 nm without changing of spectrum form and intensity of absorption on peaks.

Changing of spectrum form accompanied by essential reduction of absorption from 4 to 0.095 with the maximum on the length 243 nm takes place while diluting solution of ascorbic acid with paracetamol up to the concentration 10^{-5} mole/l (Table 4).

TABLE 4 Maximum values of absorption of ascorbic acid with paracetamol solutions (1:1) in ultra-violet field.

Concentration, mole/l	Absorption maximum	
	Wave length, nm	A
10^{-1}	237	4,651
10^{-3}	235	4,208
10^{-5}	243	0,095
10^{-7}	205	0,242

TABLE 4 *(Continued)*

Concentration, mole/l	Absorption maximum	
	Wave length, nm	A
10^{-9}	205	0,293
10^{-11}	207	0,084
10^{-13}	206	0,367
10^{-15}	205	0,186
10^{-17}	207	0,086
10^{-19}	207	0,089
10^{-21}	206	0,261
10^{-23}	207	0,105

Two peaks on lengths 205–206 nm and 270–273nm are shown on spectra of solutions with paracetamol and ascorbic acid concentrations 10^{-7}, 10^{-9},10^{-13}, 10^{-15}, and 10^{-21} mole/l. One peak on the length 207 nm is shown on spectra of solutions with paracetamol and ascorbic acid concentrations 10^{-11}, 10^{-17}, 10^{-19}, and 10^{-23} mole/l. Display of maximum on the length 270 nm occurred under conditions that absorption on maximum 205 nm was not less than 0.1.

Increase of optical density is observed in spectra of solutions with paracetamol and ascorbic acid concentrations 10^{-7}, 10^{-9}, 10^{-13}, 10^{-19}, and 10^{-21} mole/l.

As a result of this work, we by cathetometric method, have defined concentration ranges of possible display of medical compounds biological activity that forms 10^{-9} mole/l and 10^{-13}–10^{-17} mole/l for ascorbic acid and 10^{-9}, 10^{-15} of paracetamole, and also 10^{-9}, 10^{-13}–10^{-17} for the mixture of ascorbic acid–paracetamol. This allows to assume compatibility of components data in given concentration ranges.

While studying solutions in supersmall doses of ascorbic acid, paracetamol and also at their joint presence by method of electronic spectroscopy it has been found out that irregular growth of absorption is observed for solutions with ascorbic acid concentrations 10^{-9}, 10^{-15}, and 10^{-19} mole/l in comparison with more concentrated solutions and for paracetamol solutions with concentrations 10^{-9}, 10^{-13}, 10^{-15}, and 10^{-21} mole/l. But for the mixture ascorbic acid–paracetamol such interval is 10^{-9}, 10^{-13}, 10^{-15}, and 10^{-21} mole/l.

This fact allows to speak about appearance of medical compounds structures with water in this solution which differ by relatively high absorption.

Information received by method of ultra-violet spectroscopy, agrees with data of cathetometric screening both for solutions of ascorbic acid and of paracetamol.

KEYWORDS

- **Ascorbic acid**
- **Electronic spectroscopy**
- **Meniscus height changing**
- **Paracetamol solution**
- **Supersmall doses**

REFERENCES

1. Konovalov, A. I. Physico-chemical mystery of super-small doses. *Chemistry and life*, **2** 5–9 (2009).
2. Kuznetsov, P. E., Zlobin, V. A., and Nazarov, G. V. On the question about physical nature of super low concentrations action. *Heads of reports at III International Symposium Mechanisms of super-small doses action*. Moscow, p. 229 (December 3–6, 2002).
3. Chernikov, F. R. *Method for evaluation of homoeopathic preparations and its physico-chemical foundations*. Materials of Congress of homoeopathists of Russia. Novosibirsk, pp 73 (1999).
4. Pal'mina, N. P. Mechanisms of super-small doses action. *Chemistry and life*, **N2** 10 (2009).
5. Maslikovsky, M. D. *Combined analgesic-febrifugal and anti-inflammatory preparations*. 208 (1995).
6. Niyazi, F. F. and Kuvardin, N. V. Method for determining abilities of bioactive substances to display "super-small doses" effect. Patent N 2346260 of February 10 (2009).

21 Some Aspects of Chemical Physics in Polymers

G. E. Zaikov, M. D. Goldfein, and N. V. Kozhevnikov

CONTENTS

21.1 Introduction 209
21.2 Experimental........ 210
21.3 Discussion and Results 210
21.3.1 Effect of Salts in Varivalent Metals [2, 3] 210
21.3.2 Effect of Some Organic Solvents [4-6] 211
21.3.3 Homopolymerization and Copolymerization of Acrylonitrile in an Aqueous Solution of Sodium Sulfocyanide [7-10] 211
21.3.4 Stable Radicals in the Polymerization Kinetics of Vinyl Monomers [11-18] 214
21.3.5 Monomer Stabilization [19-25]
21.3.6 Some Peculiarities of the Kinetics and Mechanism of Emulsion Polymerization [24-36] 215
21.3.7 Compositions for Production of Rigid Foamed Polyurethane [25, 37, 38, 40] 217
21.3.8 Scientific Basics of the Synthesis of a High-Molecular-Weight Flocculent [24, 25, 39, 40] 218
21.4 Conclusion 218
Keywords 219
References 219

21.1 INTRODUCTION

The paper summarizes the basic results of our studies related to further development of known concepts of the chemical physics of polymers and creation of new ones in environment protection and life safety. The scientific novelty of our data is in development of the theory of radical-chain processes of polymer formation on the basis of exploring the gross kinetics of radical polymerization of vinyl monomers, establishment of the mechanisms of elementary reactions of chain initiation, propagation, and termination, estimation of reactivity parameters of monomers, initiators, and inhibitors in the conditions of homopolymerization and copolymerization in bulk, in solution, and in emulsion in both presence and absence of oxygen. The practical importance of our work includes the development of novel methods to regulate polymerization at its initial stages and at deep conversion degrees, the development of the scientific foundations of environmentally safe technologies and new effective monitoring techniques of the environment status.

Chemical physics is a most important fundamental science in the modern natural sciences [1]. Chemical physics concepts qualitatively and quantitatively explain the mechanisms of various processes, such as the reactions of oxidation, burning, and explosion, obtaining food products and drugs, oil hydrocarbon cracking, polymeric material formation, biochemical reactions underlying metabolism and genetic information transfer, and so on. Chemical physics comprises notions of the structure, properties, and reactivity of various substances, free radicals as active centers of chain processes proceeding in both mineral and organic nature, the scientific foundations of low-waste and resource-saving technologies. The design and wide usage of synthetic polymeric materials is a lead in chemistry. This results in the appearance of new environmental problems due to pollution of the environment with these materials and the wastes of their production and monomer synthesis. The presented results of our studies of the kinetics and mechanism of vinyl monomer polymerization which can proceed in quite various conditions point to their usability in the scientific justification of technological regime optimization of monomer and polymer synthesis, and in solving both local and global environmental problems.

21.2 EXPERIMENTAL

To obtain reliable experimental data and to correctly interpret them, we used such physicochemical and analytical techniques as dilatometry, viscometry, UV and IR spectroscopy, electronic paramagnetic resonance, Raman light scattering spectroscopy, electron microscopy, and gas-liquid chromatography. To analyze the properties of polymeric dispersions, the turbidity spectrum method was used, and the efficiency of flocculants was estimated gravimetrically and by the sedimentation speed of special water-suspended imitators (e.g. copper oxide).

21.3 DISCUSSION AND RESULTS

21.3.1 Effect of Salts in Varivalent Metals [2, 3]

Additives of the stearates of iron (IS), copper (CpS), cobalt (CbS), zinc (ZS), and lead (LS) within a certain concentration range were found to increase the polymerization rate of styrene and methylmethacrylate (MMA) in comparison with thermal polymerization. By initiating activity, they can be arranged as LS < CbS < ZS < IS < CpS. The decreases in the effective activation energy, the activation energy of the initiating reaction, and the kinetic reaction order with respect to monomer point to the monomer's active participation in chain initiation. IR spectroscopy data show that an intermediate monomer–stearate complex is formed and then decomposed into active radicals to initiate polymerization. The benzoyl peroxide (BP)–IS (or CpS) systems can be used for effective polymerization initiation. Concentration inversion of the catalytic properties of stearates has been found, which depends on salt concentration and conversion degree. The accelerating action efficiency decreases with increasing temperature, the BP–IS system possessing the highest initiating activity. The initiating mechanism for these systems principally differs from the redox one. It follows from our experimental data (color changes before and in the course of polymerization, the absorption and IR spectra of reactive mixtures, electron microscopy observations, etc.) that initiation occurs due to the stearate radicals formed at decomposition of the complex consisting of one BP molecule and two stearate ones.

Concentration and temperature inversion of the catalytic properties of gold, platinum, osmium, and palladium chlorides at thermal and initiated polymerization of styrene and MMA has been discovered. The mechanism of ambiguous action of noble metal salts is caused by the competition of the initiating influence of monomer complexes with colloidal metal particles and the inhibition reaction proceeding by ligand transfer.

21.3.2 Effect of Some Organic Solvents [4-6]

The influence of acetonitrile (ACN) and dimethyl formamide (DMFA) on the radical polymerization of styrene and (meth) acrylic esters initiated with azo isobutyronitrile (AIBN) or BP was studied. Basic characteristics of the gross kinetics of polymerization in solution have been found. The reaction order with respect to initiator concentration is always 0.5, which points to bimolecular chain termination. The reaction order with respect to monomer varies within rather wide limits (above or below unity) depending on the chemical nature of the solvent, monomer, and initiator. The lowest (0.83) and highest (1.6) values were found for BP-initiated polymerization of MMA in DMFA solution and for AIBN-initiated polymerization of styrene in ACN solution, respectively. Such a high value of the order with respect to monomer is caused by the abnormally low rate of AIBN decomposition in ACN, which, in turn, is explained by the donor–acceptor interaction of the alkyl and nitrile groups of the initiator with the nitrile and alkyl groups of the same solvent molecule. The influence of DMFA and ACN on each elementary stage of polymerization has also been ascertained, which manifests itself in the dependences of the rate constants of the reactions of initiation, propagation, and termination on the monomer concentration. This is caused by such factors as changes in the initiator decomposition rate, macro radical solvation with the molecules of an electron–donor solvent, diffusional-controlled chain termination, and conformational changes of macromolecules in solution.

21.3.3 Homopolymerization and Copolymerization of Acrylonitrile in an Aqueous Solution of Sodium Sulfocyanide [7-10]

When acrylonitrile (AN)-based fiber-formation polymers are obtained, a spinning solution ready for fiber formation appears as a result of polymerization in some solvent. Organic solvents or solutions of inorganic salts are used as solvents in these cases. First, a comparative study was made of the kinetics and mechanism of polymerization of AN in DMFA and in an aqueous solution of sodium sulfocyanide (ASSSC) initiated with AIBN and our newly-synthesized azonitriles (azobiscyanopentanol, azobiscyanovalerian acid, azobisdimethylethylamidoxime). The polymerization rate in ASSSC turns out to be significantly higher in comparison with the reaction in DMFA, in spite of the initiation rate in the presence of the said azonitriles being more than by 1.5 times lower than that in DMFA. The lower initiation rate in ASSSC is associated with stronger manifestation of the "cell effect" due to the higher viscosity of the water–salt solvent and the ability of water to form H-bonds (which hinders initiation). The ratio of the rate constants of chain propagation and termination ($K_p/K_o^{0.5}$) was found to be ca. tenfold higher in ASSSC than in DMFA. The molecular mass of the polymer formed is correspondingly higher. These differences are caused by the medium viscosity influence on K_o and the formation of H-bonds with the nitrile groups of the end chain of the macroradical and the monomer being added. But sodium sulfocyanide (which forms charge-transfer complexes with a molecule of AN or its radical to activate them) mainly contributes into the higher K_p.

The kinetics of AN copolymerization with methylacrylate (MA) or vinylacetate (VA) in ASSSC is qualitatively similar to AN homopolymerization in identical conditions. At the same time, the initial reaction rate, copolymer molecular mass, effective activation energy (E_{ef}), orders with respect to initiator and total monomer concentration differ. Example, the lower E_{ef} is due to the presence of a more reactive monomer (MA), and in the AN–VA system the non-end monomer units in macroradicals influence the rate constant of cross chain termination. For these binary systems, the copolymerization constants have been estimated, whose values point to a certain mechanism of chain propagation, which leads to the MA concentration in the copolymer being significantly higher than in the source mixture. The same is observed in the case of AN with VA polymerization (it it naturally that the absolute amount of AN in the final product is much higher than that of MA or VA due to its higher initial concentration in the mixture).

Obtaining synthetic PAN fiber of a nitrone type is preceded by the preparation of a spinning solution by means of copolymerization of AN with MA (or VA) and itaconic acid (IA), or acrylic acid (AA), or methacrylic acid (MAA), or methallyl sulfonate (MAS). Usually, mixtures contain 15% of AN, 5–6% of the second monomer, 1–2% of the third one, and ca. 80 wt. % of 51.5% ASSSC. When a third monomer is introduced into the reaction mixture, the process rate and the molecular mass of the copolymer decrease, which allows treating these comonomers as peculiar low-effective inhibitors. In such a case, it becomes possible to estimate the inhibition (retardation) constant which is, in essence, the rate constant of one of the reactions of chain propagation. At copolymerization of ternary monomeric systems based on AN in ASSSC, the chain initiation rate increases with the total monomer concentration. The order of the initiation reaction with respect to the total monomer concentration varies from 0.5 to 0.8 depending on the degree of the retarding effect of the third monomer. Besides, AN in the 3-component system is shown to participate less actively in the chain initiation reaction than at its homopolymerization and copolymerization with MA or VA.

Thus, the obtained results enable regulating the copolymerization kinetics and the structure of the copolymer formed, which, finally, is a way of chemical modification of synthetic fibers.

21.3.4 Stable Radicals in the Polymerization Kinetics of Vinyl Monomers [11-18]

The laboratory of chemical physics (Saratov State University) is a leading research center in Russia to conduct studies of the influence of stable radicals on the kinetics and mechanism of polymerization of vinyl monomers. In general, as is known, free radicals are neutral or charged particles with one or several uncoupled electrons. Unlike usual (short-living) radicals, stable (long-living) ones are characteristic of paramagnetic substances whose chemical particles possess strongly delocalized uncoupled electrons and sterically screened reactive centers. This is the very cause of the high stability of many classes of nitroxyl radicals of aromatic, aliphatic-aromatic, and heterocyclic series, and radical ions and their complexes.

Peculiarities of thermal and initiated polymerization of vinyl monomers in the presence of tetracyanoquinodimethane (TCQM) radical anions were investigated. TCQM radical anions are shown to effectively inhibit both thermal and AIBN-initiated polymerization of styrene, MMA, and MA in ACN and dimethylformamide solutions. Inhibition is accomplished by the recombination mechanism and by electron transfer to the primary (relative to the initiator) or polymeric radical. The electron-transfer reaction leads to the appearance of a neutral TCQM which regenerates the inhibitor in the medium of electron-donor solvents. Our calculation of the corresponding radical-chain scheme has allowed us to derive an equation to describe the dependence of the induction period duration on the initiator and inhibitor concentrations, and how the polymerization rate changes with time. The mechanism of the initiating effect of the peroxide—TCQM − system has been revealed, according to which a single-electron transfer reaction proceeds between an radical anion and a BP molecule, with subsequent reactions between the formed neutral TCQM and benzoate anion, and a benzoate radical and one more TCQM radical anion. Free radicals initiating polymerization are formed at the redox interaction between the products of the aforesaid processes and peroxide molecules. The TCQM radical anion interacts with peroxide only at a rather high affinity of this peroxide to electron (BP, lauryl peroxide): in the presence of cumyl peroxide, the radical anion inhibits polymerization only.

Iminoxyl radicals which are stable in air and are easily synthesized in chemically pure state (mainly, crystalline brightly-colored substances) present a principally new type of nitroxyl paramagnets. Organic paramagnets are used to intensify chemical processes, to increase the selectivity of catalytic systems, to improve the quality of production (anaerobic sealants, epoxy resins, and polyolefins). They have found application in biophysical and molecular-biological studies as spin labels and probes, in forensic medicinal diagnostics, analytical chemistry, to improve the adhesion of polymeric coatings, at making cinema and photo materials, in device building, in oil-extracting geophysics and flaw detection of solids, as effective inhibitors of polymerization, thermal and light oxidation of various materials, including polymers.

In this connection, systematic studies were made of the inhibiting effect of many stable mono and poly radicals on the kinetics and mechanism of vinyl monomer polymerization. The efficiency of nitroxyls as free radical acceptors has promoted their usage to explore the mechanism of polymerization by means of the inhibition technique. Usually, nitroxyls have time to only react with a part of the radicals formed at azonitrile decomposition, and they do not react at all with primary radicals at peroxide initiation. Iminoxyls have been found to terminate chains by both recombination and disproportionation in the presence of azonitriles. Inhibitor regeneration proceeds as a result of detachment of a hydrogen atom from an iminoxyl by an active radical to form the corresponding nitroso compound. The mechanism of inhibition by a nitroso compound is in addition of a propagating chain to a N = O fragment to form a stable radical again. The interaction of iminoxyls with peroxides depends on the type of solvent. For example, in vinyl monomers, induced decomposition of BP occurs to form a heterocyclic oxide (nitrone) and benzoic acid. In contrast to iminoxyls, aromatic nitroxyls in a monomeric medium interact with peroxides to form non-radical products. Imidazoline-based nitroxyl radicals possess advantages over common azotoxides which are their stability in acidic media (owing to the presence of an imin or nitrone functional group) and the possibility of complex formation and cyclometalling with no radical center involved.

21.3.5 Monomer Stabilization [19-25]

The practical importance of inhibitors is often associated with their usage for monomer stabilization and preventing various spontaneous and undesirable polymerization processes. In industrial conditions, polymerization may proceed in the presence of air, oxygen, and peroxide radicals MOO·are active centers of this chain reaction. In such cases, compounds with mobile hydrogen atoms, example phenols and aromatic amines, are used for monomer stabilization. They inhibit polymerization in the presence of oxygen only, that is are antioxidants. As inhibitors of polymerization of (meth) acrylates proceeding in the atmosphere of air, some aromatic amines known as polymer stabilizers were studied, namely, dimethyldi-(*n*-phenyl-aminophenoxy) silane, dimethyldi-(*n*-β-naphthyl aminophenoxy)silane, 2-oxy-1,3-di-(*n*-phenylamino phenoxy) propane, 2-oxy-1,3-di-(*n*-β-naphthyl aminophenoxy)propane. These compounds have proven to be much more effective stabilizers in comparison with the widely used hydroquinone (HQ), which is evidenced by high values of the stoichiometric inhibition coefficients (by 3–5 times higher than that of HQ). It has been found that inhibition of thermal polymerization of the esters of acrylic and MAAs at relatively high temperatures (100°C and higher) is characterized by a sharp increase of the induction periods when some critical concentration of the inhibitor $[X]_{cr}$ is exceeded. This is caused by that the formation of polymeric peroxides as a result of copolymerization of the monomers with oxygen should be taken into account when polymerization proceeds in air. Decomposition of polyperoxides occurs during the induction period as well and can be regarded as degenerated branching. The presence of critical phenomena is characteristic of chain branching. However, in early works describing inhibition of thermal oxidative polymerization, no degenerated chain branching on polymeric peroxides was taken into account. It follows from the results obtained that the value of critical concentration of inhibitor $[X]_{cr}$ can be one of the basic characteristics of its efficacy.

Inhibition of spontaneous polymerization of (meth) acrylates is necessary not only at their storage but also in the conditions of their synthesis proceeding in the presence of sulfuric acid. In this case, monomer stabilization is more urgent, since sulfuric acid not only deactivates many inhibitors but also is capable of intensifying polymer formation. The concentration dependence of induction periods in these conditions has a brightly expressed nonlinear character. And, unlike polymerization in bulk, decomposition of polymeric peroxides is observed at relatively low temperatures in the presence of sulfuric acid, and the values $[X]_{cr}$ of the amines studied are by ca. 10 times lower than $[HQ]_{cr}$.

Synthesis of MMA from acetone cyanohydrin is a widely spread technique of its industrial production. The process proceeds in the presence of sulfuric acid in several stages, when various monomers are formed and interconverted. Separate stages of this synthesis were modeled with reaction systems containing, along with MMA, methacrylamide and MAA, and water with sulfuric acid in various ratios. As heterogeneous as well as homogeneous systems appeared, inhibition was studied in both static and dynamic conditions. The aforesaid aromatic amines appear to effectively suppress polymerization at different stages of the synthesis and purification of MMA. Their advantages over hydroquinone are strongly exhibited in the presence of sulfuric acid in homogeneous conditions, or under stirring in biphasic reaction systems. Besides, application of polymerization inhibitors is highly needed in dynamic conditions at the stage of esterification.

The usage of monomer stabilizers to prevent various spontaneous polymerization processes implies further release of the monomer from the inhibitor prior to its processing into a polymer. Usually, it is achieved by monomer rectification, often with preliminary extraction or chemical deactivation of the inhibitor, which requires high energy expenses and entails large monomer losses and extra pollution of the environment. It would be optimal to develop such a way of stabilization, at which the inhibitor would effectively suppress polymerization at monomer storage but would almost not affect it at polymer synthesis. The usage of inhibitors low-soluble in the monomer is one of possible variants. When the monomer is stored and the rate of polymerization initiation is low, the quantity of the inhibitor dissolved could be enough for stabilization. Besides, as the inhibitor is being spent, its permanent replenishment is possible due to additional dissolution of the earlier unsolved substance. The ammonium salt of N-nitroso-N-phenylhydroxylamine (cupferron) and some cupferronates were studied as such low-soluble inhibitors. The solubility of these compounds in acrylates, its dependence on the monomer moisture degree, the influence of the quantity of the inhibitor and the duration of its dissolution on subsequent polymerization was studied. Differences in the action of cupferronates are due to their solubility in monomers, their various stability in solution, and the ability of deactivation. All this results in poorer influence of the inhibitor on monomer polymerization at producing polymer.

21.3.6 Some Peculiarities of the Kinetics and Mechanism of Emulsion Polymerization [24-36]

Emulsion polymerization, being one of the methods of polymer synthesis, enables the process to proceed with a high rate to form a polymer with a high molecular weight, high-concentrated latexes with a relatively low viscosity to be obtained, polymeric dispersions to be used at their processing without separation of the polymer from the reaction mixture, and the fire-resistance of the product to be significantly raised. At the same time, the kinetics and mechanism of emulsion polymerization feature ambiguity caused by such specific factors as the multiphase nature of the reaction system and the variety of kinetic parameters, whose values depend not so much on the reagent reactivity as on the character of their distribution over phases, reaction topochemistry, the way and mechanism of nucleation and stabilization of particles. The obtained results pointing to the discrepancy with classical concepts can be characterized by the following effects: (1) recombination of radicals in an aqueous phase, leading to a reduction of the number of particles and to the formation of surfactant oligomers capable of acting as emulsifiers, (2) the presence of several growing radicals in polymer–monomer particles, which causes the gel effect appearance and an increase in the polymerization rate at high conversion degrees, (3) a decrease in the number of latex particles with the growing conversion degree, which is associated with their flocculation at various polymerization stages, (4) an increase in the number of particles in the reaction course when using monomer-soluble emulsifiers, and also due to the formation of an own emulsifier (oligomers).

Surfactants (emulsifiers of various chemical nature) are usually applied as stabilizers of disperse systems, they are rather stable, poorly destructed under the influence of natural factors, and contaminate the environment. The principal possibility to synthesize emulsifier-free latexes was shown. In the absence of emulsifier (but in emulsion polymerization conditions) with the usage of persulfate-type initiators (e.g. ammonium persulfate), the particles of acrylate latexes can be stabilized by the ionized end groups of macromolecules. The M_nSO_4 radical ions appearing in the aqueous phase of the reaction medium, having reached some critical chain length, precipitate to form primary particles, which flocculate up to the formation of aggregates with a charge density providing their stability. Besides, due to radical recombination, oligomeric molecules are formed in the aqueous phase, which possess properties of surfactants and are able to form micelle-like structures. Then, the monomer and oligomeric radicals are absorbed by these micelles, where chains can grow. In the absence of a specially introduced emulsifier, all basic kinetic regularities of emulsion polymerization are observed, and differences are only concerned with the stage of particle generation and the mechanism of their stabilization, which can be amplified at copolymerization of hydrophobic monomers with highly hydrophilic comonomers. Increasing temperature results in a higher polymerization rate and a growth of the number of latex particles in the dispersion formed, in decreasing their sizes and the quantity of the coagulum formed, and in improved stability of the dispersion. At emulsifier-free polymerization of alkyl acrylates, the stability of emulsions and obtained dispersions rises in the monomer row: MA < ethylacrylate < butylacrylate that is the stability grows at lowering the polarity of the main monomer.

Our account of the aforesaid factors influencing the kinetics and mechanism of emulsion polymerization (in both presence and absence of an emulsifier) has enabled the influence of comonomers on the processes of formation of polymeric dispersions based on (meth) acrylates to be explained. Changes of some reaction conditions have turned out to affect the influence character of other ones. Example, increasing the concentration of MAA at its copolymerization with MA at a relatively low initiation rate leads to a decrease in the rate and particle number and an increase in the coagulum amount. At high initiation rates, the number of particles in the dispersion in the presence of MAA rises and their stability improves. The same effects were revealed for emulsifier-free polymerization of butylacrylate as well, when its partial replacement by MAA at high temperatures results in better stabilization of the dispersion, an increase in the reaction rate and the number of particles (whereas their decrease was observed in the presence of an emulsifier). Similar effects were found for AN as well, which worsens the stability of the dispersion at relatively low temperatures but improves it at high ones. Increasing the AN concentration in the ternary monomeric system with a high MAA content leads to a higher number of particles and better stability of the dispersion at relatively low temperatures as well.

Emulsion copolymerization of acrylic monomers and unconjugated dienes was studied with the aim to explore the possibility to synthesize dispersions whose particles would contain reactive polymeric molecules with free multiple C=C bonds. The usage of such latexes to finish fabrics and some other materials promotes getting strong indelible coatings. The kinetics and mechanism of emulsion copolymerization of ethylacrylate (EA) and butylacrylate (BA) with allylacrylate (AlA) (with ammonium persulfate (APS) or the APS—sodium thiosulfate system as initiators) were studied. The found constants of copolymerization of AlA with EA ($r_{AlA} = 1.05$: $r_{EA} = 0.8$) and AlA with BA ($r_{AlA} = 1.1$: $r_{BA} = 0.4$) point to different degrees of the copolymer unsaturation with AlA units. Emulsion copolymerization of multicomponent monomeric BA-based systems (with AN, MAA, and the unconjugated diene acryloxyethylmaleate (AOEM) as comonomers) was also studied. The effect of AN and MAA is described. Copolymerization with AOEM depends on reaction conditions.Example, AOEM reduces the polymerization rate in MAA-free systems. In the presence of MAA, the rate of the process at high conversion degrees increases with the AOEM concentration, this monomer promoting the gel effect due to partial chain linking in polymer–monomer particles by the side groups with C=C bonds. The unsaturation degree of diene units in the copolymer is subject to the composition of monomers, AOEM concentration, and temperature. In the BA—AOEM system, an increase of the diene concentration results in an increase in the unsaturation degree. This means that the diene radical is added to its own monomer with a higher rate than to BA ($r_{AOEM} = 7.7$). The higher unsaturation degree of diene units in the MAA-containing systems points to that the AOEM· radical interacts with MAA with a higher rate than with BA, and the probability of cyclization reduces, the degree of polymer unsaturation increases, and the diene's retarding effect upon polymerization weakens.

21.3.7 Compositions for Production of Rigid Foamed Polyurethane [25, 37, 38, 40]

In the field of polymer physico-chemistry, studies were made according to the requirements of the Montreal protocol on substances that deplete the ozone layer (1987), which demands to drastically reduce the production and consumption of chlorofluorocarbons (CFC, Freons) and even replace ozone-dangerous substances by ozone-safe ones. Our investigations dealt with the replacement of trichlorofluoromethane (Freon-11), which had been used as a foaming agent in the synthesis of rigid foam polyurethane (FPU) over a long period of time, which was thermal insulator in freezing chambers and building constructions. On the basis of our experimental dependences of the kinetic parameters of the foaming process (the instant of start, the time of structurization, and the instant of foam ending rising), the values of density and heat conductivity of pilot foam plastic samples on the concentration of the reaction mixture components and physico-chemical conditions of FPU synthesis, optimal compositions (recipes) of mixtures with Freon-11 replaced by an ozone-safe (with ozone-destruction potentials by an order of magnitude lower in comparison with Freon-11) azeotropic mixture of dichlorotrifluoroethane and dichlorofluoroethane were found. Their practical implementation requires no principal changes of known technological procedures and no usage of new chemical reagents, which is an important merit of our achievements.

21.3.8 Scientific Basics of the Synthesis of a High-Molecular-Weight Flocculent [24, 25, 39, 40]

There exists a problem of purification of natural water and industrial sewage from various pollutants, including suspended and colloid particles, associated with the growing consumption of water and its deteriorating quality (owing to anthropogenic influence). It is known that flocculants can be used for these purposes, which high-molecular-weight compounds capable of adsorption are on disperse particles to form quickly sedimenting aggregates. Polyacrylamide (PAA) is most active of them. As many countries (including Russian Federation) suffer from acrylamide (AA) deficit, we have developed modifications of the PAA flocculent synthesis by means of the usage of AN and sulfuric acid to bring about the reactions of hydrolysis and polymerization. The AN has turned out to participate in both processes simultaneously in the presence of a radical initiator of polymerization and sulfuric acid, and as AA is being formed from AN, their joint polymerization begins. Desired polymer properties were achieved at AN polymerization in an aqueous solution of sulfuric acid up to a certain conversion degree with subsequent hydrolysis of the polymerizate (a two-stage synthesis scheme) or at achieving an optimal ratio of the rates of these reactions proceeding in one stage. The influence of the nature and concentration of initiator, the sulfuric acid content temperature and reaction duration on the quantity and molecular mass of the polymer contained in the final product, its solubility in water and flocculating properties were studied. The required conversion degree at the first stage of synthesis by the two-stage scheme is determined by the concentrations of AN and the aqueous solution of sulfuric acid. At one-stage synthesis, changes in temperature and monomer amount almost equally affect the hydrolysis and polymerization rates and do not strongly affect the copolymer composition. But changes in the concentration of either acid or initiator rather strongly influence the molecular mass and composition of macromolecules, which causes extremal dependences of the flocculating activity on these factors.

21.4 CONCLUSION

New kinetic regularities at polymerization of vinyl monomers in homophase and heterophase conditions in the presence of additives of transition metal salts, azonitriles, peroxides, stable nitroxyl radicals and radical anions (and their complexes), aromatic amines and their derivatives, emulsifiers and solvents of various nature were revealed. The mechanisms of the studied processes have been established in the whole and as elementary stages, their basic kinetic characteristics have been determined. Equations to describe the behavior of the studied chemical systems in polymerization reactions proceeding in various physico-chemical conditions have been derived. Scientific principles of regulating polymer synthesis processes have been elaborated, which allows optimization of some industrial technologies and solving most important problems of environment protection.

KEYWORDS

- **Aqueous solution of sodium sulfocyanide**
- **Chemical physics**
- **Dimethyl formamide**
- **Methylmethacrylate**
- **Tetracyanoquinodimethane**

REFERENCES

1. Semenov, N. N. Chemical Physics. *Chemical Physics on the Threshold of the 21 Century*. Nauka, Moscow, pp. 5–9 (1996).
2. Goldfein, M. D. *Kinetics and mechanism of radical polymerization of vinyl monomers*. Saratov Univ. Press, Saratov, p. 139 (1986).
3. Goldfein, M. D., Kozhevnikov, N. V., and Trubnikov, A. V. *Kinetics and regulation mechanism of polymer formation processes*. Saratov Univ. Press, Saratov, p. 178 (1989).
4. Kozhevnikov, N. V., Gayvoronskaya, S. I., and Leontieva, L. T. In *Kinetics and mechanism of radical and polymerization processes*. Saratov Univ. Press, Saratov, pp. 85–93 (1973).
5. Stepukhovich, A. D., Kozhevnikov, N. V., and Leontieva, L. T. *Polymer Science, Series A*. **16**(7), 1522–1529 (1974).
6. Kozhevnikov, N. V. and Stepukhovich, A. D. *Polymer Science, Series A*. **21**(7), 1593–1599 (1979).
7. Goldfein, M. D., Kozhevnikov, N. V., and Rafikov, E. A. et al. *Polymer Science, Series A*. **17**(10), 2282–2287 (1975).
8. Goldfein, M. D., Rafikov, E. A., and Kozhevnikov, N. V. et al. *Polymer Science, Series A*. **19**(2), 275–280 (1977).
9. Goldfein, M. D., Rafikov, E. A., and Kozhevnikov, N. V. et al. *Polymer Science, Series A*. **19**(11), 2557–2562 (1977).
10. Goldfein, M. D. and Zyubin, B. A. *Polymer Science, Series A*. **32**(11), 2243–2263 (1990).
11. Stepukhovich, A. D. and Kozhevnikov, N. V. *Polymer Science, Series A*. **18**(4), 872–878 (1976).
12. Kozhevnikov, N. V. and Stepukhovich, A. D. *Polymer Science, Series A*. **22**(5), 963–971 (1980).
13. Kozhevnikov, N. V. Proceedings of Russian Higher-Educational Establishments. *Chemistry and Chemical Technology*, **30**(4), 103–106 (1987).
14. Rozantsev, E. G., Goldfein, M. D., and Trubnikov, A. V. *Advances in Chemistry*, **55**(11), 1881–1897 (1986).
15. Rozantsev, E. G., Goldfein, M. D., and Pulin, V. F. *Organic Paramagnets*. Saratov Univ. Press, Saratov, p. 340 (2000).
16. Rozantsev, E. G. and Goldfein, M. D. *Oxidation Communications*, Sofia. **31**(2), 241–263 (2000).
17. Rozantsev, E. G. and Goldfein, M. D. *Polymers Research Journal*. Nova Science Publishers, New York, **2**(1), 5–28 (2008).
18. Rozantsev, E. G. and Goldfein, M. D. *Chemistry and Biochemistry*. From Pure to Applied Science, New Horizons, New York. **3**, 145–169 (2009).

19. Goldfein, M. D., Kozhevnikov, N. V., and Trubnikov, A. V. *Polymer Science, Series B.* **25**(4), 268–271 (1983).
20. Goldfein, M. D. and Gladyshev, G. P. *Advances in Chemistry*, **57**(11), 1888–1912 (1988).
21. Goldfein, M. D., Kozhevnikov, N. V., and Trubnikov, A. V. *Russian Chemical Industry*. No 1. 20–22 (1989).
22. Simontseva, N. S., Kashanova, T. T., and Goldfein M. D. *Russian Plastics*, (12). 25–28 (1989).
23. Goldfein, M. D., Gladyshev, G. P., and Trubnikov, A. V. *Polymer Yearbook*, (13). 163–190 (1996).
24. Kozhevnikov, N. V., Goldfein, M. D., and Kozhevnikova, N. I. *Journal of the Balkan Tribological Association*, Sofia, **14**(4), 560–571 (2008).
25. Kozhevnikov, N. V., Kozhevnikova, N. I., and Goldfein, M. D. Proceedings of Saratov University. *New Series. Chemistry, Biology, Ecology*, **10**(2), 34–42 (2010).
26. Kozhevnikov, N. V., Goldfein, M. D., Zyubin, B. A., and Trubnikov, A. V. *Polymer Science, Series A,* **33**(6), 1272–1280 (1991).
27. Goldfein, M. D., Kozhevnikov, N. V., and Trubnikov, A. V. *Polymer Science, Series A*, 33(10), 2035–2049 (1991).
28. Goldfein, M. D., Kozhevnikov, N. V., and Trubnikov, A. V. *Polymer Yearbook*, (12), 89–104 (1995).
29. Kozhevnikov, N. V., Zyubin, B. A., and Simontsev, D. V. *Polymer Science, Series A.* **37**(5), 758–763 (1995).
30. Kozhevnikov, N. V., Goldfein, M. D., and Terekhina, N. V. *Russian Chemical Physics*, **16**(12), 97–102 (1997).
31. Kozhevnikov, N. V., Terekhina, N. V., and Goldfein, M. D. Proceedings of Russian Higher-Educational Establishments. *Chemistry and Chemical Technology*, **41**(4), 83–87 (1998).
32. Kozhevnikov, N. V., Goldfein, M. D., and Trubnikov, A. V. *Inter. Journal Polymer Mater*, **46**, 95–105 (2000).
33. Kozhevnikov, N. V., Goldfein, M. D., and Trubnikov, A. V. *Preparation and Properties of Monomers: Polymers and Composite Materials.* Nova Science Publishers, New York, pp. 155–163 (2007).
34. Kozhevnikov, N. V., Goldfein, M. D., Trubnikov, A. V., and Kozhevnikova, N. I. *Journal of the Balkan Tribological Association*, Sofia, **13**(3), 379–386 (2007).
35. Kozhevnikov, N. V., Kozhevnikova, N. I., and Goldfein, M. D. Proceedings of Russian Higher-Educational Establishments. *Chemistry and Chemical Technology*, **53**(2), 64–68 (2010).
36. Kozhevnikov, N. V., Goldfein, M. D., and Kozhevnikova, N. I. *Journal of Characterization and Development of Novel Materials*, **2**(1), 53–62 (2011).
37. Goldfein, M. D. and Kozhevnikov, N. V. *Problems of Regional Ecology.*, (4). 92–95 (2005).
38. Goldfein, M. D. Proceedings of Saratov University. *New Series. Chemistry, Biology, Ecology*, **9**(2), 79–83 (2009).
39. Kozhevnikov, N. V., Goldfein, M. D., and Kozhevnikova, N. I. *Modern Tendencies in Organic and Bioorganic Chemistry: Today and Tomorrow*. Nova Science Publishers, Inc. New York, pp. 379–384 (2008).
40. Goldfein, M. D., Ivanov, A. V., and Kozhevnikov, N. V. *Fundamentals of General Ecology, Life Safety and Environment Protection.* Nova Science Publishers, p. 210 (2010).

22 A Note on Antioxidant Properties

G. E. Zaikov, N. N. Sazhina, A. E. Ordyan, V. M. Misin

CONTENTS

22.1 Introduction 221
22.2 Experimental 222
22.3 Discussion and Results 224
22.4 Conclusion 238
Keywords 238
References 238

22.1 INTRODUCTION

Definition of the total phenol type antioxidant content and activity of compounds with respect to oxygen and its radicals in various alcohol drinks by two electrochemical methods.

The total content of phenol type antioxidants (AO) and their activity with respect to oxygen and its radicals are measured by two operative electrochemical methods: ammetric and voltammetric in samples more than 100 various alcohol drinks. Results of measurements show good (80–90%) correlation of these methods for dry red and white wines. Use of indicated methods allows revealing forged alcoholic production.

For a long time salutary influence of dry grape wines, cognacs and other qualitative alcoholic drinks on human health is known, certainly at their moderate consumption. It is connected, basically, with presence in them of natural AO, the main things from which are polyphenols. According to some researches [1, 2], polyphenols of grapes, in particular bioflavonoids, play a main role in display of "the French paradox", consisting that among the regularly consuming red dry grape wines population, lower susceptibility to cardiovascular and oncological diseases [1, 2] is revealed. Therefore, the sufficient attention is given to research antioxidant activity of various wines and their components *in vitro* and *in vivo*, despite a considerable quantity of already available works is paid now. Authors of the book [3] result the wide review of these works in which structure of different alcoholic drinks, the total antioxidant content and their biological activity are investigated by various methods. Basically, these methods are chemical methods and liquid chromatography methods [4, 5]. The aim of the present work was research of the total antioxidant activity of various alcoholic drinks by two electrochemical methods: ammetric and voltammetric and comparison of the received results.

22.2 EXPERIMENTAL

Objects of research were dry, semidry, dessert, fortified red and white wines, cognacs, liquors and infusions received from manufactures or their distributors at an exhibition in Moscow in 2009.

Used ammetric method allows defining the total phenol type antioxidant content in investigated samples [3, 6]. The essence of this method consists in measurement of the electric current arising at oxidation of investigated substance on a surface of a working electrode at certain values of electric potential (0–1.3). At such values of potential there is an oxidation only OH-groups of natural phenolic AO (R-OH). The electrochemical oxidation proceeding under scheme $R–OH \rightarrow R–O^{\bullet} + e^- + H^+$, can be used as model at measurement of free radical absorption activity. Capture of free radicals is carried out according to reaction $R–OH \rightarrow R–O^{\bullet} + H^{\bullet}$. Both reactions include rupture of same communication O–H. In this case ability to capture of free radicals by flavanoids or other polyphenols can be measured by oxidability of these substractions on a working electrode of ammetric detector [7]. The registered signal (the area under a current curve) is compared to a signal received in the same conditions for the sample of comparison with known concentration. Gallic acid (GA) was used as such sample in the present work. Measurement error of the total phenol type antioxidant content taking into account reproducibility of results was 10%. Measurement time of one sample makes 10–15 min.

The voltammetric method uses the process of oxygen electroreduction (ER O_2) as modeling reaction. It proceeds at the working mercury film electrode (MFE) in several stages with formation of the reactive oxygen species (ROS), such as O_2^- and HO_2 [8]:

$$O_2 + e^- \rightleftharpoons O_2^- \quad (1)$$

$$O_2^- + H^+ \rightleftharpoons HO_2 \quad (2)$$

$$HO_2 + H^+ + e^- \rightleftharpoons H_2O_2 \quad (3)$$

$$H_2O_2 + 2H^+ + 2e^- \rightleftharpoons 2H_2O \quad (4)$$

For determination of total antioxidant activity it is offered to use the first wave of ER O_2 corresponding to stages (1)–(3) when on a surface of a MFE active oxygen radicals and hydrogen peroxide are formed. It should be noted that AO of various natures were divided into 4 groups according to their mechanisms of interaction with oxygen and its radicals (Table 1) [8, 9].

TABLE 1 Groups of biological active substances (BAS) divided according mechanisms of interaction with oxygen and its radicals.

N Group	1 group	2 group	3 group	4 group
Substance names	Catalyze, phtalocyanines of metals, humic acids.	Phenol nature substances, vitamins A, E, C, B, flavonoids.	N, S, Se-containing substances, amines, amino acids.	Superoxide dismutase (SOD), porphyry metals, cytochrome C
Influence on ER O_2 process	Increase in current of ER O_2, potential shift in negative area	Decrease of current of ER O_2, potential shift in positive area	Decrease of current of ER O_2, potential shift in negative area	Increase in current of ER O_2, potential shift in positive area
The prospective electrode mechanism.	EC* mechanism with the following reaction of hydrogen peroxide disproportion and partial regeneration of molecular oxygen.	EC mechanism with the following chemical reaction of interaction of AO with active oxygen radicals	CEC mechanism with chemical reactions of interaction of AO with oxygen and its active radicals	EC* mechanism with catalytic oxygen reduction via formation of intermediate complex.

*The note: E = electrode stage of process, C = chemical reaction.

Kinetics criterion K is used as an antioxidant activity criterion of the investigated substances. It reflects quantity of oxygen and their radicals which have reacted with AO (or their mixes) in a minute:

For the second and third groups of AO: $K = \frac{C_0}{t}(1 - \frac{I}{I_0})$, μmol/l×min, (5)

For the firth and four groups of AO: $K = \frac{C_0}{t}(1 - \frac{I_0}{I})$, μmol/l×min, (6)

where, I, I_0 = limited values of the current of oxygen electroreduction, accordingly, at presence I and at absence I_0 of AO in the supporting electrolyte, C_0 = initial concentration of oxygen in the supporting electrolyte (μmol/l), that is solubility of oxygen in supporting electrolyte under normal conditions, t = time of interaction of AO with oxygen and its radicals, min.

To implement this method automated device "AOA" (Ltd. "Polyant", Tomsk, Russia) was used [9]. A phosphate buffer solution of 10 ml with known initial concentration of molecular oxygen was used as the background electrolyte, in which doses of the test samples (50–150 ml) were added. That concentration of investigated wine samples in volume of a buffer solution was corresponded to concentration of the samples entered into a cell of the ammetric device; wines were diluted in 50–300 times before introduction in an ammetric cell. The voltammetric method has good sensitivity and is simple and cheap. However, as in any electrochemical method of this type, the scatter of instrument readings at given measurement conditions is rather high (up to a factor of 1.5–2.0). Therefore, each sample was tested 4–5 times, and results were averaged. The maximum standart deviation of kinetic criterion K from average value for all investigated samples was 30%.

22.3 DISCUSSION AND RESULTS

Table 2 shows results of measurements of the total phenol type antioxidant content in units of GA concentration (C, mg/l GA), obtained by ammetric, and kinetic criterion (*K*, μmol/l×min), characterizing the total antioxidant activity with respect to oxygen and its radicals, for more than 100 samples of alcoholic drinks.

For clarity, diagrams (Figures 1 and 2) presents comparative measuring data of the total AO content in some typical samples of drinks from Table 2 for dry, semi-dry, semi-sweet wines and champagnes (Figure 1), and fortified wines and cognacs (Figure 2).

Diagrams show that dry and semi-dry red wines have the highest values of the total phenol AO content (from 200 to 430 mg/l GA), that is essential more than dry and semi-dry white wines (from 50 to 200 mg/l GA). The phenolic content of red semi-sweet wines also dominate compared with semi-sweet white wines and champagnes. In work [10] the total content of phenols has been measured in dry red wines by the Folin-Ciocalteau method with GA as a standard. Values of the phenol content have appeared in 2–5 times more, than in the present work. Apparently, it is possible to explain that besides phenols, with a reactant of Folin-Ciocalteau can react other substances

TABLE 2 Results of measurements of the total phenol type antioxidant content (C, mg/l GA), and kinetic criterion (K, mmol/l×min).

№ of sample	The company, country	Name, type of wine, wine age.	C, mg/l GA	*K*, mmol/l×min
1	Navarra, Spain	Palacio de OTAZU red semi-dry, 2001	328.1	1.20
2	Navarra, Spain	Palacio de OTAZU red semi-dry, 2003	302.3	1.21
3	Navarra, Spain	Palacio de OTAZU red semi-dry, 2006	277.7	1.05
4	Lancelot, Italy	Castello del Sono, red semi-dry, 2009	261.1	1.07
5	Launcelot, France	Cabernet Sauvignon, red semi-dry, 2008	254.6	0.90
6	Legenda Kryma, Moldova	Sauvignon de Purcari, dry white, 2005	82.5	0.57
7	Batitu Classic, Chile	Sauvignon Blanc, a dry white, 2009	74.2	0.58
8	Vega Libre, Spain	Malbeo, dry white, 2008	83.5	0.61
9	Batitu Classic, Chile	Chardonnay, dry white, 2008	122.5	0.85
10	Batitu Reserv, Chile	Cabernet Sauvignon, dry red, 2007	245.5	0.77
11	Batitu Classic, Chile	Cabernet Sauvignon, dry red, 2008	217.4	0.80
12	Batitu Reserve, Chile	Merlot, dry red, 2007	236.0	1.09
13	Negru De Purcaru, Moldova	Cabernet Sauvignon, dry red, 2003	209.2	0.95
14	Batitu Classic, Chile	Merlot, dry red, 2008	206.2	1.03
15	Inkerman, Crimea	Aligote, dry white, 2006	86.0	0.57

TABLE 2 *(Continued)*

№ of sample	The company, country	Name, type of wine, wine age.	C, mg/l GA	*K*, mmol/l×min
16	Legenda Kryma	Aligote, dry white, 2006	127.3	0.88
17	Koktybel, Crimea	Aligote, dry white, 2006	83.8	0.47
18	Inkerman, Crimea	Cabernet Kachinske, dry red, 2006	199.9	0.97
19	Inkerman, Crimea	Cabernet Coptobe, dry red, 2009	312.5	0.95
20	Legenda Kryma	Cabernet, dry red wine, 2009	254.8	0.88
21	Legenda Kryma	Muscat Sater, white semi-sweet, 2009	115.1	0.6
22	Legenda Kryma	Muscat Sater, semi-sweet red, 2009	153.3	0.91
23	Koktybel, Crimea	Monte Rouge, semi-sweet red, 2008	206.5	0.95
24	Cherny Polkovnik, Crimea	Coupage, red special, 1998	192.4	0.71
25	Koktybel, Crimea	Cahors, red special, 2008	174.3	0.84
26	Cherny Doktor, Crimea	Solnechnaya Dolina, red special, 2003	196.1	0.98
27	Parteniyskaya Dolina, Crimea	Tauris, red port, 2009	99.2	1.19
28	Magarach, Crimea	Chernovoy Magarach, red port, 2005	180.1	1.48
29	Krymskoe Shampanskoe Sevastopol Winery, Crimea	semi-sweet champagne , 2009	155.9	0.62

TABLE 2 *(Continued)*

№ of sample	The company, country	Name, type of wine, wine age.	C, mg/l GA	*K*, mmol/l×min
30	Sevastopol Winery, Crimea	Sparkling white muscat, 2009	84.4	0.52
31	Artemevkoe Shampanskoe, Crimea	Brut champagne white, 2009	92.1	0.52
32	Novy Svet, Crimea	Brut champagne, 2005	132.9	0.82
33	Novy Svet, Crimea	Brut champagne white, 2006	90.7	0.50
34	J. Bouchon, Chile	Sauvignon Blanc, dry white, 2009	60.9	0.32
35	Grand, France	Chateau Grand Xilon, dry white, 2008	98.8	0.65
36	Vin De Bordeux, France	Chateau Grand Xilon, dry red, 2005	257.7	0.95
37	Krevansky combine Ararat, Armenia	Noah, cognac, 2002	29.2	0.91
38	Hardy, France	V.S.O.P., Cognac, 2002	59.6	0.51
39	Chaten Le Don, France	Bordeaux, dry red, 2006	324.9	1.04
40	Sadylly, Azerbaijan	dry white (coupage)	83.8	0.52
41	Matrasa, Azerbaijan	dry red (coupage)	244.0	0.94
42	Baku, Azerbaijan	semi-sweet red (coupage)	251.2	0.78
43	Ganuzh, Azerbaijan	semi-sweet red (coupage)	238.1	0.80

TABLE 2 *(Continued)*

№ of sample	The company, country	Name, type of wine, wine age.	C, mg/l GA	*K*, mmol/l×min
44	Baku, Azerbaijan	Cognac, 2001	18.3	0.79
45	Moscva, Azerbaijan	Cognac, 1998	26.3	0.76
46	Shirvan, Azerbaijan	Cognac, 1995	45.3	0.96
47	Apsheron, Azerbaijan	Ordinary cognac, 2005	5.5	0.33
48	Port wine Adam, Azerbaijan	Port wine, 2008	63.6	1.29
49	Martini Vermouth, Azerbaijan	Vermouth white, 2009	28.7	0.50
50	Dar Bogov, Amtel	Ginseng infusion, 2009	5.3	0.22
51	Vilage Saint Deni, France	Blanc Mouellux, white semi-sweet, 2009	58.7	0.41
52	Maison De La Fer, France	Blanc Mouellux, white semi-sweet, 2009	69.8	0.50
53	Baron Du Rua, France	Blanc Mouellux, white semi-sweet, 2009	57.3	0.39
54	Vilage Saint Deni, France	Rouge Mouellux, semi-sweet red, 2009	185.8	-
55	Maison De La Fer, France	Rouge Mouellux, semi-sweet red, 2009	160.0	0.75
56	Baron Du Rua, France	Rouge Mouellux, semi-sweet red, 2009	196.2	-
57	Severnaya Zvezda, St. Petersburg	Cognac, 2006	23.1	0.70
58	Severnaya Zvezda, St. Petersburg	Cognac, 2004	33.2	0.75

TABLE 2 *(Continued)*

№ of sample	The company, country	Name, type of wine, wine age.	C, mg/l GA	K, mmol/l×min
59	Generalissimus, St. Petersburg	Cognac, 2006	19.3	0.70
60	Chateau Du Razo, France	Sauvignon, dry white, 2007	75.8	0.53
61	Russkaya Loza, Krasnodar	Sauvignon, dry white, 2007	103.6	0.65
62	Russkaya Loza, Krasnodar	Chardonnay, dry white, 2007	103.3	0.61
63	Russkaya Loza, Krasnodar	Muscat, dry white, 2007	125.1	0.75
64	Russkaya Loza, Krasnodar	Cabernet dry red, 2007	271.0	0.91
65	Chateau Del Ain, France	Bordeaux, dry red, 2007	289.1	0.99
66	Carmen, Chile	Merlot, dry red, 2007	347.0	1.21
67	Carmen, Chile	Carmenere, dry red, 2006	321.6	1.01
68	Norton, Argentina	Malbec, dry red, 2004	274.0	0.88
69	Lo Tengo, Argentina	Malbec, dry red, 2007	277.2	1.23
70	Saint-Anac, France	Moellux Dor De Alix, white semi-sweet, 2007	86.5	0.60
71	Cocoa	Liquor, liqueur, 2009	7.6	0.41
72	Strawberries	Liquor, 2009	75.0	0.83

TABLE 2 *(Continued)*

№ of sample	The company, country	Name, type of wine, wine age.	C, mg/l GA	*K*, mmol/l×min
73	Mint	Liquor, 2009	6.2	0.43
74	Taso Real, Spain	Gresalino, white semi-sweet, 2008	93.1	0.50
75	El Toril, Mexico	Tequila, 2009	4.8	0.60
76	Vina Sutil, Chile	Chardonnay, dry white, 2009	104.9	0.66
77	Vismos, Moldova	White Agate, cognac, 2004	69.7	0.82
78	Barton & Guestier, France	Sauvignon Blanc, dry white, 2009	136.0	0.88
79	Barton & Guestier, France	Bordeaux dry white wine, 2008	105.5	0.67
80	Vina Sutil, Chile	Cabernet dry red wine, 2006	276.2	0.94
81	Barton & Guestier, France	Merlot, dry red, 2008	290.4	1.05
82	Chinese fruit, Rotor House, China	Fruit wine, plum, 2007	94.0	-
83	Algeria, Algeria	Medea, Rose, 2008	172.4	-
84	ONCV, Algeria	Rose de la Hitijia, Rose, 2009	172.5	-
85	Algeria, Algeria	Medea, dry red, 2008	287.0	1.01
86	Vinariatiganga, Moldova	Fiteasca Regala, dry white, 2007	86.9	0.59
87	Lambouri, Cyprus	Lambouri, dry white, 2007	102.7	0.91

TABLE 2 *(Continued)*

№ of sample	The company, country	Name, type of wine, wine age.	C, mg/l GA	*K*, mmol/l×min
88	Stretto, Russia	Stratton, dry red, 2006 -	312.6	1.15
89	Lambouri, Cyprus	Maratheptico, dry red wine, 2006	323.3	1.06
90	Mavrud, Bulgaria	Fsenovgrad, dry red, 2004	350.3	1.08
91	Vicariya Tsyganka , Moldova	Cabernet Sauvignon, dry red wine, 2006	259.7	0.89
92	Carmello Pati, Argentina	Cabernet Sauvignon, dry red wine, 2002	317.9	1.02
93	Carmello Pati, Argentina	Cabernet Sauvignon, dry red wine, 1999	289.7	0.87
94	Tokaji, Hungary	Furmint, dry white, 2005	137.5	1.02
95	Tokaji, Hungary	Furmint, white semi-sweet 2003	155.6	0.71
96	Tokaji, Hungary	Furmint, white semi-sweet 2002	185.9	-
97	Tokaji, Hungary	Kuve white sweet, 2006	164.5	-
99	Tokaji, Hungary	Aszu 5 putonyami, white sweet, 2003	222.9	-
100	Tokaji, Hungary	Aszu 3 putonyami, white sweet, 2000	236.7	-
101	Dalvina, Macedonia	Vranec, dry red wine, 2008	412.0	1.50
102	Dalvina, Macedonia	Merlo , dry red wine, 2008	408.1	1.45
133	Dalvina, Macedonia	Cabernet Sauvignon, dry red, 2008	399.5	1.21

TABLE 2 *(Continued)*

№ of sample	The company, country	Name, type of wine, wine age.	C, mg/l GA	*K*, mmol/l×min
104	Dalvina, Macedonia	Rhein Risling, dry white, 2008	115.8	0.95
105	Dalvina, Macedonia	Astraton, dry white, 2008	90.8	0.53
106	Skovin, Macedonia	Makedonsko Belo, dry white, 2008	210.1	1.39
107	Skovin, Macedonia	Chardonnay, dry white, 2008	153.2	0.95
108	Skovin, Macedonia	Syrah Cabernet, dry red, 2007	427.1	1.50
109	Skovin, Macedonia	Kale, dry red wine, 2007	410.0	1.46
110	Skovin, Macedonia	Makedonsko Crveno, dry red, 2008	346.7	1.07
111	Zonte's Foot step, Australia	Cabernet, dry red, 2008	428.5	1.35
112	Zonte's Foot step, Australia	Shiraz, dry red wine, 2008	415.0	1.28
113	Woop Woop, Australia	Shiraz, dry red wine, 2008	321.7	0.99
114	Ries, Australia	Merlot, red semi-dry, 2009	370.3	1.02
115	Farrese, Italy	Trebyano a'Arbutstso Farnese, dry white, 2008	128.9	0.89
116	Isla Negra, Chile	Cabernet Sauvignon Merlot, red semi-dry, 2009	410.4	1.32

presenting in wines , such as restoring sugars, proteins and amino acids [11]. The ammetric method excludes it. Smaller values of the total phenol content in the red wines, received in the present experiments, can be explained, besides, by not total oxidation of these phenols at passage through ammetric cell [6].

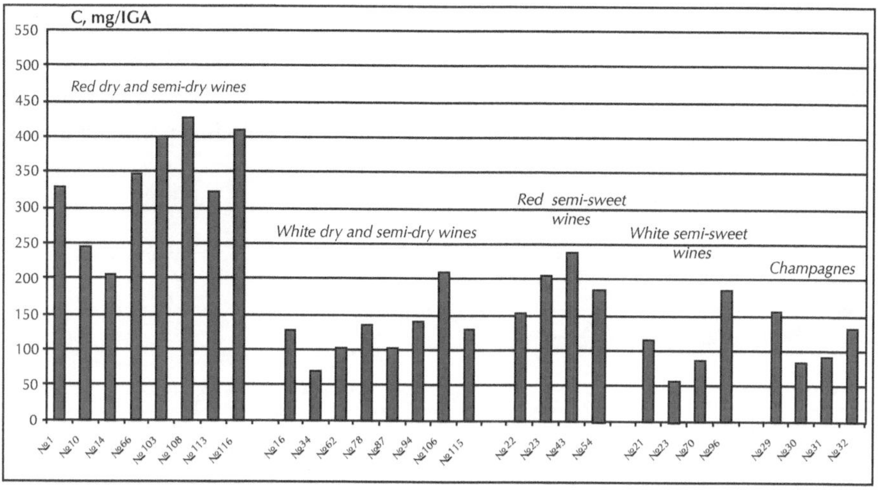

FIGURE 1 The total phenol type antioxidant content (C, mg/l GA) in samples of dry, semi-dry and semi-sweet wines.

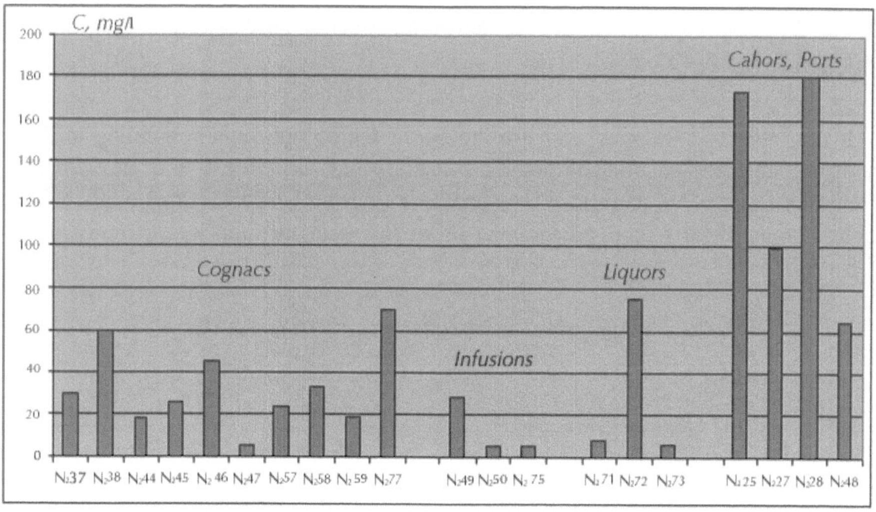

FIGURE 2 The total phenol type antioxidant content (C, mg/l GA) in samples of fortified wines, infusions, liquors and cognacs.

Red port wines and cahors have the lead positions between fortified wines (Figure 2, sample №25, 28). The level of phenol compounds in cognacs was significantly lower (20–70 mg/l GA). At this level the cognac sample № 47 is sharply behind (5.5 mg/l GA). A similar results for infusion of ginseng (№ 50), for a tequila (№ 75), cocoa and peppermint liqueurs (№ 71, № 73) were observed. The same minor values of C were recorded for 40% ethanol. Relatively low values of phenol AO content in these samples may indicate that they are probably falsified.

With regard to the total activity of drink AOs with respect to oxygen and its radicals, different samples show different mechanisms of its influence on the ER O_2 process and the reduction current form. Figure 3 shows a typical voltammograms (VA-grams) for all red dry, semi-dry, semi-sweet, special wines, including ports and cahors.

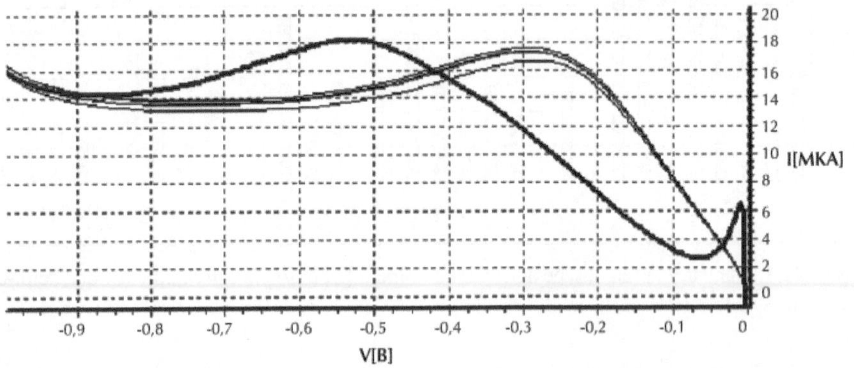

FIGURE 3 Typical VA-grams of ER O_2 current for red wines in absence of sample (the left curve) and presence (the right curves).

It is visible that the mechanism of action of the polyphenols containing in these wines, is characteristic for compounds of 2nd groups in Table 1 and the values of potential shift of EV O_2 current limit increase with growth of the total phenol content. For this group of substances (R-OH) the following mechanism of interaction with the ROS is offered [8].

$$O_2 + e^- \rightleftarrows O_2^{\cdot -} + \text{R-OH} \rightleftarrows HO_2 + \text{R=O} \quad (7)$$

$$HO_2^{\cdot} + \text{R-OH} \rightleftarrows H_2O_2 + \text{R=O} \quad (8),$$

where, R=O = oxidative form of the AO (R–O$^{\cdot}$). According to the theory of electrode processes, an indicator of thermodynamics of electrochemical processes is the potential of a half wave of a EV O_2 current which is described by the classical Nernst equation having for electrode process (7) at temperature T = +25°C the following kind [9]:

$$E_{O2/O2}^{\cdot-} = E_{O2/O2}^{\cdot-} + 0{,}059 \cdot \lg(C_{O2}/C_{O2}^{-}) \quad (9)$$

If superoxide anion (O_2^-) reacts in reaction (7) with an antioxidant (R-OH), CO_2^- decreases, the potential increases according to (9) and the maximum of EV O_2 current haves shift in positive area. The values of kinetic criterion K arise with increasing of the total phenol AO content, reaching values of 1.3–1.5 μmol/l×min for samples of red wines (Table 2, № 66, 101, 108). Figure 4 shows the correlation between K and C for 40 samples of dry and semi-dry red wines from Table 2. The correlation coefficient was r = 0.8212.

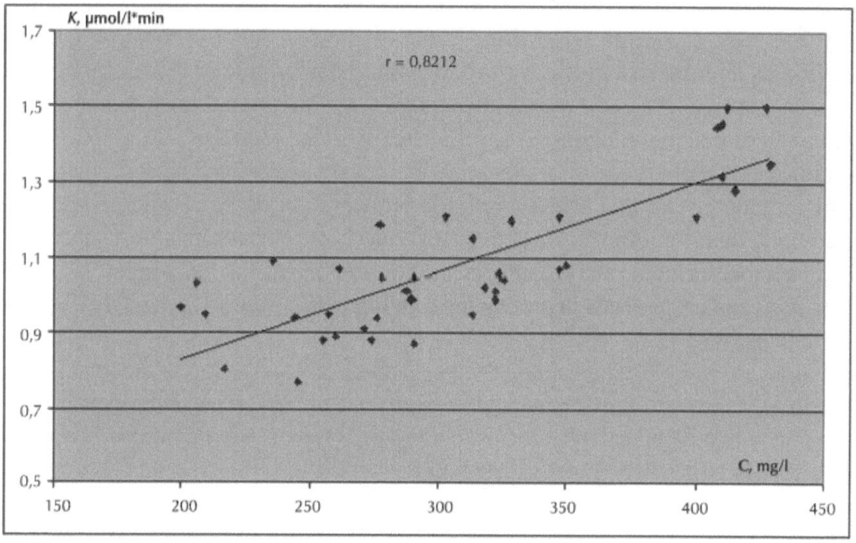

FIGURE 4 Correlation dependence between the kinetic criterion K and the total AO content C for dry and semi-dry red wines (r = 0.8212).

It is known from numerous studies [3] of red grape wines, that in they more than 100 polyphenolic compounds, including flavonoids and polyphenols non-flavonoid nature were identified [12]. The main contribution to their AO activity make compounds of following groups: anthocyans, catechins, oligomeric procyanidins, flavonols, stilbenes, glucosides and so on. [3]. Anthocyans (anthocyanidins and anthocyanins) are the main pigments of grape and cause the red wine color [12]. They have a wide spectrum of biological activity for human–increase elasticity of blood vessels and improve visual acuity [13]. In addition, anthocyans affect capillary permeability, as well as the hematopoietic function of bone marrow [14]. Catechins and procyanidins, whose content is more than 90% of the total polyphenols of grape and wine [15], are the most activity AO [16]. Some catechins, for example epicatechin and epigallokatehingallat, can induce apoptosis in cancer cells [17]. Procyanidins brake oxidation of low-viscosity lipoproteins in blood, preventing cardiovascular disorders [18]. The high AO activity have in wine also phenolic acids, particularly gallic and caffeic, and flavonols: quercetin, rutin, myricetin, reducing blood cholesterol level [19]. The above and other phenol compounds [3] contribute mainly their part in high values of the total content of AO and their activity for red wines.

Obtained in this study VA-grams of ER O_2 current for dry and semi-dry white wines are not much different from those for red wines (Figure 3), however, potential maximum shift of working current relatively to the background is slight, indicating a lower level of the total phenol AO content in white wines. The range of kinetic criterion values K for investigated white wines was from 0.4 to 1.0 mmol/l×min. Correlation between the kinetic criterion and the total phenol AO content, built for 24 samples of dry and semi-dry white wines from Table 2, showed better correlation between two methods than for red wines, with r = 0.9061. Comparative analysis of white and red wines, made in [20, 21], shows a smaller (5–10 times) total polyphenol content in white wines than in red wines. It is associated with the fact, that in fermentation of grape juice skins and seeds of grapes, rich by anthocyans, catechins and other polyphenol compounds are included. During the fermentation of white wine, grape juice is used only, therefore much less active AO: catechins, epicatechins, GA, quercetin, rutin, cyanidin, miricitin, and proanthocyanidins are found in white wines [20]. Inclusion of skins and grape seeds in production of white wines significantly increases their antioxidant activity [21].

With regard to fortified alcoholic drinks, in addition to reducing of the total phenol AO content (Figure 2), they show another character of VA-grams and mechanism of interaction with oxygen and its radicals. On Figure 5 typical VA-gram of a EV O_2 current for the investigated cognacs is presented.

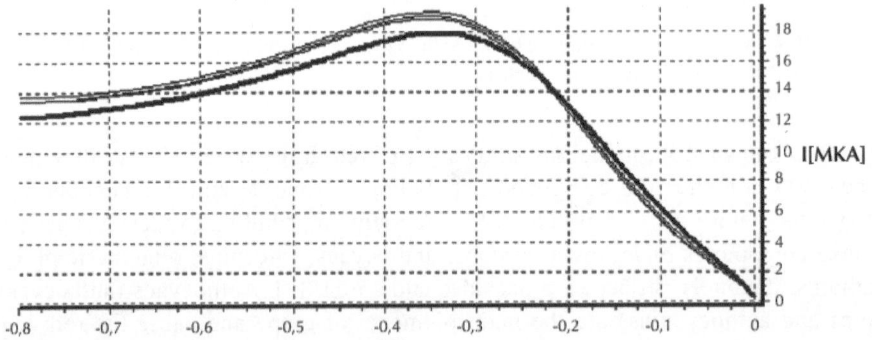

FIGURE 5 Typical VA-grams of ER O_2 current for cognacs in absence of sample (the bottom curve) and presence (the top curves).

Such behavior of a current is characteristic for compounds of 1st group in Table 1. For these compounds increase in current of ER O_2 results from influence of their catalytic properties on EV O_2 kinetics thanks to reaction of hydrogen peroxide disproportion and partial regeneration of O_2 (10)–(12) [9]:

$$O_2 + e^- + H^+ \rightleftarrows HO_2 \quad (10)$$

$$HO_2^{\cdot} + e^- + H^+ \rightleftarrows H_2O_2 \quad (11)$$

$$2H_2O_2 \xrightarrow{\text{catalyst}} 2H_2O + O_2 \quad (12)$$

Values of kinetic criterion K, calculated under the formula (6), have appeared for cognacs high enough (from 0.5 to 1.0 mmol/l×min), despite of the low total phenol AO content (Figure 2). Any essential correlation between the AO content and values K for cognacs was not observed. For the sample of cognac №47 value K has made 0.33 mmol/l×min and character of VA-grams did not correspond to typical VA-grams for cognacs (Figure 5), and reminded VA-gram for 40% ethanol (Figure 6), on which background and working currents practically merge. The same basic kind VA-gram had ginseng infusion (№50), tequila (№75), and liquors №71 and №73. It once again gives the reason to suspect their falsification. Strawberry liquor (№72) behaved like wines, showing decrease of ER O2 current and potential shift in positive area (Figure 3), also as vermouth (№49) and ports.

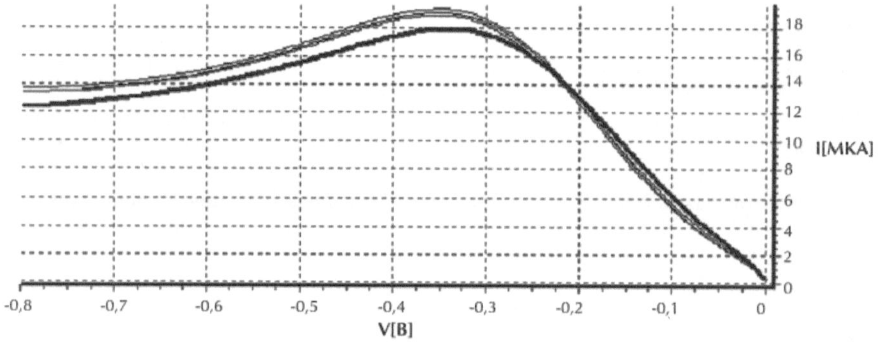

FIGURE 6 Typical VA-gram of ER O_2 current for 40% ethanol in absence of sample (the top curve) and presence (the bottom curves).

Quality cognacs contain AO, which are contained in brandy spirit and also tannins, a lignin, reducing sugars, oils, pitches, and enzymes getting to a ready product due to the extraction of them by spirit from wood of oak butts during "aging or endurance" of cognac. Cognac alcohol becomes golden color and is filled with woody-vanilla aromas due to the extraction of timber oksiaromatic acids, the main ones are lilac, ellagic, vanillic, as well as vanillin and lilac aldehyde [3]. Apparently, some enzymes of cognac and aromatic compounds in which include phthalocyanines, being effective catalysts of electrochemical reduction of oxygen, and determine the dominant mechanism of ER O_2 and character of VA-grams. Thanks to presence of tannins in cognacs, they help from gastro enteric frustration. Cognac also expands vessels of a brain and raises an organism tone. However medical properties of cognac are shown, if it is used only in small doses (30–50 ml).

22.4 CONCLUSION

Thus, use of two operative electrochemical methods allows you to determine quickly and relatively cheaply the total content of AO and their activity with respect oxygen and its radicals in various grades of wines and cognacs and to identify low-quality production. The results of the present work received for more than 100 samples of alcoholic drinks, show rather high (80–90%) correlation of these methods for dry, semi-dry red and white wines. These methods can be applied for quality control of wines and other alcoholic drinks.

KEYWORDS

- **Electrochemical methods**
- **Gallic acid**
- **Mercury film electrode**
- **Reactive oxygen species**
- **Voltammograms**

REFERENCES

1. Renaud, S. and De Longeril, M. Wine, alkogol, platelets and the French paradox for coronary heart disease. *Lancet*, **339** 1523–1526 (1992).
2. Perez, D. D., Strobel, P., Foncea, R., Diez, M. S., Vasquez, L., UrQuiaga, I., Castillo, O., Cuevas, A., San Martin, A., and Leignton, F. Wine, diet, antioxidant defenses and oxidative damage. *Ann. NY Acad. Sci.*, **957** 136–145 (2002).
3. Yashin, A. Ya., Yashin, Ya. I., Ryzhnev, V. Yu., and Chernousova, N. I. *Natural antioxidants*. The total content in foodstuff and their influence on health and human aging. Moscow, Translit, pp. 92–103 (2009).
4. Dolores Rivero-Perez, M., Muniz Pilar, and Gonzalez Sanjose Maria, L. Antioxidant Profile of Red Wines Evaluated by Total Antioxidant Capacity, Scavenger Activity, and Biomarkers of Oxidative Stress Methodologies. *J. Agric. Food Chem.*, **55** 3–8 (2007).
5. Alonso Angeles, M., Dominguez Cristina, Guillen Dominico, A., and Barroso Carmelo, G. Determination of Antioxidant Power of Red and White Wines by a New Electrochemical Method and Its Correlation with Polyphenolic Content. *J. Agric. Food Chem.*, **50** 3112–3115 (2002).
6. Yashin, A. Ya. Inject-flowing system with ammetric detector for selective definition of antioxidants in foodstuff and drinks. *Russian chemical magazine*, **LII**(2) 130–135 (2008).
7. Peyrat Maillard, M. N., Bonnely, S., and Berset, C. Determination of the antioxidant activity of phenolic compounds by coulometric detection. *Talanta*, **51** 709–714 (2000).
8. Korotkova, E. I., Karbainov, Y. A., and Avramchik, O. A. Investigation of antioxidant and catalytic properties of some biological-active substances by voltammetry. Anal. *and Bioanal. Chem.*, **375**(1–3) 465–468 (2003).

9. Korotkova, E. I. *Voltammetric method of determining the total AO activity in the objects of artificial and natural origin*. Doctoral thesis. Tomsk (2009).
10. Roginsky, V., De Beer, D., Harbertson, J. F., Kilmartin, P. A., Barsukova, T., and Adams, D. O. The antioxidant activity of Californian red wines does not correlate with wine age. *J. Sci. Food Agric.*, **86** 834–840 (2006).
11. Korenman, I. M. Method of organic compounds determination. *Photometric analysis. M.* (1970).
12. De Villiers, A., Vanhoenacker, G., Majek, P., and Sondra, P. Determination of anthocyanins in wine by direct injection liquid Chromatography-diode array detection-mass spectrometry and classification of wines using discriminant analysis. *J. Chromat. A.*, **1054** 194–204 (2004).
13. Kong, J. M., Chia, L. S., Goh, N. K., Chia, T. F., and Brouillard, R. Analysis and biological activities of anthocyanins. *Phytochemistry*, **64**(5) 923–933 (2003).
14. Lila, M. A. Anthocyanins and Human Health: An *in vitro* Investigative Approach. *J. Biomedicine and Biotechnology*, (5) 306–313 (2004).
15. Gachons, C. P. and Kennedy, J. A. Direct Method for Determining Seed and Skin Proanthocyanidin Extraction into Red Wine. *J. Agric. Food Chem.*, **51** 5877–5881 (2003).
16. Bagchi, D., Bagchi, M., Stohs, S. J., Das, D. K., Ray, S. D., Kuszynski, C. A., Joshi, S. S., and Pruess, H. G. Free radicals and grape seed proanthocyanidn extract: Importance in human health and disease prevention. *Toxicology*, **148** 187–197 (2000).
17. Ahmad, N., Gupta, S., and Mukhtar, H. Green tea polyphenol epigallocatechin-3-gallate differentially modulates nuclear factor kB in cancer cells versus normal cells. *Archives of Biochemistry and Biophysics*. **376** 338–346 (2000).
18. Bagchi, D., Sen, C. K., Ray, S. D., Dipak, K., Bagchi, M., Preuss, H. G., and Vinson, J. A. Molecular mechanisms of cardioprotection by a novel grape seed proanthocyanidin extract. *Mutation Research*, **523** 87–97 (2003).
19. Chen, C. K. and Pace-Asciak, C. R. Vasorelaxing activity of resveratrol and quercetin in isolated rat aorta. *Gen. Pharmacol*, **27** 363–366 (1996).
20. Simonetty, P., Pietta, P., and Testolin, G. Polyphenol Content and Total Antioxidant Potential of Selected Italian Wines. *J. Agric. Food Chem.*, **45** 1152–1155 (1997).
21. Fuhrman, B., Volkova, N., Soraski, A., and Aviram, M. White wine with red wine-like properties: Increased extraction of grape skin polyphenols improves the antioxidant capacity of the derived white wine. *J. Agric. Food Chem.*, **49** 3164–3168 (2001).

23 Antioxidant Activity of Various Biological Objects

G. E. Zaikov, N. N. Sazhina, E. I. Korotkova, and V. M. Misin

CONTENTS

23.1 Introduction 241
23.2 Experimental Part 242
23.2.1 Ammetric Method for Determining the Total Content of Antioxidants 242
23.2.2 Voltammetric Method for Determining the Total Activity of Antioxidants with Respect to Oxygen and its Radicals 243
23.3 Discussion and Results 246
23.3.1 Juices of Medicinal Plants 246
23.3.2 Water Extract of Mint 248
23.3.3 Extracts of Tea, Vegetative Additives and their Mixes 250
23.3.4 Plasma of Human Blood 253
23.4 Conclusion 254
Keywords 254
References 254

23.1 INTRODUCTION

Definition of the total phenol antioxidant content and activity of compounds with respect to oxygen and its radicals in various biological objects by two electrochemical methods and analysis of obtained results.

A comparison of the total content of antioxidants (AO) and their activity with respect to oxygen and its radicals in juice and extracts of herbs, extracts of a tea and also in human blood plasma was carried out in the present work by use of two operative electrochemical methods: ammetric and voltammetric. Efficiency of methods has allowed studying dynamics of AO content and activity change in same objects during time. Good correlation between the total phenol antioxidant content in the studied samples and values of the kinetic criterion defining activity with respect to oxygen and its radicals is observed.

For the last quarter of the century there was a considerable quantity of works devoted to research of the activity of AO in herbs, foodstuff, drinks, biological liquids, and other objects. It is known that the increase in activity of free radical oxidation processes in a human organism leads to destruction of structure and properties of lipid membranes. There is a direct communication between the superfluous content of free radicals in an organism and occurrence of dangerous diseases [1, 2]. The AO are class of biologically active substances which remove excessive free radicals, decreasing the lipid oxidation. Therefore a detailed research of the total antioxidant activity of various biological objects represents doubtless interest.

At present, there are a large number of various methods for determining the total AO content and also their activity with respect to free radicals in foodstuffs, biologically active additives, herbs and preparations, biological liquids and other objects [3]. However, it is impossible to compare the results obtained by different methods, since they are based on different principles of measurements, different modeling systems, and have different dimensions of the antioxidant activity index. In such cases, it is unreasonable to compare numerical values, but it is possible to establish a correlation between results obtained by different methods.

Ones from the simplest methods for study of antioxidant activity of various biological objects are electrochemical methods, in particular, ammetry, and voltammetry. A comparative analysis of the AO content and their activity in juice and extracts of herbs, extracts of tea and vegetative additives, and also in plasma of human blood is carried out in present work. Operability of methods has allowed studying also dynamics of change of the AO content and their activity in same objects during time.

23.2 EXPERIMENTAL PART

23.2.1 Ammetric Method for Determining the Total Content of Antioxidants

The essence of the given method consists in measurement of the electric current arising at oxidation of investigated substance on a surface of a working electrode at certain potential. An oxidation of only OH—groups of natural phenol type AO (R-OH) there is at this potential. The electrochemical oxidation proceeding under scheme R–OH → R–O$^{\bullet}$ + e^- + H^+ can be used under the assumption of authors [4], as model for measurement of free radical absorption activity which is carried out according to equation R–OH → R–O$^{\bullet}$ + H$^{\bullet}$. Both reactions include the rupture of the same bond O–H. In this case, the ability of same phenol type AO to capture free radicals can be measured by value of the oxidizability of these compounds on a working electrode of the ammetric detector [4].

Ammetric device "TsvetJauza-01-AA" in which this method is used, represents an electrochemical cell with a glassy-carbon anode and a stainless steel cathode to which a potential 1,3 V is applied [5]. The analyte is introduced into eluent by a special valve. As the analyte pass through the cell the electrochemical AO oxidation current is recorded and displayed on the computer monitor. The integral signal is compared to the signal received in same conditions for the comparison sample with known concentration. Quercetin and gallic acid (GA) were used in work as the comparison sample.

The root-mean-square deviation for several identical instrument readings makes no more than 5% [5].The error in determination of the AO content including the error by reproducibility of results was within 10%. The method involves no model chemical reaction, and measurement time makes 10–15 min.

23.2.2 Voltammetric Method for Determining the Total Activity of Antioxidants with Respect to Oxygen and its Radicals

The voltammetric method uses the process of oxygen electroreduction (ER O_2) as modeling reaction. This process is similar to oxygen reduction in tissues and plant extracts. It proceeds at the working mercury film electrode (MFE) in several stages with formation of the reactive oxygen species (ROS), such as O_2^- and HO_2 [6]:

$$O_2 + e^- \rightleftarrows O_2^{\cdot-} \qquad (1)$$

$$O_2^{\cdot-} + H^+ \rightleftarrows HO_2^{\cdot} \qquad (2)$$

$$HO_2 + H^+ + e^- \rightleftarrows H_2O_2 \qquad (3)$$

$$H_2O_2 + 2H^+ + 2e^- \rightleftarrows 2H_2O \qquad (4)$$

For determination of total antioxidant activity it is used the ER O_2. It should be noted that AO of various natures were divided into 4 groups according to their mechanisms of interaction with oxygen and its radicals (Table 1) [7].

The first group of substances increased ER O_2 current according mechanism (5)–(7):

$$O_2 + e^- + H^+ \rightleftarrows HO_2^{\cdot} \qquad (5)$$

$$HO_2^{\cdot} + e^- + H^+ \rightleftarrows H_2O_2 \qquad (6)$$

$$2H_2O_2 \xrightarrow{\text{catalyst}} 2H_2O + O_2 \qquad (7)$$

For the second group of the AO we suppose following mechanism of interaction of AO with the ROS (8):

TABLE 1 Groups of biological active substances (BAS) divided according mechanisms of interaction with oxygen and its radicals.

N Group	1 group	2 group	3 group	4 group
Substance names	Catalyze, phtalocyanines of metals, humic acids.	Phenol nature substances, vitamins A, E, C, B, flavonoids.	N, S-containing substances, amines, amino acids.	Superoxide dismutase(SOD), porphyry metals, cytochrome C
Influence on ER O_2 process	Increase of ER O_2 current, potential shift in negative area	Decrease of ER O_2 current, potential shift in positive area	Decrease of ER O_2 current, potential shift in negative area	Increase in ER O_2 current, potential shift in positive area
The prospective electrode mechanism.	EC* mechanism with the following reaction of hydrogen peroxide disproportion and partial regeneration of molecular oxygen.	EC mechanism with the following chemical reaction of interaction of AO with active oxygen radicals	CEC mechanism with chemical reactions of interaction of AO with oxygen and its active radicals	EC* mechanism with catalytic oxygen reduction via formation of intermediate complex.

*The note: E = electrode stage of process, C = chemical reaction.

$$O_2 + e^- \overset{k_0}{\rightleftarrows} O_2^{\cdot -} + \text{R-OH} + H^+ \overset{k_1^*}{\rightleftarrows} H_2O_2 + \text{R–O}^{\cdot} \tag{8}$$

The third group of the BAS decreased ER O_2 current via the following mechanism (9)–(11):

$$O_2 \longrightarrow O_2^S \xrightarrow[k_1]{+RSH} HO_2 + \bar{e} \overset{k_{R1}}{\rightleftarrows} HO_2^{\bullet -} + RS^{\bullet} \tag{9}$$

$$HO_2^{\bullet -} + RSH \rightleftarrows H_2O_2 + RS^{\bullet} \tag{10}$$

$$RS^{\bullet} + RS^{\bullet} \rightleftarrows RS - SR \tag{11}$$

For the fourth group of substance we could suggest mechanism with catalytic oxygen reduction *via* formation of intermediate complex similar by SOD (12).

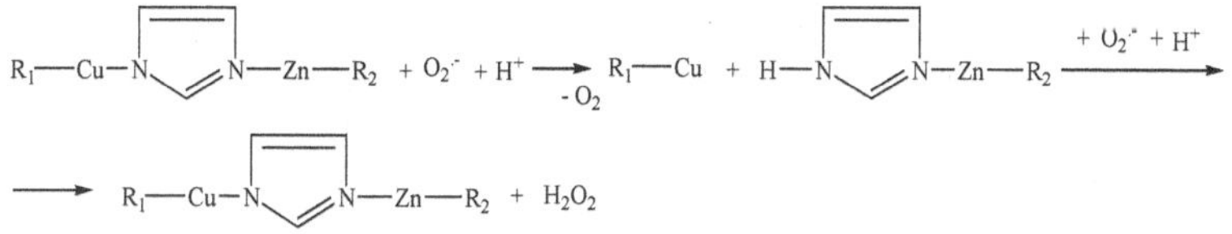

(12)

For voltammetric study of the total antioxidant activity of the samples automated voltammetric analyzer "Analyzer of TAA" (Ltd. "Polyant" Tomsk, Russia) was used. As supporting electrolyte the 10 ml of phosphate buffer (pH = 6.76) with known initial concentration of molecular oxygen was used [7]. The electrochemical cell (V = 20 ml) was connected to the analyzer and consisted of a working MFE, a silver-silver chloride reference electrode with KCl saturated ($Ag|AgCl|KCl_{sat}$) and a silver-silver chloride auxiliary electrode. The investigated samples (10–500 ml) were added in cell.

Criterion *K* is used as an antioxidant activity criterion of the investigated substances:

For the second and third groups of AO: $K = \frac{C_0}{t}(1 - \frac{I}{I_0})$ μmol/l×min, (13)

For the firth and four groups of AO: $K = \frac{C_0}{t}(1 - \frac{I_0}{I})$, μmol/l×min, (14)

where, I, I_0 = limited values of the ER O_2 current, accordingly, at presence and at absence of AO in the supporting electrolyte, C_0 = initial concentration of oxygen (μmol/l), that is solubility of oxygen in supporting electrolyte under normal conditions, t = time of interaction of AOs with oxygen and its radicals, min.

This method of research has good sensitivity. It is simple and cheap. However, as in any electrochemical method of this type, the scatter of instrument readings at given measurement conditions is rather high (up to a factor of 1.5–2.0). Therefore, each sample was tested 3–5 times, and results were averaged. The maximum standard deviation of kinetic criterion *K* for all investigated samples was within 30%.

23.3 DISCUSSION AND RESULTS

23.3.1 Juices of Medicinal Plants

In the present work juices pressed out from different parts of various medicinal plants, such as basket plant or golden tendril (Callisia fragrans), Moses-in-the-cradle (Rhoeo spatacea), Dichorisandra fragrants (Dichorisandra fragrans), Blossfelda kalanchoe (Kalanchoe blossfeldiana), air plant (Kalanchoe pinnatum) and devil's backbone (Kalanchoe daigremontiana) were investigated [8]. To preserve the properties of the juices, they were stored in refrigerator at –12°C and unfrozen to room temperature immediately before the experiment. In both methods before being poured into the measuring cell, the test juice was diluted 100-fold. A quercetin was used as the sample of comparison. Results of measurement of the total content of AO and their activity with respect to oxygen and its radicals in juice of investigated plants, received by above described methods, are presented in Table 2. For samples 1, 4, 7, and 12 where the small content of phenol type AO is observed, voltamperograms (VA-grams) look like, characteristic for substances of the third group of Table 1 (Figure 1). Other samples show the classical phenolic mechanism as substances of the second group (Figure 2), and have high values of the AO content and kinetic criterion *K*, especially the sample 10.

TABLE 2 The total content of AOs in the juice samples and their kinetic criterion K.

Sample no.	Plant name	Content of antioxidantsC, mg/l	*K*, μmol/(l min)
1	Juice from *Callisia fragrans* (golden tendril) leaves	63,6	0,72
2	Juice from *Callisia fragrans* (golden tendril) lateral sprouts (4–5 mm in diameter)	279,6	1,81
3	Juice from *Rhoeo spathacea* (Moses_in_the_ cradle) leaves	461,2	2,45
4	Juice from *Callisia fragrans* leaves	73,2	0,83
5	Juice from *Callisia fragrans* stalks (1–2 mm in diameter)	119,3	1,23

TABLE 2 *(Continued)*

Sample no.	Plant name	Content of antioxidantsC, mg/l	*K*, µmol/(l min)
6	Juice from kalanchoe (*Kalanchoe blossfeldiana*) bulblets	251,3	1,78
7	Juice from *Dichorisandra fragrans* stalks (6 mm in diameter)	38,41	0,52
8	Juice from *Dichorisandra fragrans* leaves	179,1	0,77
9	Juice from *Kalanchoe pinnata* (air plant) leaves	201,5	1,99
10	Juice from *Kalanchoe daigremontiana* (devil's backbone) leaves with bulblets	742,4	4,12
11	Juice from *Kalanchoe daigremontiana* stalks (5 mm in diameter)	142,5	1,47
12	Juice from *Callisia fragrans* herb + 20% ethanol	70,2	0,88

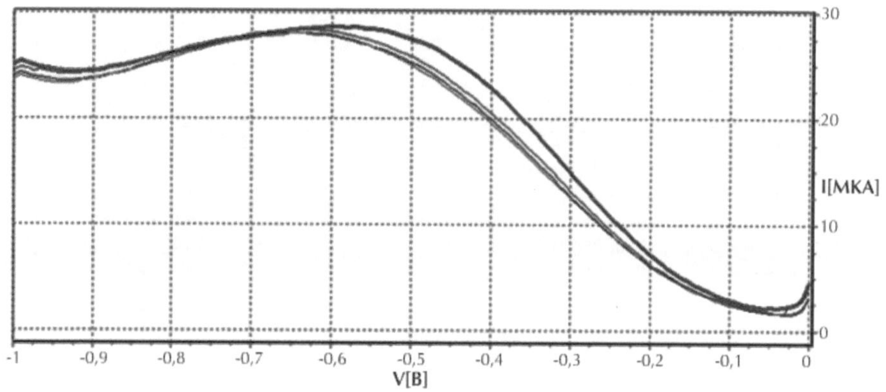

FIGURE 1 Typical VA-grams of the ER O2 current for compounds from the third group in Table1. The upper curve is the background current in AOs absence; left curves were recorded in AOs presence.

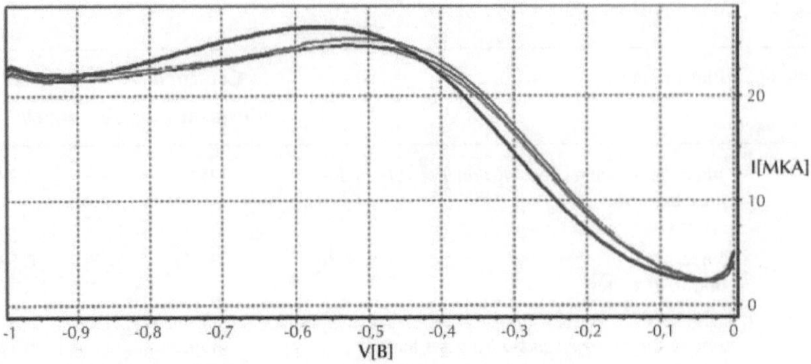

FIGURE 2 Typical VA-grams for AOs of the second substances group in Table 1. The upper curve is the background current.

Correlation dependence between the kinetic criterion *K* and the total AOs content is presented on Figure 3. The results of measurements spent for juices of medicinal plants, show high correlation (r = 0.96) between these methods. The explanation of received results is resulted in [8].

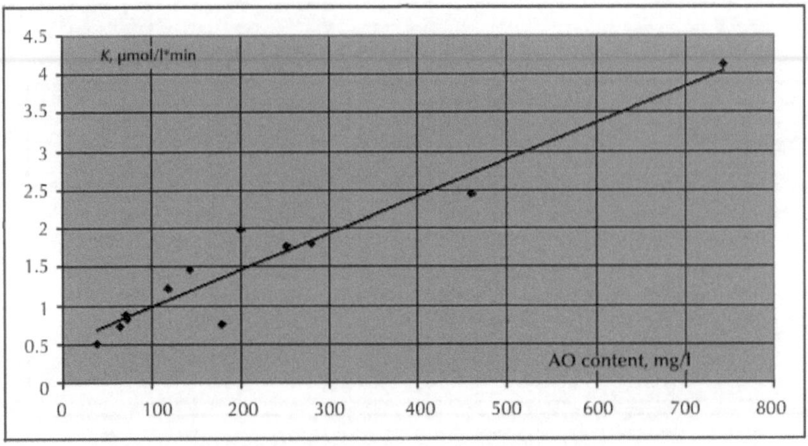

FIGURE 3 Correlation dependence between the kinetic criterion K and the total AOs content (r = 0.96).

23.3.2 Water Extract of Mint

The purpose of the present work is to measure the total antioxidant activity of a water mint extract by voltammetric method and to study the mechanisms of influence of mint components on the process of ER O_2. Concurrently measurements of the total phenol AO content were carried out by an ammetric method [9]. The object of the research was water extract of the mint peppery (Mentha piperita). Dry herb was grinded in a mortar till particles of the size 1–2 mm. Further this herb (0.5 g) was immersed into 50 ml of distilled water with T = 95°C and held during 10 min without thermostating. Then the extract was carefully filtered through a paper filter and if necessary diluted before measurements.

The VA grams of the ER O_2 current have been received in various times after mint extraction. It has appeared that the fresh extract (time after extraction t = 5 min) "works" on the mechanism of classical AO, reducing a current maximum and shifting its potential in positive area (Figure 2). However, approximately through t = 60 min, character of interaction of mint components with oxygen and its radicals changes, following the mechanism, characteristic for substances of the 4th group in Table 1 (Figure 4). Transition from one mechanism to another occurs approximately during t = 30 min after extraction and the kinetic criterion *K* becomes thus close to zero. At the further storage of an extract *in vitro* character of VA-grams essentially does not change, and the kinetic criterion caused by other mechanism, grows to values 2.0 μmol/l·min during 3 hr after extraction.

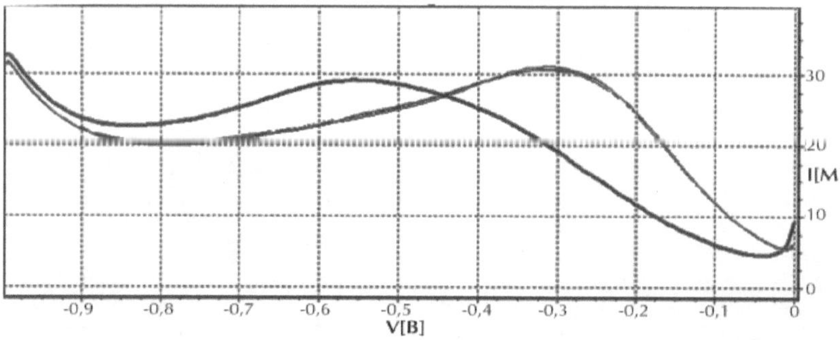

FIGURE 4 VA-grams of the ER O2 current in the absence (left curve) and in the presence (right curves) of mint extract at t = 60 min after mint extraction.

In parallel for the same mint extract the registration of the total phenol AO content C in mg of GA per 1 g of dry mint has been made during extract storage time t by ammetric method. Dependence C on t testifies to notable falling of C after extraction of a mint extract (approximately 20% for 2 hr of storage). It is possibly explained by destruction of unstable phenol substances contained in an extract. Therefore, apparently, VA-grams character and *K* values during the first moment after extraction could be established as influence of classical phenol AOs on the ER O_2 process. The general character of ER O_2 is defined already by the mint substances entering into 4th group of the Table 1. Probably, various metal complexes as mint components are dominated causing increase of ER O_2 current and catalytic mechanism of ER O_2. More detailed statement of experimental materials given in [9].

23.3.3 Extracts of Tea, Vegetative Additives and their Mixes

In the given section results of measurements of the total AO content and activity by two methods in water extracts of some kinds of tea and vegetative additives are presented [10]. These parameters were measured also in extracts of their binary mixes to study possible interference of mixes components into each other. Objects of research were water extracts of three kinds of tea (Chinese green tea Eyelashes of the Beauty», grey tea with bergamot "Earl grey tea" and black Ceylon tea "Real"), mint peppery (*Mentha piperita*) and the dry lemon crusts. Ten extracts of binary mixes of the listed samples with a different weight parity of components have been investigated also. Preparation of samples and extraction spent in the same conditions, as for mint (the previous section). The GA was used as the comparison sample in an ammetric method. Efficiency of this method has allowed tracking the dynamics of AO content change in investigated samples directly after extraction. On Figure 5 dynamics of total phenol AO content change for 5 samples is shown. The most considerable content C decrease is observed in tea extracts (20–25%) that, possibly, as well as for mint, is explained by destruction of the unstable phenol substances in extracts (katehins, teaflavins, tearubigins, etc.). For extract of lemon crusts the total AO content is much less and practically does not change during first minutes after extraction.

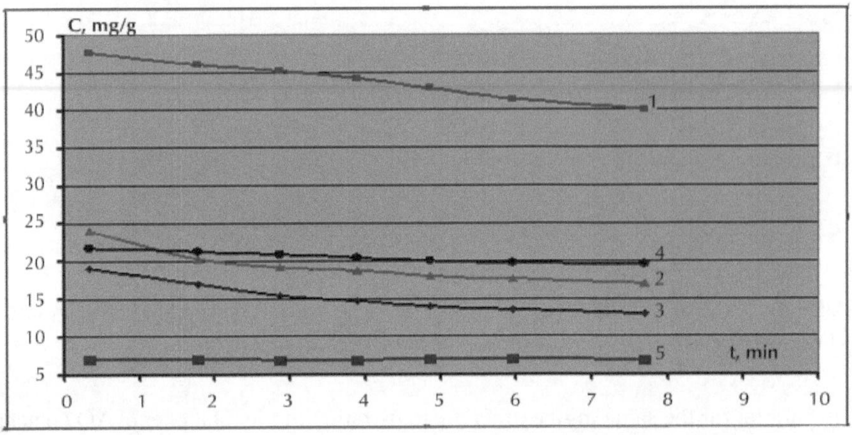

FIGURE 5 Dynamics of the total AO content C change (in units of gallic acid) during extract storage time t: 1 = green Chinese tea, 2 = grey tea with bergamot, 3 = black Ceylon tea, 4 = mint, 5 = lemon crusts (for lemon crusts C was increased in 5 times).

Measurement results of the total AO content for extracts of tea and additives (Figure 6(a)) and extracts of their mixes in a different parity (Figure 6(b)) are presented. The AO content in mixes (c) is calculated under the additive AO content contribution of mix components, taken from Figure 6(a) according to their parity.

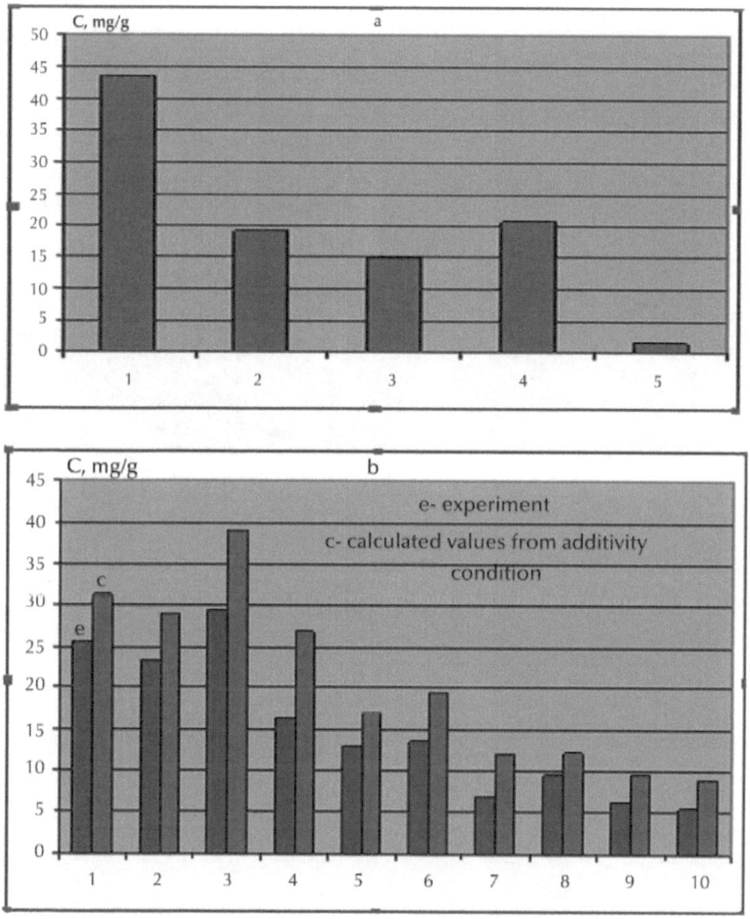

FIGURE 6 The total AO content C (in units of gallic acid): (a) in extracts of: 1 = green Chinese tea, 2 = grey tea with bergamot, 3 = black Ceylon tea, 4 = mint, 5 = lemon crusts.
(b) in extracts of tea and additives mixes: 1 = tea 1 + tea 2 (1:1), 2 = tea 1 + tea 3 (1:1), 3 = tea 1 + mint (4:1), 4 = tea 1 + lemon crusts (3:2), 5 = tea 2 + tea 3 (1:1), 6 = tea 2 + mint (4:1), 7 = tea 2 + lemon crusts (3:2), 8 = tea 3 + mint (4:1), 9 = – tea 3 + lemon crusts (3:2), 10 = mint (4:1) + lemon crusts (2:3). In brackets the parity between components of mixes is specified.

As to mixes of tea and additives (Figure 6(b)) the measured values of the AO content in extracts of investigated mixes (e) have considerable reduction in comparison with the additive contribution of the phenol AO content of mix components, that is observed their strong antagonism. Especially it is considerable for extracts of tea with lemon crusts mixes.

The measured values of the total AO activity with respect to oxygen and its radicals *K* are presented for extracts of tea, additives (Figure 7(a)) and their mixes (Figure 7(b)).

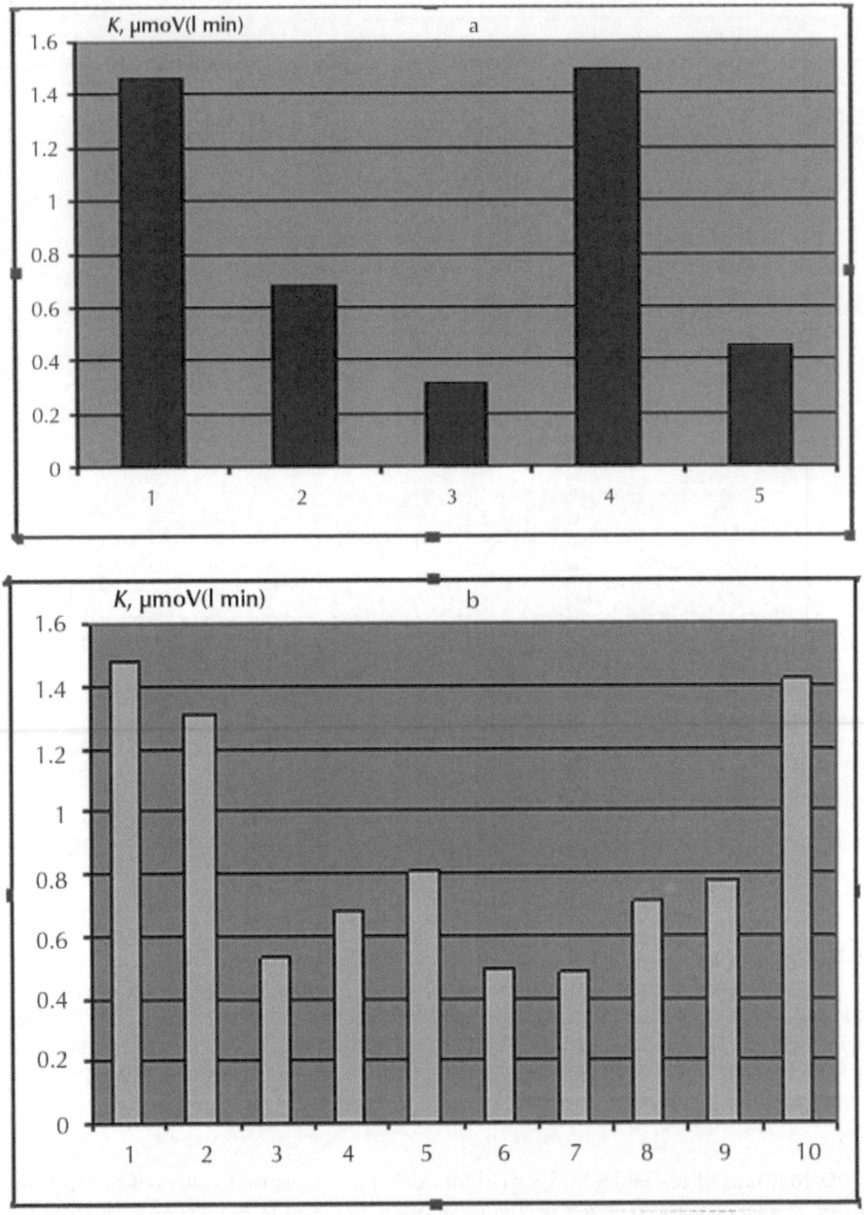

FIGURE 7 The total AO activity with respect to oxygen and its radicals K: (a) in extracts of tea and additives: 1 = green Chinese tea, 2 = grey tea with bergamot, 3 = black Ceylon tea, 4 = mint, 5 = lemon crusts, (b) in extracts of tea and additives mixes: 1 = tea 1 + tea 2 (1:1), 2 = tea 1 + tea 3 (1:1), 3 = tea 1 + mint (4:1), 4 = tea 1 + lemon crusts (3:2), 5 = tea 2 + tea 3 (1:1), 6 = tea 2 + mint (4:1), 7 = tea 2 + lemon crusts (3:2), 8 = tea 3 + mint (4:1), 9 = – tea 3 + lemon crusts (3:2), 10 = mint (4:1) + lemon crusts (2:3). In brackets the parity between components of mixes is specified.

For all samples of extracts, except an extract of the lemon crusts, dominating character of interaction of extract components with oxygen and its radicals has not phenolic, but the catalytic nature. This interaction proceeds on the mechanism (12), characteristic for substances of 4th group in Table 1. Lemon crusts extract "works" on the mechanism of classical AO (8). Unlike values of the phenolic AO content measured by an ammetric method, the kinetic criterion has appeared maximum not only for extract of green tea, but also for mint extract, minimum–for extract of black tea. In spite of the fact that the phenol AO content in extracts of tea and mint has appeared much more, than in lemon crusts, final total activity of tea and mint extracts is defined not by phenol type substances, but, apparently, various metal complexes, present in them, and catalyze of proceeding chemical processes. Considerable shift of the ER O_2 current maximum potential in positive area attests in favor of enough high content of phenol substances in teas and mint extracts.

For the activity of mixes extracts deviations of the measured values *K* from the values calculated on additivity (here are not presented) are big enough and are observed both towards reduction, and towards increase. For this method, apparently, the additively principle does not "work" since activity of mix components has the different nature and the mechanism of interaction with oxygen and its radicals. Activity of mixes is defined not only chemical interactions between substances, but diffusion factors of these substances to an electrode and so on. Activity of tea mixes extracts changes weaker in comparison with activity of separate tea extracts. The possible explanation of the received results is presented in [10].

23.3.4 Plasma of Human Blood

A human blood plasma is a difficult substance for researches. Its antioxidant activity is defined, mainly, by presence in it of amino acids, uric acid, vitamins E, C, glucose, hormones, enzymes, inorganic salts, and also intermediate and end metabolism products. The total activity of blood plasma is integrated parameter characterizing potential possibility of AO action of all plasma components, considering their interactions with each other. The purpose of this work was measurement of total activity with respect to oxygen and its radicals of blood plasma of 30 persons simultaneously with measurement of the total AO content in plasma. Blood plasma has been received by centrifuging at 1,500 r/min of blood of 30 patients from usual polyclinic with different age, sex and pathology. It is necessary to notice that for the majority of plasma samples VA-grams were stable for 3–4 identical measurements and resulted on 2nd group of Table 1 (Figure 2). The ER O_2 current potential shift was small (0–0.03 V) that testifies to presence in plasma of small phenol substances content. For some blood plasma samples (4 samples from 30) VA-gram character was corresponded to substances of 4th group in Table1. For studying of correlation of the results received by two methods, values of *K*, measured during 30 min after plasma defrosting have been selected for the samples having VA-grams of 2nd group. On Figure 9 these values are presented together with corresponding measured values of the total phenol AO content C, spent also during 30 min after plasma defrosting. Correlation of received results with factor r = 0.81 is observed. It means that in blood plasma of many patients there are phenol substances which define dominant processes of plasma components interaction with oxygen and its radicals. Results of this work were published in [11].

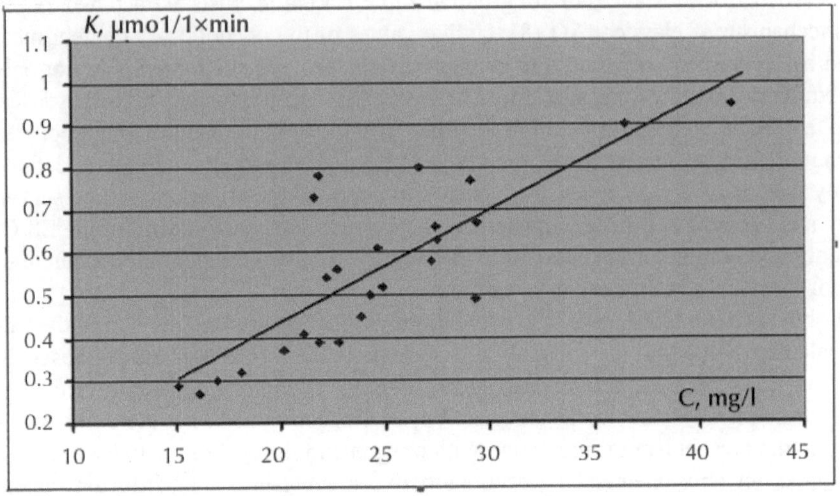

FIGURE 8 Correlation dependence between the kinetic criterion K and the total AO content C (in units of gallic acid) for samples of blood plasma, having VA-grams of 2nd type. R = 0.81.

23.4 CONCLUSION

Use of two operative electrochemical methods realized in devices "TsvetJauza-01-AA" and "Analyzer of TAA" allows quickly and cheaply to define the total content of AO and their activity with respect to oxygen and its radicals in various biological objects. Results of the present work show good correlation of these methods, and the specified devices can be applied widely in various areas.

KEYWORDS

- **Antioxidants**
- **Electrochemical methods**
- **Free radicals**
- **Mercury film electrode**
- **Voltammetric method**

REFERENCES

1. *Study of Synthetic and Natural Antioxidant in vivo and in vitro*. E. B. Burlakova (Ed.), Nauka, Moscow (1992).
2. Vladimirov, Yu. A. and Archakov, A. I. *Lipid peroxidation in biologicall membranes*. Nauka, Moscow (1972).
3. Roginsky, V. and Lissy, E. Review of methods of food antioxidant activity determination. *Food Chemistry*, **92** 235–254 (2005).

4. Peyrat_Maillard, M. N., Bonnely, S., and Berset, C. Determination of the antioxidant activity of phenolic compounds by coulometric detection. *Talanta*, **51** 709–714 (2000).
5. Yashin, A. Ya. Inject-flowing system with ammetric detector for selective definition of antioxidants in foodstuff and drinks. *Russian chemical magazine*, **LII**(2) 130–135 (2008).
6. Korotkova, E. I., Karbainov, Y. A., and Avramchik, O. A. Investigation of antioxidant and catalytic properties of some biological-active substances by voltammetry. *Anal. and Bioanal.Chem.*, **375**(1–3) 465–468 (2003).
7. Korotkova, E. I. *Voltammetric method of determining the total AO activity in the objects of artificial and natural origin*. Doctoral thesis. Tomsk (2009).
8. Misin, V. M. and Sazhina, N. N. Content and Activity of Low_Molecular Antioxidants in Juices of Medicinal Plants. *Khimicheskaya Fizika*, **29**(9) 1–5 (2010).
9. Natalia Sazhina, Vyacheslav Misin, Elena Korotkova. Study of mint extracts antioxidant activity by electrochemical methods. *Chemistry and Chemical Technology*, **5**(1) 13–18 (2011).
10. Misin, V. M., Sazhina, N. N., and Korotkova, E. I. Measurement of tea mixes extracts antioxidant activity by electrochemical methods. *Khim. Rastit. Syr'ya*, (2) 137–143 (2011).
11. Sazhina, N. N., Misin, V. M, and Korotkova, E. I. The comparative analysis of the total content of antioxidants and their activity in the human blood plasma. *Theses of reports of 8th International conference "Bioantioxidant"*, Moscow, pp. 301–303 (2010).

Index

A

Acetic acid, 16, 21
Acrylic acid (AA), 85
Acrylonitrile, homopolymerization and copolymerization
  acrylic acid (AA), 212
  aqueous solution of sodium sulfocyanide (ASSSC), 211
    polymerization rate, 211–212
  cell effect, 212
  chain propagation, 212
  charge-transfer complexes, 212
  copolymer molecular mass, 220
  cross chain termination.
    rate constant of, 212
  fiber-formation polymers, 211
  inorganic salts, 211
  itaconic acid (IA), 212
  methacrylic acid (MAA), 212
  methallyl sulfonate (MAS), 212
  methylacrylate (MA), 212
  organic solvents, 211
  retarding effect, 212
  sodium sulfocyanide, 212
  ternary monomeric systems, 212
  vinylacetate (VA), 212
Acryloxyethylmaleate (AOEM), 90
Ammonium persulfate (APS), 90
Amonton's friction law, 181
Antioxidant properties
  amino acids, 233
  ammetric cell, 224
  ammetric detector, 222
  anthocyans, 235
  biological active substances (BAS) groups, 223
  biological activity, 221
  cardiovascular disease, 221
  catechins, 235
  chemical methods, 222
  classical Nernst equation, 234
  correlation coefficient, 235
  electric potential, 222
  electrochemical methods, 221
  electrochemical oxidation, 222
  electrode processes theory, 234
  electrolyte, 224
  free radical absorption activity
    measurement of, 222
  French paradox, 221
  hematopoietic function, 235
  hydrogen peroxide, 222
  kinetics criterion, 224
  lilac aldehyde, 237
  mercury film electrode (MFE), 222
  OH-groups, 222
  oncological diseases, 221
  oxygen electroreduction, 222
  oxygen radicals, 222
  phenolic content, 224
  phenol type antioxidant content, 233
    measurements of, 225–232
  phosphate buffer solution, 224
  phthalocyanines, 237
  procyanidins, 235
  reactive oxygen species (ROS), 222
  superoxide anion, 235
  timber oksiaromatic acid, 237
  woody-vanilla aromas, 237
Aromatic tetracarboxylic acids, 8
Ascorbic acid and paracetamol high-diluted solutions
  anaesthetic effect, 200
  analgesic-antipyretic action, 200
  anti-inflammatory preparations, 200
  antitumorous agents, 200
  bioactive compounds, 200
  bioactive substances, 199
  biological activity, 201
  cathetometric method, 200
  classical methods, 200
  electronic spectroscopy method, 200
  equivalent lowering, 201
  febrifugal effect, 200
  high biological activity, 199

illusory concentrations, 199
medical compounds structures, 206
meniscus height, 200
growth of, 201–203
increase, 202
reduction, 202
values of, 201–203
neupeptides, 200
neutropic preparations, 200
one-component solution, 200
optical density value
reduction, 203
paracetamol concentration
reduction of, 204
physico-chemical ones, 199
physico-chemical properties, 200
polymodal dependence effect concentration, 201
pretherapeutic content, 199
radioprotectors, 200
reverse osmosic, 200
sharp respiratory illnesses, 200
spectrophotometer Cary 100, 202
supersmall concentrations, 200
supersmall doses (SSD) effect, 199
physico-chemical bases, 199
two-component solutions, 200
ultra-violet field
absorption values, 204–206
ultra-violet spectra, 202
ultra-violet spectroscopy method, 206
UV-visible Spectrophotometer, 202
wide absorption band, 205
Average Fiber Diameter (AFD), response surfaces
applied voltage, 162–163
solution concentration, 162
polymer chain entanglements, 162
tip to collector distance, 163
volume flow rate, 163

B

Benzoyl peroxide (BP), 84
Bioadditives, motor fuel
acetone condensation, 26
acetone vapor, 30
4-acetoxymethyl -2,2-dimethyl -1,3-dioxolane
solketal mixture, 28
additive structure, 26
alcohol-gasoline blends, 25
aluminum, 30
biodiesel production, 26
carbonyl component, 31
carbonyl compounds, 26, 29
catalytic effect, 31
catalytic systems, 30
cyclic ketals (dioxolanes), 29
cyclohexanone, 31
dihydroxyacetone, 26
2,2-dimethyl-1 ,3-dioxolane
ethyleneglycole, 28
solketal mixture, 28
ethanol-gasoline blends, 25, 31
ethylene glycol, 31
fuel compositions
phase stability of, 25
separation temperatures of, 26
glycerol
condensation of, 26
ketalization, 30
mixture of, 26–27
product of, 26
glycerol–acetone mixture, 31
hydrocarbon fuels, 25, 29
hydrochloric acid, 30
hydroxyl groups, 26
4-hydroxymethyl -2,2-dimethyl -1,3-dioxolane, 26
acetic anhydride, 27
amberlyst, 27
analysis, 28
chromatograms, 27
2,2-dimethyl-1,3-dioxolane-4-methanol, 27
ethyleneglycol dimethacrylate, 27
glycerol ketalization, 28
hydroxymethyl group, 27
KU-2 acid, 27
magneto sensitive sulfocationite, 27
polystyrene, 27
solketal–glycerol mixtures, 27
sulfated oxides, 27

sulfated zirconium oxide, 28
sulfo groups, 27
sulfuric acid, 27
temperature programming mode, 28
zeolite, 27
zirconium hydroxide, 28
4-Hydroxymethyl-2-methyl-2-ethyl-1,3-dioxolane
methylethylketone, 28
solketal mixture, 28
4-Hydroxymethyl -2,2-pentamethylene -1,3-dioxolane
cyclohexanone, 28
solketal mixture, 28
ketal additives properties, 30
ketalyzation process, 31
metal particles, 30
methyl ethyl ketone, 31
natural polyols, 29
nonpolar hydrocarbon media, 29
octane number, 26, 29
phase stability, 26
phosphoric acid, 30
polyatomic alcohols, 26
polymeric sulfoacids, 30
semiketals formation, 30–31
silicon oxides, 30
solketal, 26
titanium, 30
Biological objects, antioxidant activity
ammetric method
ammetric detector, 242
ammetric device, 242
electrochemical cell, 242
electrochemical oxidation, 242
free radical absorption activity, 242
gallic acid (GA), 242
glassy-carbon anode, 242
root-mean-square deviation, 242
electrochemical methods, 241
free radical oxidation processes, 242
Human Blood Plasma
amino acids, 253
plasma components, 254
uric acid, 253
human blood plasma, 241
lipid membranes
structure and properties, 242
medicinal plants, 246–249
tea extracts, 250–253
vegetative additives, 250–253
voltammetric method
biological active substances (BAS), 244
electrochemical cell, 245
kinetic criterion, 246
mercury film electrode (MFE), 243
oxygen electroreduction process, 243
reactive oxygen species (ROS), 243
voltammetric analyzer, 245
Bis(oxymethyl) phosphonic acid, watering and melamine effect
Arabidopsis thaliana, 192
bioenergetic indices, 195
biological membranes, 188
cell
components, 188
membranes, 188
metabolism, 188
digalactosyl-diacyl- glycerol, 188
eicosanic acid, 193
eicosanoic acid, 192
Fatty acid methyl esters (FAMEs), 189
fatty acids
one-step methylation, 189
saturated, 188
unsaturated, 188
unsaturation coefficient, 194
linoleic acid
relative content, 192
linolenic acid, 193
lipid–protein relation, 188, 192
materials, 189
metabolic processes, 188
mitochondria
electron-transport chain, 188
energy metabolism, 188
isolation of, 189
rate of, 189
respiratory chain, 195
mitochondrial energy metabolism, 188

mitochondrial membranes, 188
  chemical properties, 195
  lipid phase, 195
  LPO products fluorescence spectra, 191
  physical properties, 195
  physicochemical state, 192
monogalactosyl-diacyl-glycerol, 188
NAD-dependent substrate oxidation
  rate of, 193–194
NAD-dependent substrates oxidation, 188
NADH-dehydrogenase complex, 195
organelle membranes, 188
oxidative phosphorylation efficiency, 192–193
pea seedlings, 191
pea seeds germination, 189
  acidic methanolysis, 189
  chloroform, 190
  chromato-mass-spectrometry, 189
  electron impact ionization regime, 190
  emission wavelength, 190
  epicotyls, 189
  excitation wavelength, 190
  FAME analysis, 190
  flame-ionization detector, 190
  fluorescent method, 190
  germinated seeds, 189
  helium carrier gas rate, 190
  Hewlett-Packard-6890 spectrophotometer, 189
  hexanol elution, 189
  HP-5972 mass-selective detector, 189
  4-hydroxynonenals, 190
  lipid peroxidation (LPO) activity, 190
  methanol, 190
  methanolysis process, 189
  pentadecane, 189
  Schiff's bases, 190
  soap solution, 189
  thin layer chromatography method, 189
physiological indices, 195
plant growth regulator, 188
plant metabolism, 188
plant resistance, 193
reagents, 190
  acetyl chloride, 190
  FA-free BSA, 190
  glutamate, 190
  Hepes, 190
  malate, 190
  potassium carbonate, 190
  sucrose, 190
respiratory control rate (RCR), 192
statistics, 190
  standard deviations, 190
  statistica v. 6 software, 190
stearic acid, 188
water deficit, 188
Bruster Vertgame's law, 45
Butylacrylate (BA), 90

C

chitosan (CHT)
  acetic acid, 16
  carbohydrates concentration, 18
  chains, 18
    dependence of, 18
  destruction, 17
  enzymatic destruction, 15
  hydrolysis
    problem of, 15
  hydrolysis process, 15
  intermolecular hydrogen bonds, 18
  intrinsic viscosity
    dependence of, 16
  solutions, 16
  supermolecular structure, 16
Conjugated polymer
  general properties, 2
Cross-linked polymers, susceptibility
  detection operator, 71
  electromagnetic anisotropy's property, 71
  electromagnetic radiation, 71
  electromagnetic susceptibility equilibrium elastic coefficient
    experimental values, 72
  hardeners molar amounts ratio, 71

hyperelastic state, 69
intermolecular interaction's level growth, 71
mathematical model
BrusterVertgame's law, 70
electromagnetic perception, 70
electromagnetic susceptibility elastic coefficients, 70
equilibrium elastic coefficient, 70
inductive capacity, 70
mathematical formalization, 70
polymeric dielectric tensor, 70
polymeric meshe's cross-site chains, 70
polymer model, 70
relaxation operator, 70
shear pliability coefficients, 70
strain electromagnetic susceptibility, 71
weighting coefficient, 71
mechanical scale, 72
optical anisotropy curve, 72
optical double refraction, 71
photocreep curves, 72
polarization optical method, 73
radioparency coefficient, 69
shear pliability
operators, 72
relaxation spectrums, 72
shear tensions, 71
strain electromagnetic anisotropy, 72
strain electromagnetic susceptibility
relaxation spectrums, 72
weighting coefficient
experimental values, 72
Cross-linked polymers with nanoscale cross-site chains
acetone, 77
1-aminohexane(AH), 77
bisphenol- A (DGEBA), 77
cross-site mesh's chains
equipotential fractional operator, 76
local conformation mobility, 76
dense polymer meshes, 77
diglycocide ether, 77
fiberglass product technology, 75
glass transition temperatures
experimental values, 80
glassy state, 75, 77
hardened reactionary mixtures, 77
heredity's theory equations, 75
hexamethylenediamine (HMDA), 77
homogeneous topological structure, 76
hyperelastic state, 77
experimental values, 80
hyperelastic state constant, 78
equilibrium shear pliability, 78
hardeners molar ratio, 78
intermolecular interaction, 78
mesh points concentration, 78
polymerization number growth, 78
mathematical model
equilibrium volumetric pliability strain tensor, 76
shear pliability, 76
shear tension, 76
Volter's operator, 76
matrix assessment
strain properties value, 75
model constants, 77
multimodal distribution, 76
nanofragment cross-linked matrix, 75
pliability momentary, 75
polymer mesh, 75
reiteration freezing defrosting cycles, 77
a-relaxation times spectrums properties
cross-linked topology, 78
isothermal creep curves, 78
least squares method, 78
molecular polymer structure relaxation spectrum, 79
structure glass transition temperatures, 78
spring-damper models, 75
thermal prehistory influence, 77
thermal properties, 77
heat expansion coefficients, 77
structure glass transition temperature, 77
thermomechanical curves, 80
transition region, 77
viscoelastic module, 75

viscoelastic pliability, 80
weighting coefficient, 76, 78
experimental values, 80

D

Densely cross-linked polymers, viscoelastic and electromagnetic properties
Bruster Vertgame's equation, 51
cross-links viscoelastic pliability, 35
different physical states, 50
electromagnetic radiation, 51
electromagnetic susceptibility's
elastic coefficient, 52
equilibrium coefficient, 53–54
epoxy amine polymers, 35
experimental models' physical states, thermal properties
solid expansion study, 53
thermal expansion's coefficients, 53
experimental objects
1-aminohexane (AH), 52
diglycide ether of bisphenol-A (DGEBA), 52
hexamethylenediamine (HMDA), 52
methods of study, 53
molar ratio, 52
restoration, 52
synthesis, 53
fabrication projection, 49
fractional exponential operator, 52
glass transition temperature values, 57
glassy and hyperelastic states operator, 51
hyperelastic state's constant, 53
input analog conception, 51
mathematical description, 57
mathematical model's constants, 47
mesh's points
co-operative mobility, 52
modeling
thermomechanical course, 47
thermo optical experiments' course, 47
optical anisotropy
freezing process, 58
optimum cross-linked polymer matrix, 49
pliability's momentary constituent, 50
polymer system's, 50
radioparence's coefficient
isotropy, 50
radioparent fiberglass product, 35
relaxation operator, 52
shearing modulus' spectrum, 51
stable radioparence coefficient, 49–50
strain electromagnetic anisotropy, 51
strain electromagnetic susceptibility's, 51
relaxation operator, 52
temperature time realization, 50
temperature time realizations, 47
thermomechanical course
experiment, 56
prediction of, 56–57
thermomechanical rating, 36
thermooptical curves, 36
thermooptical experiments course
cooling rate, 57
curves' trend, 56
dependence, 58
prediction of, 56–57
topological structure, experimental objects
cross-linked structure, 36
mesh pointsa nitrogen atom, 36
polymer meshes fragment, 36
statistical parameters, 36
transition region, 51
viscoelastic properties module's, 50
weighting coefficients, 54
Dimethyldi-(n-b-naphthyl aminophenoxy) silane, 87
Dimethyldi-(n-phenyl-aminophenoxy) silane, 87

E

Electrochromic conjugated polymers
applications, 1
Electromagnetic susceptibility's equilibrium elastic coefficient
Avogadro's constant, 63
deformation's tensor independent components, 62

experimental objects
topological structure of, 64
glassy state, 62
glassy state theoretical rating, 62
heat expansion's coefficient, 62
hyperelastic state, 62
polymers deformation, 63
inductive capacity tensor, 62
polymer topology, 62
repeated mesh's fragment
Van der Waals's volume of, 63
structure glass transition temperature, 62
theoretical calculation of, 63
unstressed polymer
inductive capacity of, 63
Electromagnetic susceptibility's equilibrium elastic coefficient, 62–64
Electron-rich triarylamines, 2
Electrophotographic (EPG) method, 2
Electrospun fiber mat
Contact Angle (CA)
applied voltage, 164
collector distance, 165
process parameters optimum values, 166
solution concentration, 163–164
volume flow rate, 165
morphology, 166
relationship between AFD and CA, 166
interfibrillar distance, 166
nanoscale fibers, 166
surface roughness, 166
variation of CA with AFD, 166
structure, 166
Electrospun fibers
fiber diameter distributions, 154–157
Emulsion polymerization
acrylic monomers
copolymerization, 90
acryloxyethylmaleate (AOEM), 90
BA-AOEM system, 90
polymerization rate, 90
alkyl acrylates, 90
allylacrylate (AlA), 90
classical concepts, 89
initiation rates, 90
micelles, 89
monomer radicals, 89
oligomeric radicals, 89
polymeric dispersions, 90
polymerization rate, 89
polymer synthesis, 89
synthesize emulsifier
principal possibility, 89
ternary monomeric system, 90
unconjugated dienes, 90
Emulsion polymerization, kinetics and mechanism
acrylic monomers, 217
alkyl acrylates, 216
allylacrylate (AlA), 217
ammonium persulfate (APS), 217
BA-AOEM system, 217
butylacrylate (BA), 217
emulsifier-free latexes, 216
ethylacrylate (EA), 217
gel effect, 216
hydrophilic comonomers, 216
hydrophobic monomers, 216
kinetic parameters, 216
latex particles, 216
monomer-soluble emulsifiers, 216
oligomeric molecules, 216
oligomeric radicals, 216
polymeric dispersions, 215
polymer–monomer particles, 216
unsaturation degree, 217
Enzymatic destruction
acetate buffer solutions, 23–22
chitosan (CHT) film, 21
CHT film coating, 21
CHT solutions, 22
intrinsic viscosity of, 22
physico chemical characteristics, 23
properties of, 22
enzymatic preparation
concentration, 22
solution, 22
enzymatic stability, 23
enzymatic transformation, 22
Huggins constant, 22
macromolecular chains, 21
polar polymers, 22

polymeric material, 21
polymer solution, 22
reversible agglomeration process, 22
supermolecular solution structure, 23
thermodynamic quality, 23
thermo-modification, 23
Ethylacrylate (EA), 90

F

Fatty acid composition. See Bis(oxymethyl) phosphonic acid, watering and melamine effect
Field assisted thermo dissociation (FATD), 5
Fluorene
analogous derivatives, 1
Folin-Ciocalteau method, 224

G

Glycoside bond
hydrolytic rupture of, 17

H

High-molecular-weight flocculent synthesis
scientific basics
AN polymerization, 91
flocculants, 91
hydrolysis rates, 91
one-stage synthesis, 91
polymerization rates, 91
Howell hanging fiber method, 131
Hydroquinone (HQ), 87

I

Inclined plane method, 131
Itaconic acid (IA), 85

L

Light emitting diodes, 1
liquid chromatography methods, 222

M

Materials
aromatic tetracarboxylic dianhydrides, 9
3,3',4,4'-benzophenone tetracarboxylic dianhydride, 9
N,N,-bis(4-aminohenyl)-N', 8
bis(dibenzylideneacetone) palladium, 9
1,1'-bis(diphenylphosphino)ferrocene (DPPF), 9
N'-bis[4-(2-phenyl-2-isopropyl)phenyl]-1, 8
bis[4-(2-phenyl-2-isopropyl)phenyl] amine, 8–9
bis(4-tertbutylphenyl) amine, 8
3,3',4,4'-bitetracarboxylic dianhydride, 9
cesium fluoride (ACROS), 8–9
N,N-dimethylacetamide, 9
dimethyl sulfoxide (DMSO), 9
3,3',4,4'-diphenyl sulfone-tetracarboxylic dianhydride, 9
4-fluoro-nitro- benzene (ACROS), 8–9
4,4'-hexafluoroisopropylidenediphathalic dianhydride, 9
hydrazine monohydrate (MERCK), 9
iodobenzene, 9
N-methyl-2-pyrrolidinone (NMP), 9
N-N, dimethylformamide (DMF), 151
4,4'-oxydiphthalic anhydride, 9
p-fluoronitrobenzene, 8
1,4-phenylene diamine, 8
polyacrylonitrile (PAN) powder, 151
pyridine, 9
sodium hydride, 8–9
sodium tert-butoxide, 9
Mathematical formalism, constants
equilibrium constatnts
abstract linear polymer parameter, 42
atomic refractions, 42
Avogadro's constant, 40
cross-linked polymer parameter, 42
density expression, 40
electromagnetic anisotropy, 41, 43
electromagnetic susceptibility's equilibrium coefficient, 41, 43
equilibrium shear pliability dependence, 41
experimental models' topological structure, 44

front-factor on n3f, 41
glass transition temperature, 39, 43
hyperelastic state's constant, 39, 42
inductive capacity's, 41
linear law, 40
mesh's points' frequency, 40
molar masses of elementary links, 40
molecular package's coefficient, 40
polymer density, 39
strainless experimental models, 42
thermal expansion's coefficients, 43
topological structure, 41
universal gas constant, 39
Van der Waals's volume, 42
thermal constants
dispersion interaction, 38
experimental models, 39
glass transition temperature, 37–38
intermolecular interaction, 37
linear fragments, 37
oxygen atom, 37
partial volumetric coefficients, 38
SimkhiBoyer's equation, 39
thermal expansion's coefficient, 38
Van der Waals's volume, 37
weighting coefficients
average experimental value, 46
BrusterVertgame's law, 45
glassy state experimental models, 46
glassy state shear pliability, 44
inductive capacity's tensor, 44
oligomer average value, 46
shear pliability's relaxation operator, 44
strain electromagnetic anisotropy, 45
theoretical values, 46
thermal expansion's coefficient, glassy state, 46
Medicinal plants
air plant, 246
basket plant, 246
Blossfelda kalanchoe, 246
classical phenolic mechanism, 246
devil's backbone, 246
Dichorisandra fragrants, 246
kinetic criterion, 246–247
Moses-in-the-cradle, 246
Methacrylic acid (MAA), 85
Methallyl sulfonate (MAS), 85
Monomer stabilization
acetone cyanohydrin, 215
acrylic
esters of, 214
aforesaid aromatic amines, 215
ammonium salt, 215
aromatic amines, 214
dimethyldi-(n-b-naphthyl aminophenoxy)silane, 214
dimethyldi-(n-phenyl-aminophenoxy) silane, 214
hydroquinone (HQ), 214
inhibition, 215
usage of, 215
N-nitroso-N-phenylhydroxylamine, 215
2-oxy-1,3-di-(n-b-naphthyl aminophenoxy)propane, 214
2-oxy-1,3-di-(n-phenylamino phenoxy) propane, 214
polymeric peroxides
formation of, 214
polymer stabilizers, 214
polyperoxides
decomposition of, 214
stoichiometric inhibition coefficients, 214
sulfuric acid, 214–215
thermal oxidative polymerization
inhibition of, 214
thermal polymerization
inhibition of, 214
Monomer synthesis
N,N,-bis(4-aminohenyl)-N',N'-bis[4-(2-phenyl-2-isopropyl) phenyl]-1,4-phenylene-diamine, 11–12
N,N,-bis(4-nitrophenyl)-N',N'-bis[4-(2-phenyl-2-isopropyl) phenyl]-1,4-phenylene-diamine, 10–11
bis[4-(2-phenyl-2-isopropyl)phenyl]-4-aminophenylamine
hydrazine monohydrate, 9–10
bis[4-(2-phenyl-2-isopropyl)phenyl]-4-nitrophenylamine
4-fluoro-nitrobenzene, 9

reaction mixture, 9
silica gel column chromatography, 9
sodium hydroxide, 9
IR spectrum, 12

N

Nanofibers morphological analysis
Analysis of Variance (ANOVA), 158
average fiber diameter (AFD), 159
electrospun fiber mat, 160
regression analysis, 161
regression equation, 159
responses, 158
chain mobility, 153
electrospinning process, 153
electrospun fiber mat
predicted versus actual plot, 161–162
p-values, 161
electrospun fibers
fiber diameter distributions, 153
electrospun nanofiber
electric field strength, 153
electrostatic repulsive force, 153
fiber diameter distribution, 153
fiber morphology, 153
PAN solution, 153
SEM image, 153–157
Nanofragment cross-linked polymers, mathematical model
compliance's constants, 171
cross-linked epoxy amine polymers, 169
epoxy amine polymers topological structure
elastic effective mesh points, 170
elementary units, 170
meshes' structure, 170
methylene groups, 170
nitrogen atom, 170
repeated fragment of, 170
tetramethylene units, 170
experimental objects, thermomechanical behavior, 179
chemical structure, 179
thermomechanical curve, 179
fiberglass products, 169
hyperelastic state's constant
elastic effective mesh points, 173
elementary links, 173
equilibrium shear pliability, 175
front factor's values, 175
molar mass, 174
molecular packing's coefficient, 174
repeated mesh's fragment, 174
mechanical tensions, 169
optimal cross-linked polymer matrix, 169
physical state, 169
pliability, 169
a-relaxation times spectrum, parameters, 179
molecular mass distribution, 179
strain properties, 169
thermal expansion
coefficients, 171
glass transition temperature, 171
glassy state, 172
hyperelastic state, 172
intermolecular dipole-dipole interaction, 171
linear chains, 171
oxygen atom, 171
repeated fragment, 171
Simkhi-Boyer's equation, 172
structural glass transition temperature, 173
Vander Waals's volume, 171
weighting coefficient
Bruster-Vertgame's law, 176
dielectric permeability's tensor, 176
epoxide resins, 178
glass transition state's shear pliability, 176
hypothetical linear polymer, 178
inductive capacity, 176
polymer's topology, 178
relaxational operator, 176
square refractive index, 176
strain electromagnetic anisotropy, 176
strain electromagnetic susceptibility, 176
transverse strain, 176

unstressed polymer, 177
valence angles deformation, 177
N-nitroso-N-phenylhydroxylamine, 88

O

Organic solvents effect
acetonitrile (ACN)
influence of, 211
alkyl
donor–acceptor interaction, 211
azo isobutyronitrile (AIBN), 211
bimolecular chain termination, 211
diffusional-controlled chain termination, 211
dimethyl formamide (DMFA)
influence of, 211
electron-donor solvent, 211
initiator decomposition rate, 211
macro radical solvation, 211
nitrile groups
donor–acceptor interaction, 211
Organic thin film transistors, 1
Oscillation Model
constrained yarns, 137
crimp angle, 134
discriminating equations, 137
elastic modulus, 134
equivalent spring constant, 137
evaluate, 134
fabric deformation data, 134
flexible string vibration, 134
force balance model
discriminating equations, 134
empirical factors, 134
linear density, 134
modified warp yarns modulus, 137
parameters, 134
adhesive interactions, 133
Coulombs law, 132–133
different zones, 132
predictive equations, 133
pullout design, 132
pullout force, 133
slippage causes, 133
static friction resistance, 132
stick-slip motion, 133
yarn pullout test, 132
partial differential equation, 133
polyester fabrics, 137
predicts, 137
principals, 137
textile structures, 137
vibration part
equation, 133
2-Oxy-1,3-di-(n-b-naphthyl aminophenoxy)propane, 87
2-Oxy-1,3-di-(n-phenylamino phenoxy) propane, 87

P

p-conjugated polymers, 1
Photoelectric characteristics
acceptor fragments EA, 5
band absorption edge, 4–5
Charge Transfer Complexes (CTC), 4
bands, 5
charge-transfer mobility, 1
dark conductivity
sharp increase, 7
distinct spectral peaks, 5
electrophotographic (EPG) method, 2
electrophotographic study
equivalent optical density, 3
excitation photon energy, 3
film optical density, 3
kinetics, 3
maximum potential, 3
PES values comparison, 3–4
PI films, 4
potential dark decay rate, 3
potential photo discharging rate, 3
Potential photoinduced decay, 3
strong absorption, 3
weak absorption, 3
excimers formation, 1
Fluorene, 1
geminate recombination Onsager model, 6
homopolyimides synthesis
chemical cyclodehydration, 2
polymer solution, 2
stirred solution, 2
long wavelength absorption band edge
energy position, 5

novel conjugated polymers, 8
optoelectronic devices, 2
p-conjugated polymers, 1
PES spectral
comparison, 4
photochemical transformation, 8
photo generation quantum, 5
field dependence, 6
photostimulated current (PSC), 6
photostimulated generation, 8
photostimulation, 5
PI films
PES of, 4
PI series, 4
polyimides (PIs), 2
potential applications, 1
stabilized cation-radicals (D+), 5
stable cation-radicals, 6
accumulation, 7
observation of PSC, 8
synthesized PI, 2
technological applications, 2
unsatisfied thermal stability, 1
UV region
b values, 6
Photoshop software, 123
Photovoltaic's, 1
Plain woven polymeric fabrics, internal mechanical properties
Amonton's friction law, 181
Amonton's Law, 184
analytical approach
densely weaved fabric, 182
fabric deformation angle, 184
fabric modulus, 184
force-balance model, 184
friction coefficient, 183
friction force, 183
image processing method, 184
pullout force, 182
shear stresses effects, 183
stresses, 182
weave angle, 182
yarn pullout test, 182
analytical model, 181, 184
correlation coefficient calculation, 182
fabric
mechanical parameters, 181
fabric mechanical behavior
structural characteristics, 181
force-balance model, 182
image processing method, 181
internal mechanical properties
plain woven fabrics, 184
mechanical parameters, 184
mechanistic model, 181
parameters, 181
Pullout test, 181
weave angle variations, 184
yarn-to-yarn friction coefficient, 183
Polyacrylamide (PAA)
flocculent synthesis, 91
radical initiator, 91
Polymer, chemical physics
analytical techniques, 210
benzoyl peroxide (BP)–IS systems, 210
colloidal metal particles, 211
effective activation energy, 210
high-molecular-weight flocculent
acrylamide (AA) deficit, 218
one-stage synthesis, 218
polyacrylamide (PAA), 218
monomer–stearate complex, 210
noble metal salts, 211
novel methods, 210
physicochemical techniques, 210
polymeric dispersions properties
analyze, 210
polymeric material formation, 210
polymer synthesis, 210
radical-chain processes theory, 209
radical polymerization, 209
salts effects in varivalent metals, 210
methylmethacrylate (MMA), 210
stearates
catalytic properties, 210–211
styrene
polymerization rate, 210
technological regime optimization, 210
thermal polymerization, 210
turbidity spectrum method, 210
vinyl monomers, 209
polymerization, 210

water-suspended imitators, 210
polymer deformation's tensor
a-relaxation times, 51
compression deformation's tensor, 51
equilibrium shear pliability, 50
equilibrium volumetric pliability, 50
fractional exponential operator, 50
homogeneous topological structure, 50
Krokener's tensor, 50
legacy conceptions, 50
mesh's chemical structure, 50
polymeric meshe's crosssite chains
conformational mobility, 50
shear deformation's tensor, 51
shear pliability's relaxation operator, 50
strainless polymer object
inductive capacity, 51
transverse strain's tensor, 50
weighting coefficien, 50
Polymer enzyme system, 17
Polymeric nanofibers
aerosol filtration properties, 142
aerosol particle obstruction, 141
aerosol particle protection, 142
airborne particles, 142
air resistance, 141
applied voltage, 152
approximation error, 153
average fiber diameter (AFD), 151
Brunauer-Emmett- Teller (BET) surface area, 142
bulk nanofiber mat
moisture transfer, 145
bulky nanofiber mat
characteristics, 145–146
transport properties, 145–146
characterization
electrospun fiber mat, 151
gold-sputtered electrospun fibers, 151
PCTV vision software, 151
chemical composition, 151
coating weight, 142
contact angle (CA) measurements, 151
set up, 152
conventional fiber spinning systems, 150
data acquisition system, 143
design-expert software, 153
determination coefficient, 153
diffusion cell, 143
electric field force, 150
electrospinning, 141
before, 143
after, 143
apparatus, 151
homogenous PAN/DMF solution, 151
PAN powder, 151
parameters, 150
set up, 150, 151
electrospinning apparatus, 150
electrospinning method, 142
electrospinning process, 150
electrospun fiber mat, 141–142, 151, 152
electrospun layers, 142
electrospun nanofibers, 142
ambient conditions, 151
diameter, 150
processing conditions, 151
solution properties, 150
electrospun PAN nanofiber mat, 151
electrospun polylacticacid (PLA) membranes
pore size distribution, 142
fibrous materials, 141
filter media, 141
filter performance, 141
filtration efficiency, 141
fluid jet, 141
geometrical microstructure, 151
guarded hot plate, 143
high barrier performance, 142
high efficiency particulate air (HEPA), 142
linear coefficients, 153
linear model, 153
liquid phase, 151
mats gas transport properties, 142
mechanical force, 150
mechanical properties, 141
microstructure measurement software, 143

moisture transfer measurement, 144
moisture vapor transport, 142
nanofiber density, 143
nanofiber mat
  air permeability, 142, 145
  air permeability versus bulk porosity, 146
  bulk porosity, 142–143, 145
  density, 145–146
  electrospinning system, 142
  heat transfer, 142
  modified electospinning system, 143
  moisture transfer, 142, 145
  properties, 142
  relative humidity, 146
nanofibrous materials, 150
nanowebs, 141
nonwoven mat, 141, 150
optimal Conditions determination, 166
  spinning distance, 166
polymer solution, 141
  concentration, 142
  surface tension forces, 150
polymer solution concentration, 152
protective clothing applications, 141–142, 150
quadratic coefficients, 153
quadratic model, 152–153
regression analysis, 153
response surface methodology, 151
second order interaction coefficients, 153
Shirley permeation analyzer, 143
solid phase, 151
solid surfaces wettability, 151
substrates, 141
superhydrophilic surface, 151
superhydrophobic surface, 151
textile materials
  moisture transport behavior, 142
thermal comfort, 142
tissue engineering scaffolds, 150
ultra fine fibers, 141
ultra thin mat, 141
volume flow rate, 152
water vapor diffusion, 147

Polymer molecular mass, changes
  alkaline deacetylation, 15
  chitosan (CHT) hydrolysis, 15
  enzymatic destruction, 16
    acetate buffer, 16
    polymer solution, 16
  enzymatic preparation, 15
  ferricyanide method, 16
  hydrolysis process, 17
  intrinsic viscosity, 15
  polymer molecular weight, 16
  viscosimetric method, 15
Polymers role in technologies and environment
  accelerating action efficiency, 84
  activation energy, 84
  analytical techniques, 83–84
  aqueous solution of sodium sulfocyanide (ASSSC)
    acrylonitrile (AN)-based fiber formation, 85
    AN copolymerization kinetics, 85
    AN homopolymerization, 85
    binary systems, 85
    cell effect, 85
    copolymerization, 85
    homopolymerization, 85
    initiation reaction, 86
    lower initiation rate, 85
    methylacrylate (MA), 85
    monomer, 86
    organic solvents, 85
    polymerization rate, 85
    sodium sulfocyanide, 85
    stable radicals, 86
    synthesized azonitriles, 85
    ternary monomeric systems, 86
    three-component system, 86
    vinylacetate (VA), 85
  benzoyl peroxide (BP), 84
  BP–IS system, 84
  chemical physics, 83
  concentration inversion, 84
  emulsion polymerization
    kinetics and mechanism, 89
  environmental problems, 83
  experimental data, 83

free radicals, 83
intermediate monomer-stearate complex, 84
IR spectroscopy data, 84
methylmethacrylate (MMA), 84
monomer stabilization
acetone cyanohydrin, 88
acrylates, 87
aforesaid aromatic amines, 88
ammonium salt, 88
biphasic reaction systems, 88
copolymerization, 87–88
critical phenomena, 88
inhibitors low-soluble, 88
inhibit polymerization, 87
MMA synthesis, 88
polymer stabilizers, 87
polyperoxides decomposition, 88
spontaneous polymerization, 88
spontaneous polymerization processes, 88
stoichiometric inhibition coefficients, 87
sulfuric acid, 88
thermal polymerization, 87
noble metal salts
ambiguous action mechanism, 84
organic solvents
acetonitrile (ACN), 84
alkyl groups, 85
azoisobutyronitrile (AIBN), 84
basic characteristics, 84
BP initiated polymerization, 84
dimethyl formamide (DMFA), 84
effect of, 84
nitrile groups, 85
rate constants, 85
reaction order, 84
physicochemical techniques, 83–84
polymerization rate, 84
spinning solution, 85
stearates
catalytic properties, 84
synthetic PAN fiber, 85
synthetic polymeric materials, 83
technological regime optimization, 83
varivalent metals
effect of salts, 83–84
vinyl monomer polymerization
kinetics and mechanism, 83
Pullout test
Adobe Premier software, 124
fabric deformation, 118
fabrics
mechanical behavior, 118
tear resistance, 118
Fast-Dizzle software, 124
frictional force, 118
process, 124
Taylor's work, 118

Q

Quantumchemical calculation, update
a-cyclopropyl-2,4-dimethylstyrene, 95
charges on atoms, 98
electronic molecular structure, 97
geometric molecular structure, 97
optimized bond lengths, 98
valence corners, 98
a-cyclopropyl-p-ftorstyrene, 95
charges on atoms, 99
electronic molecular structure, 99
geometric molecular structure, 99
optimized bond lengths, 99
valence corners, 99
a-cyclopropyl-p-izopropylstyrene, 95
charges on atoms, 96–97
electronic molecular structure, 96
geometric molecular structure, 96
optimized bond lengths, 96–97
valence corners, 96–97
electronic structures, 95
geometric structures, 95
quantum-chemical method, 95
standard gradient method, 95

R

a-Relaxation times' spectrum parameters
connection of, 55
cross-linked polymers
confirms validity, 55

glass transition temperature approximately, 55
temperature function, 54–55
cross-site chains, 47
molar ratio, 56
fluctuating unconfined space's thermal dependence, 55
isothermal creep's, 54
photocreep curves, 54
relaxation spectrum, 54
structural sensitive parameter, 56
temperature function, 47
theoretical rating, 47
Response Surface Methodology (RSM)
central composite design (CCD), 152
mathematical technique, 152
optimization process, 152
quadratic surface, 152
second-order designs, 152
statistical technique, 152
Rigid foamed polyurethane
production compositions
azeotropic mixture, 91
chlorofluorocarbons, 91
montreal protocol, 91
optimal compositions, 91
physicochemical conditions, 91
polymer physicochemistry, 91
trichlorofluoromethane (Freon-11), 91
Rigid foamed polyurethane production, compositions
chlorofluorocarbons, 217
dichlorofluoroethane, 218
dichlorotrifluoroethane, 218
foaming process, 218
montreal protocol, 217
ozone layer, 217
ozone-safe ones, 217
physico-chemical conditions, 218
trichlorofluoromethane, 217

S

Shear pliability
relaxation spectrums, 65
Silver nanoparticles
Ag-gelatin immobilized matrix, 108
alkaline water solutions, 103
blue light-filter, 105–108
compact-packed crystal structure, 115
complex forming reagent
threshold concentration, 112
complex forming reagents, 104, 109
crystal lattice, 112
elemental silver
nuclear particles, 112
structure, 114
XRD-pattern, 113–114
elementary cell parameter, 112, 115
ethanediamine-1,2, 104
face-centered lattice, 112
Fm3m group of symmetry, 112
gelatin-immobilized chemical compounds, 109
gelatin-immobilized matrix, 103
elemental silver, 104
gelatin layers
development of, 103
intermolecular cavities, 110
gelatin mass, 110
gelatin matrix, 103
gelatin structure, 110–111
hydroxo complex, 110
3-(2-hydroxyethyl)-3-azapenthanediol-1,5, 104
inorganic reagents, 105
inter-plane distances, 105
ligand ratio, 109
Macbeth TD504 photometer, 105
molar concentrations, 108–109
negative charged acid ligands, 110
non-charged ligands, 110
novel phase structure, 112
optical density Ag-GIM, 105
organic reagents, 105
photochemical reduction, 103
polymer layer, 109
polynuclear particles, 112
potassium hexacyanoferrate, 104
potassium hydroxide, 104
potassium thiocyanate, 104
Powder Cell program, 105
proteolytic enzymes, 104
redox process, 109

reflexes intensities, 105
re-precipitation process, 112
silver bromid, 112
silver halide photographic emulsions, 103
silver hexacyanoferrate, 103
reduction process, 105–108
sodium trioxocarbonate, 104
sodium trioxosulphidosulphate, 104
spectrometer D8 Advance, 105
standard software package EVA, 105
thiocyanate anion, 108
tin chloride, 104
transformation process, 105
X-ray diffraction method, 105
X-ray film Structurix D-10, 103
X-ray powder diffraction patterns, 112
X-ray radiation, 104
XRD spectra, 105
SLEE MAINZ microtome device model cut 4055, 123
Strain electromagnetic anisotropy, mathematical model constants
theoretical determination of densely cross-linked polymers
electromagnetic susceptibility's equilibrium elastic coefficient, 62–64
weighting coefficient, 65
Strain electromagnetic anisotropy, modeling change of experimental objects
influence of constant load, 65
optical anisotropy
freezing curve of, 66
thermooptic curve, 66
Strain electromagnetic susceptibility relaxation
practical hints on recovery of, 61
intensive polymer's state, 61
radioparence coefficient, 61, 66
radioparent fiberglass product, 66
a-relaxation times, spectrum and parameters, 65
strain electromagnetic anisotropy, mathematical model constants (see Strain electromagnetic anisotropy, mathematical model constants)
Supermolecular polymer structure, 18, 21
Suzuki coupling reaction, 8

T

Technologies and environment protection, polymers role. See Polymers role in technologies and environment
Tetracyanoquinodimethane (TCQM), 86
Textile performance, mechanics
adhesion formation, 125
after first junction rupture, 125
analytical model, 132
bleached cotton fabric, 135
complex deformations, 117
correlation coefficient, 129
cross sectional microscopic image, 135
dynamic friction coefficient, 125
dynamic lateral forces, 125
dynamic normal load, 125
elastic modulus, 135
equivalent spring constant, 136
fabric deformation angle, 136
fabric displacements, 125
fabrics
along warp direction, 124
characteristics, 135
deformation angle, 135
failure, 118
geometrical properties, 118
internal mechanical characteristics, 119
mechanical behavior, 117
mechanical properties, 118, 127, 137
Structural properties, 127, 137
structure, 118
tensile specifications, 124
warp crimp angle, 135
yarn pullout test, 119
force balance model, 128, 132
force displacement diagram, 127
frictional interactions, 131
friction coefficient, 125
geometrical analysis, 118
geometrical presumptions, 117
Howell hanging fiber method, 131
image processing method, 135
inclined plane method, 131

mathematical model
  Amounton's law, 121
  cross section schematic, 120
  fabric construction before yarn pulling, 120
  fabric deformation angle, 123
  fabric modulus, 123
  force balance model efficiency, 122
  force component, 122
  force propagation, 122
  free body diagram, 122
  friction coefficient, 121–122
  lateral forces, 122–123
  lateral strain, 120–121
  main forces, 121
  normal load, 122–123
  plain woven fabrics, 123
  pullout force, 121
  resin hardener FK20, 123
  SLEE MAINZ microtome device, 123
  weave angle, 120–121
maximum deformation angle, 135
maximum static situation, 127
mechanistic model, 118
modified lateral forces, 126
motic microscope, 123
open-end spun wefts, 135
oscillation
  behavior, 125
  method, 127
  model, 118
Pearson product correlation coefficient, 128
physical models, 118
physical study, 131
plain woven fabrics, 135
  cross section, 136
problem plan
  geometrical details, 120
  mechanical specifications, 120
  pullout force, 119
  stresses, 119
  U frame, 119
  in yarn pullout test, 119
pulling load, 125
pulling velocity, 135
pullout force displacement versus analytical method, 128, 138
pullout test, 118
ringspun yarns, 135
samples characteristics, 123
scoured cotton fabric, 135
screening speed, 127
semi empirical model, 118
smoother filaments, 127
spring junction assumptions, 118
static friction force, 125
stick-slip motion, 125
strong physical hypotheses, 131
tensiometer device, 126
textured polyester multi filaments, 135
twist friction (capstan) method, 131
two-dimensional continuum sheets, 117
warp direction, 136
weft yarns, 135
yarn modulus Ey, 126
yarn pullout force displacement profile, 126
yarn pullout process, 118
yarns frictional behavior, 125
yarn-to-yarn friction, 118, 131
yarn-to-yarn resistant, 125
Zwick tensiometer, 135
Triarylamine-based polymers, 2

V

Vinyl monomers
  iminoxyl radicals, 86
  organic paramagnets, 87
  polymerization kinetics
    imidazoline-based nitroxyl radicals, 87
    iminoxyls interaction, 87
    initiated polymerization, 86
    nitroxyls efficiency, 87
    polymeric radical, 86
    poly radicals, 87
    radical chain scheme, 86
    regularities, 92
    stable mono, 87
    stable radicals, 86
    TCQM radical anions, 86
    thermal peculiarities, 86

Vinyl monomers polymerization kinetics, stable radicals
benzoic acid, 214
catalytic systems, 213
cumyl peroxide, 213
electron-transfer reaction, 213
heterocyclic oxide, 214
iminoxyl radicals, 213
induction period duration, 213
influence of, 213
nitroxyl
efficiency of, 213
nitroxyl paramagnets, 213
nitroxyl radicals, 213
organic paramagnets, 213
paramagnetic substances
characteristic of, 213
polymeric radical, 213
radicalchain scheme, 213
tetracyanoquinodimethane (TCQM) radical anions, 213
voltammetric method, 222

W

Weighting coefficient, 65. see also Strain electromagnetic susceptibility relaxation
Wine
dry, 233
fortified, 234
red, 233
red grape, 235
red port, 234
semi-sweet, 233
white
fermentation of, 236